Cardiovascular Proteomics

METHODS IN MOLECULAR BIOLOGY™

John M. Walker, SERIES EDITOR

METHODS IN MOLECULAR BIOLOGY™

Cardiovascular Proteomics

Methods and Protocols

Edited by

Fernando Vivanco

Department of Immunology
Fundación Jiménez Díaz
and
Department of Biochemistry and Molecular Biology I
Universidad Complutense
Madrid, Spain

HUMANA PRESS ✳ TOTOWA, NEW JERSEY

© 2007 Humana Press Inc.
999 Riverview Drive, Suite 208
Totowa, New Jersey 07512

humanapress.com

This publication is printed on acid-free paper. ∞
ANSI Z39.48-1984 (American Standards Institute)

Permanence of Paper for Printed Library Materials.
Cover illustration: Fig. 2, Chapter 21, "Isolation and Culture of Adult Mouse Cardiac Myocytes," by Timothy D. O'Connell, Manoj C. Rodrigo, and Paul C. Simpson

Production Editor: Amy Thau

Cover design by Patricia F. Cleary

For additional copies, pricing for bulk purchases, and/or information about other Humana titles, contact Humana at the above address or at any of the following numbers: Tel.: 973-256-1699; Fax: 973-256-8341; E-mail: orders@humanapr.com; or visit our Website: www.humanapress.com

Photocopy Authorization Policy:

Printed in the United States of America. 10 9 8 7 6 5 4 3 2 1

eISBN: 1-59745-214-9

ISSN: 1064-3745

Library of Congress Cataloging-in-Publication Data

Cardiovascular proteomics : methods and protocols / edited by Fernando Vivanco.
   p. ; cm. -- (Methods in molecular biology, ISSN 1064-3745 ; 357)
   Includes bibliographical references and index.
   ISBN-13: 978-1-58829-535-4 (alk. paper)
   ISBN-10: 1-58829-535-4 (alk. paper)
   1. Cardiovascular system--Diseases--Molecular aspects. 2. Proteomics.
 3. Biochemical markers. 4. Two-dimensional electrophoresis.
 I. Vivanco, Fernando 1948-   . II. Series: Methods in molecular biology
 (Clifton, N.J.) ; v. 357.
   [DNLM: 1. Heart. 2. Proteomics--methods. 3. Biological Markers
 --analysis. 4. Electrophoresis, Gel, Two-Dimensional--methods.
 5. Spectrometry, Mass, Matrix-Assisted Laser Desorption-Ionization
 --methods.   W1 ME9616J v.357 2006 / WG 200 C2678 2006]
 RC669.9.C388 2006

   616.1'07--dc22

                                                            2006007132

Preface

Over the past few years, the power and potential of proteomics has become widely recognized. The use of proteomics for the study of complex diseases is increasing and is particularly applicable to cardiovascular disease, the leading cause of death in developed countries. The ability to investigate the complete proteome provides a critical tool toward elucidating the complex and multifactorial basis of cardiovascular biology, especially disease processes. Proteomics involves the integration of a number of technologies with the aim of analyzing all the proteins expressed by a biological system in response to various stimuli under different pathophysiological conditions. The proteomic approach offers the ability to evaluate simultaneously the changes in protein expression and cell signaling pathways in response to such conditions as atherosclerosis, cardiac hypertrophy, stroke, or heart failure.

Cardiovascular Proteomics: Methods and Protocols covers many of the above aspects of the proteomic approach in the cardiovascular field. This volume takes the reader through the complete process of proteomic analysis, from the obtention of specific heart proteins (troponin I) to the new techniques of identifying risk biomarkers of atherome plaque rupture, analyzing the secretome of explanted endarterectomies cultured in vitro, or the application of phage display techniques to decipher the molecular diversity of blood vessels. Determining global changes in the protein expression levels of cardiac myocytes in response to states of ischemia, hypertension, hypertrophy, heart failure, or infarction may disclose new molecules directly involved in these pathologies. Thus, detailed protocols for the isolation and the short- and long-term culture of adult mouse cardiac myocytes have been included.

Cardiovascular Proteomics: Methods and Protocols covers not only the most recent advances of separating proteins by two-dimensional electrophoresis (2-DE; zoom gels, large formats, cup-loading, basic pH ranges) and 2D difference gel electrophoresis, but also the newly developed strategies of liquid chromatography coupled to mass spectrometry (LC-MS) or the SELDI-TOF approach to searching for biomarkers of stroke in human serum or of hypertension in the serum of animal models. A key requirement for analysis of the serum proteome is the depletion of the most abundant proteins (mainly albumin and immunoglobulin G) in order to detect lower abundance proteins that might prove to be potential biomarkers of disease. Thorough descriptions of the most effective methods (immunoaffinity substraction, delipidation combined with specific precipitation) to deplete these proteins have been included.

In recent years the proteome (and secretome) of the most relevant cellular elements of the cardiovascular system has begun to be depicted with the concomitant construction of 2-DE maps and databases. The proteomic strategies and protocols to study the proteome of endothelial, arterial smooth muscle cells, foam cells, circulating blood monocytes, and platelets constitute an essential part of *Cardiovascular Proteomics: Methods and Protocols*. These databases will provide a consistent basis for comparative studies of protein expression levels in these cells from healthy and patient subjects.

Special mention is given to two research areas of high current interest: the analysis of subproteomes and the characterization of posttranslational modifications. *Cardiovascular Proteomics: Methods and Protocols* includes protocols for subcellular fractionation and the obtention of several organelles, cytosol, membranes, and so on, and particularly of heart mitochondria and the analysis of its proteome (comprised of more than 1000 polypeptides). The methods for the purification and characterization of the proteins in others subproteomes (myofilaments, caveolae from endothelial cells, phosphoproteome of human platelets) are also described in detail. The analysis and characterization of complex associations of proteins are central aims of proteomics. A good example is included in *Cardiovascular Proteomics: Methods and Protocols* with the description of the proteomic analysis of the subunit composition of complex I (ubiquinone oxidoreductase) from bovine heart mitochondria. There are numerous examples of cardiovascular functions whose molecular pathways are mediated through posttranslational processes, such as phosphorylation and glycosylation. Several protocols are included illustrating these topics (identification of targets of phosphorylation in mitochondria, analysis of phosphorylated isoforms of HSP-27 in atherosclerotic plaques, identification of *S*-nitrosylated proteins in endothelial cells).

The application of proteomics to cardiovascular disease holds great promise and offers exciting advances towards predictive, preventive, and personalized medicine. It is conceivable that the analysis of a simple plasma sample will provide unique prognostic information about an individual's risk of atherothrombotic disease or heart failure. Similarly, proteomic analysis of a myocardial biopsy specimen may provide useful prognostic information in patients with unexplained heart failure or in cardiac transplant recipients. It is our intent that our book will contribute to this important task.

We hope that readers will find *Cardiovascular Proteomics: Methods and Protocols* a useful snapshot of the currently available technologies, a starting point for evaluating the applicability of these techniques to their own research. The editor is especially grateful to all the contributing authors for the time and effort they have put into writing their chapters and particularly to the Methods in Molecular Biology series editor, John Walker, for his continuous advice and support through the editorial process.

Fernando Vivanco

Contents

Contributors

LAURE ALLARD • *Biomedical Proteomics Research Group, Department of Structural Biology and Bioinformatics, School of Medicine, Geneva University, Geneva, Switzerland*

WADIH ARAP • *Department of Genitourinary Medical Oncology, The University of Texas M. D. Anderson Cancer Center, Houston, TX*

MARIA G. BARDERAS • *Department of Immunology, Fundación Jiménez Díaz, Madrid, Spain*

ELISABETTA BOERI-ERBA • *Protein Research Group, Department of Biochemistry and Molecular Biology, University of Southern Denmark, Odense, Denmark*

DIANE E. BOVENKAMP • *Department of Medicine, Johns Hopkins University, Baltimore, MD*

JOE CARROLL • *Dunn Human Nutrition Unit, The Medical Research Council, Cambridge, United Kingdom*

DAWN R. CHRISTIANSON • *Department of Genitourinary Medical Oncology, The University of Texas M. D. Anderson Cancer Center; and Graduate School of Biomedical Sciences, The University of Texas Health Science Center, Houston, TX*

JUDITH COPPINGER • *Department of Clinical Pharmacology, Royal College of Surgeon's in Ireland, Dublin, Ireland*

VERÓNICA M. DARDÉ • *Department of Immunology, Fundación Jiménez Díaz, Madrid, Spain*

SANDRINE DELBOSC • *INSERM U698, Cardiovascular Haemostasis, Bioengineering, and Remodeling, CHU X, Bichat, Paris, France*

ROSARIO DE NICOLÁS • *Research Unit, Renal and Vascular Pathology Laboratory, Fundación Jiménez Díaz, Madrid, Spain*

MICHAEL J. DUNN • *Proteome Research Centre, Conway Institute of Biomolecular and Biomedical Research, University College Dublin, Dublin, Ireland*

ANNABELLE DUPONT • *INSERM U508, Institut Pasteur de Lille, Lille, France*

MARI-CARMEN DURÁN • *Department of Immunology, Fundación Jiménez Díaz, Madrid, Spain; and Department of Biochemistry and Molecular Biology, Protein Research Group, University of Southern Denmark, Odense, Denmark*

JESÚS EGIDO • *Renal and Vascular Pathology Laboratory, Fundación Jiménez Díaz, Universidad Autónoma, Madrid, Spain*

ANDREW EMILI • *Program in Proteomics and Bioinformatics, Banting and Best Department of Medical Research, University of Toronto, Toronto, Ontario, Canada*

IAN M. FEARNLEY • *Dunn Human Nutrition Unit, The Medical Research Council, Cambridge, United Kingdom*

DESMOND J. FITZGERALD • *Department of Molecular Medicine, Conway Institute, University College Dublin, Dublin, Ireland*

MARTINA FOY • *Department of Clinical Pharmacology, Royal College of Surgeon's in Ireland, Dublin, Ireland*

QIN FU • *Department of Medicine, Johns Hopkins University, Baltimore, MD*

JULIO GÁLLEGO-DELGADO • *Renal and Vascular Pathology Laboratory, Fundación Jiménez Díaz, Universidad Autónoma, Madrid, Spain*

JESÚS GONZÁLEZ-CABRERO • *Research Unit, Renal and Vascular Pathology Laboratory, Fundación Jiménez Díaz, Madrid, Spain*

ROBERTA A. GOTTLIEB • *Department of Molecular and Experimental Medicine, The Scripps Research Institute, La Jolla, CA*

ANTHONY O. GRAMOLINI • *Banting and Best Department of Medical Research, University of Toronto, Toronto, Ontario, Canada*

CHITTIBABU GUDA • *Gen*NY*sis Center for Excellence in Cancer Genomics, and Department of Epidemiology and Biostatistics, State University of New York at Albany, Rensselaer, NY*

PURNIMA GUDA • *Gen*NY*sis Center for Excellence in Cancer Genomics, and Department of Epidemiology and Biostatistics, State University of New York at Albany, Rensselaer, NY*

MARIANA EÇA GUIMARÃES DE ARAÚJO • *Division of Cell Biology, Innsbruck Biocenter, Innsbruck, Austria*

DONAL F. HARNEY • *Department of Clinical Pharmacology, Royal College of Surgeon's in Ireland, Dublin, Ireland*

TIMOTHY A. HAYSTEAD • *Department of Pharmacology and Cancer Biology, Duke University Medical Center, Durham, NC*

DENIS HOCHSTRASSER • *Biomedical Proteomics Research Group, Department of Structural Biology and Bioinformatics, School of Medicine, Geneva University, Geneva, Switzerland*

ANTON J. G. HORREVOETS • *Department of Medical Biochemistry, Academic Medical Center, University of Amsterdam, Amsterdam, The Netherlands*

LUKAS ALFONS HUBER • *Division of Cell Biology, Innsbruck Biocenter, Innsbruck, Austria*

HIRONORI ISHIKAWA • *Genomic Science Laboratories, Sumitomo Pharmaceuticals, Osaka, Japan*

OLE N. JENSEN • *Protein Research Group, Department of Biochemistry and Molecular Biology, University of Southern Denmark, Odense, Denmark*

MASAHARU KANAOKA • *Genomic Science Laboratories, Sumitomo Pharmaceuticals, Osaka, Japan*

LESLEY A. KANE • *Department of Biological Chemistry, Johns Hopkins University, Baltimore, MD*

THOMAS KISLINGER • *Program in Proteomics and Bioinformatics, Banting and Best Department of Medical Research, University of Toronto, Toronto, Ontario, Canada*

SHINICHI KOJIMA • *Genomic Science Laboratories, Sumitomo Pharmaceuticals, Osaka, Japan*

SANTIAGO LAMAS • *Fundación Centro Nacional de Investigaciones Cardiovasculares Carlos III (CNIC); Centro de Investigaciones Biológicas— Consejo Superior de Investigaciones Científicas (CSIC); and Instituto "Reina Sofia" de Investigaciones Nefrológicas, Madrid, Spain*

ALBERTO LÁZARO • *Renal and Vascular Pathology Laboratory, Fundación Jiménez Díaz, Universidad Autónoma, Madrid, Spain*

PIERRE LESCUYER • *Biomedical Proteomics Research Group, Department of Structural Biology and Bioinformatics, School of Medicine, Geneva University, Geneva, Switzerland*

PETER LIU • *Banting and Best Department of Medical Research, University of Toronto, Toronto, Ontario, Canada*

DAVID H. MACLENNAN • *Banting and Best Department of Medical Research, University of Toronto, Toronto, Ontario, Canada*

PATRICIA B. MAGUIRE • *Department of Molecular Medicine, Conway Institute, University College Dublin, Dublin, Ireland*

WIM MARTINET • *Division of Pharmacology, University of Antwerp, Wilrijk, Belgium*

ANTONIO MARTÍNEZ-RUIZ • *Fundación Centro Nacional de Investigaciones Cardiovasculares Carlos III (CNIC); Centro de Investigaciones Biológicas— Consejo Superior de Investigaciones Científicas (CSIC); and Instituto "Reina Sofia" de Investigaciones Nefrológicas, Madrid, Spain*

JOSE L. MARTÍN-VENTURA • *Vascular Research Department, Fundación Jiménez Díaz, Madrid, Spain*

SEBASTIÁN MAS • *Department of Immunology, Fundación Jiménez Díaz, Madrid, Spain*

OLIVIER MEILHAC • *INSERM U698, Cardiovascular Haemostasis, Bioengineering, and Remodeling, CHU X, Bichat, Paris, France*

JEAN-BAPTISTE MICHEL • *INSERM U698, Cardiovascular Haemostasis, Bioengineering, and Remodeling, CHU X, Bichat, Paris, France*

SHABAZ MOHAMMED • *Protein Research Group, Department of Biochemistry and Molecular Biology, University of Southern Denmark, Odense, Denmark*

IRINA NEVEROVA • *Department of Physiology, Queen's University, Kingston, Ontario, Canada*

TIMOTHY D. O'CONNELL • *Cardiovascular Research Institute, Department of Medicine and South Dakota Health Research Foundation, The University of South Dakota, Sioux Falls, SD*

JULIO I. OSENDE • *Cardiology Service, Hospital Universitario "Gregorio Marañón," Madrid, Spain*

MICHAEL G. OZAWA • *Department of Genitourinary Medical Oncology, The University of Texas M. D. Anderson Cancer Center, Houston, TX*

RENATA PASQUALINI • *Department of Genitourinary Medical Oncology, The University of Texas M. D. Anderson Cancer Center, Houston, TX*

FLORENCE PINET • *INSERM U744, Public Health and Molecular Epidemiology of Aging-Related Diseases, Institut Pasteur de Lille, Lille, France*

MAYTE POZO • *Research Unit, Renal and Vascular Pathology Laboratory, Fundación Jiménez Díaz, Madrid, Spain*

ALEXEY V. PSHEZHETSKY • *Department of Medical Genetics, St. Justine Hospital, Côte St. Catherine, Montreal, Quebec, Canada*

MANOJ C. RODRIGO • *Cardiology Division, San Francisco VA Medical Center, and Department of Medicine and Cardiovascular Research Institute, The University of California at San Francisco, San Francisco, CA*

YAO-CHENG RUI • *Department of Pharmacology, School of Pharmacy, Secondary Military Medical University, Shanghai, China*

JUN SAKAI • *Genomic Science Laboratories, Sumitomo Pharmaceuticals, Osaka, Japan*

JEAN-CHARLES SANCHEZ • *Biomedical Proteomics Research Group, Department of Structural Biology and Bioinformatics, School of Medicine, Geneva University, Geneva, Switzerland*

HIDESHI SATOH • *Discovery Research Laboratories I, Sumitomo Pharmaceuticals, Osaka, Japan*

PAUL C. SIMPSON • *Cardiology Division, San Francisco VA Medical Center, and Department of Medicine and Cardiovascular Research Institute, The University of California at San Francisco, San Francisco, CA*

RICHARD R. SPRENGER • *Department of Medical Biochemistry, Academic Medical Center, University of Amsterdam, Amsterdam, The Netherlands*

BRIAN A. STANLEY • *Department of Physiology, Queen's University, Kingston, Ontario, Canada; and Department of Medicine, Johns Hopkins University, Baltimore, MD*

SHANKAR SUBRAMANIAM • *San Diego Supercomputer Center, University of California, San Diego, La Jolla, CA; and Departments of Bioengineering, Chemistry, and Biochemistry, University of California, San Diego, La Jolla, CA*

JENNIFER E. VAN EYK • *Department of Physiology, Queen's University, Kingston, Ontario, Canada; and Departments of Medicine, Biological Chemistry, and Biomedical Engineering, Johns Hopkins University, Baltimore, MD*

FERNANDO VIVANCO • *Department of Immunology, Fundación Jiménez Díaz; and Department of Biochemistry and Molecular Biology I, Proteomics Unit, Universidad Complutense, Madrid, Spain*

JOHN E. WALKER • *Dunn Human Nutrition Unit, The Medical Research Council, Cambridge, United Kingdom*

ANNE A. WOOLDRIDGE • *Department of Pharmacology and Cancer Biology, Duke University Medical Center, Durham, NC*

KIERAN WYNNE • *Department of Clinical Pharmacology, Royal College of Surgeon's in Ireland, Dublin, Ireland*

SETSUKO YAMAMOTO • *Discovery Research Laboratories I, Sumitomo Pharmaceuticals, Osaka, Japan*

PENG-YUAN YANG • *Department of Pharmacology, School of Pharmacy, Secondary Military Medical University, Shanghai, China*

PENG-YUAN YANG • *Department of Chemistry, Fudan University, Shangai, China*

YAN-LING YU • *Department of Chemistry, Fudan University, Shanghai, China*

I

ANALYSIS OF THE HEART PROTEOME

1

Two-Dimensional Polyacrylamide Gel Electrophoresis for Cardiovascular Proteomics

Michael J. Dunn

Summary

The majority of cardiovascular proteomic investigations reported to date have employed two-dimensional gel electrophoresis (2DE) with immobilized pH gradients to separate the sample proteins, combined with quantitative computer analysis to detect differentially expressed proteins and mass spectrometry technologies to identify proteins of interest. In spite of the development of novel gel-free technologies, 2DE remains the only technique that routinely can be applied to parallel quantitative expression profiling of large sets of complex protein mixtures, such as those represented by cardiovascular cell and tissue lysates. This chapter details a procedure for large-format 2DE, and its variations, that has been successfully applied in cardiovascular proteomic research.

Key Words: Expression profiling; human heart; human saphenous vein; immobilized pH gradients; laser capture microdissection; mass spectrometry; myocardium; smooth muscle; two-dimensional polyacrylamide gel electrophoresis.

1. Introduction

The majority of cardiovascular proteomic investigations reported to date have employed two-dimensional gel electrophoresis (2DE) with immobilized pH gradients (IPGs) to separate the sample proteins, combined with quantitative computer analysis to detect differentially expressed proteins and mass spectrometry (MS) technologies to identify proteins of interest (*1*).

In spite of the development of novel gel-free technologies based on the use of liquid chromatography (LC) and MS (**ref.** *2*; *see* Chapter 2), 2DE remains the only technique that routinely can be applied to parallel quantitative expression profiling of large sets of complex protein mixtures, such as whole cell and tissue lysates (*3*). 2DE involves the separation of solubilized proteins in the first

From: *Methods in Molecular Biology, vol. 357: Cardiovascular Proteomics: Methods and Protocols*
Edited by: F. Vivanco © Humana Press Inc., Totowa, NJ

dimension according to their charge properties (isoelectric point [p*I*]) by iso-electric focusing (IEF) under denaturing conditions. This is then followed by sodium dodecyl sulfate-polyacrylamide gel electrophoresis (SDS-PAGE), where proteins are separated according to their relative molecular mass (M_r), the second dimension. Because the charge and mass properties of proteins are essentially independent parameters, this orthogonal combination of charge (p*I*) and size (M_r) separations results in the sample proteins being distributed across the two-dimensional gel profile. In excess of 5000 proteins can readily be resolved simultaneously (~2000 proteins routinely), with 2D allowing detection of <1 ng of protein per spot. Furthermore, a map of intact proteins is provided, reflecting changes in protein expression level, individual isoforms, or posttranslational modifications *(3)*.

2. Materials

Prepare all solution from analytical-grade reagents (except where otherwise indicated) using ultrapure (18.2 MΩ) water.

1. 18 cm IPG Immobiline DryStrip, pH 3.0–10.0, NL gel strips (GE Healthcare, Amersham, UK; *see* **Notes 1** and **2**).
2. IPG Immobiline DryStrip reswelling tray (GE Healthcare; *see* **Note 3**).
3. Multiphor II horizontal flat-bed electrophoresis unit (GE Healthcare; *see* **Note 4**).
4. Immobiline DryStrip kit for Multiphor II (GE Healthcare; *see* **Note 5**).
5. Power supply capable of providing an output of 3500 V (*see* **Note 6**).
6. MultiTemp III thermostatic circulator (GE Healthcare).
7. IEF electrode strips (GE Healthcare) cut to a length of 110 mm.
8. Urea (Ultrapure, Invitrogen, Paisley, UK; *see* **Note 7**).
9. Solution A: 9.5 *M* urea (50 mL): Dissolve 60 g of urea and make up to 50 mL in ultrapure water. Deionize the solution by adding 1 g Amberlite MB-1 monobed resin (Sigma-Aldrich, UK) and stirring for 1 h. Filter the solution using a sintered glass filter.
10. Solution B: Sample lysis buffer: 9.5 *M* urea, 2% (w/v) CHAPS, 1% (w/v) dithiothreitol (DTT), 0.8% (w/v) 2D Pharmalyte, pH 3.0–10.0 (GE Healthcare; *see* **Note 8**). Add 1 g CHAPS (*see* **Note 9**), 0.5 g DTT, 1 mL of Pharmalyte, pH 3.0–10.0, and 5 tablets of protease inhibitors (Complete Mini Protease Inhibitor Cocktail tablets, Roche Diagnostics Ltd., Lewes, UK).
11. Solution C: 8 *M* urea solution (40 mL): Dissolve 19.3 g of urea in 25.6 mL of ultrapure water. Deionize the solution by adding 0.5 g Amberlite MB-1 monobed resin (Sigma-Aldrich) and stirring for 1 h. Filter the solution using a sintered glass filter.
12. Solution D: Reswelling solution: 8 *M* urea, 0.5% (w/v) CHAPS, 0.2% (w/v) DTT, 0.2% (w/v) 2D Pharmalyte pH 3.0–10.0. Add 60 mg DTT, 150 mg CHAPS, and 150 μL Pharmalyte pH 3.0–10.0, to 29.7 mL of solution C.
13. Immobiline DryStrip Cover Fluid (GE Healthcare).
14. Solution E: Electrolyte solution for both anode and cathode: ultrapure water.

15. Solution F: Equilibration buffer (100 mL): 6 *M* urea, 30% (w/v) glycerol, 2% (w/v) SDS, 50 m*M* Tris-HCl buffer, pH 8.8. Add 36 g urea, 30 g glycerol, and 2 g SDS to 3.3 mL 1.5 *M* Tris-HCl buffer, pH 8.8, and make up to 100 mL with ultrapure water.
16. Solution G: DTT stock solution: Add 200 mg DTT to 1 mL ultrapure water. Prepare immediately before use.
17. Solution H: bromophenol blue solution: Add 30 mg bromophenol blue to 10 mL 1.5 *M* Tris-HCL buffer, pH 8.8.

3. Methods

1. Sample preparation: Samples of cardiovascular cells, either primary isolates or cultured in vitro, should be washed several times by suspension in phosphate-buffered saline (PBS) to remove any proteins (e.g., serum) from the culture medium, and finally washed in isotonic sucrose (0.35 *M*) to remove any salts that can interfere with the IEF dimension. The cell pellet is then solubilized by suspension in a small volume of lysis buffer (solution B). Samples of intact cardiovascular tissues should be disrupted while still frozen by crushing between two cooled metal blocks (small samples) or by grinding in a mortar cooled with liquid nitrogen (larger samples). The resulting powder is then suspended in a small volume of lysis buffer (solution B). The final protein concentration of the samples should be about 10 mg/mL. Protein samples should be used immediately or stored frozen at –80°C (*see* Chapters 3–5).
2. Rehydration of IPG gel strips with the protein sample: Dilute an aliquot of each sample containing an appropriate amount of protein (*see* **Note 10**) with solution D to a total volume of 450 µL (*see* **Note 11**). Pipet each sample into one groove of the reswelling tray. Peel off the protective cover sheets from the IPG strips and insert the IPG strips (gel side down) into the grooves. Avoid trapping air bubbles. Cover the strips with 1 mL of DryStrip Cover Fluid, close the lid, and allow the strips to rehydrate overnight at room temperature.
3. Preparation of IEF apparatus: Ensure that the strip tray, template for strip alignment, and electrodes are clean and dry. Set the thermostatic circulator at 20°C (*see* **Note 12**) and switch on at least 15 min prior to starting the IEF separation. Pipet a few drops of DryStrip Cover Fluid onto the cooling plate and position the strip tray on the plate. This film of DryStrip Cover Fluid, which has excellent thermal conductivity properties and a low viscosity, allows for good contact between the strip tray and the cooling plate. Pipet a few drops of DryStrip Cover Fluid into the tray and insert the IPG strip alignment guide.
4. After rehydration is complete, remove the IPG strips from the reswelling tray, rinse them briefly with deionized water, and place them, gel side up, on a sheet of water-saturated filter paper. Wet a second sheet of filter paper with deionized water, blot it slightly to remove excess water, and place on the surface of the IPG strips. Blot them gently for a few seconds to remove excess rehydration solution in order to prevent urea crystallization on the surface of the gel during IEF.
5. IPG IEF dimension: Place the IPG gel strips side by side in the grooves of the alignment guide of the strip tray, which will take up to 12 strips (*see* **Note 13**). The basic

end of the IPG strips must be at the cathodic side of the apparatus. Wet the electrode wicks with about 0.5 mL of the electrode solution (solution E) and remove excess liquid with a tissue. Place the electrode wicks on top of the strips as near to the gel edges as possible. Position the electrodes and press them down onto the electrode wicks. Fill the strip tray with DryStrip Cover Fluid to protect the IPG strips from the effects of the atmosphere.

6. IEF running conditions: Run the IPG IEF gels at 0.05 mA per strip, and 5 W limiting. For the higher protein loads used for micropreparative runs, it is recommended to limit the initial voltage to 150 V for 30 min (75 Vh) and then 300 V for 60 min (300 Vh). Continue IEF with maximum settings of 3500 V, 2 mA, and 5 W until constant focusing patterns are obtained. The precise running conditions required depend on the pH gradient, the separation distances used, and the type of sample being analyzed (*see* **Note 14**).

7. After completion of IEF, remove the gel strips from the apparatus. Freeze the strips in plastic bags and store them at −80°C if they are not to be used immediately for the second-dimension separation.

8. Equilibration of IPG gel strips: Equilibrate IPG gel strips with gentle shaking for 2 × 15 min in 10 mL equilibration buffer (solution F). Add 500 μL/10 mL DTT stock (solution G) and 30 μL/10 mL bromophenol blue stock (solution H) to the first equilibration solution. Add 500 mg iodoacetamide per 10 mL of the second equilibration solution (final concentration iodoacetamide 5% [w/v]).

9. SDS-PAGE dimension: The second dimension SDS-PAGE separation is carried out using a vertical SDS-PAGE system (*see* **Note 15**) and the normal Laemmli buffer system (*4*). The gels can be either of a suitable constant percentage concentration of polyacrylamide or of a linear or nonlinear polyacrylamide concentration gradient. We routinely use 1.5-mm thick 12% T SDS-PAGE gels (26 cm × 20 cm). No stacking gel is used.

10. Rinse the equilibrated IPG gel strips with deionized water and blot them on filter paper to remove excess liquid.

11. Apply the IPG gel strips to the SDS-PAGE gels by filling the space in the cassette above the separation gel with upper reservoir buffer and gently slide the strips into place. Good contact between the tops of the SDS gels and the strips must be achieved and air bubbles must be avoided. Cement the strips in place with 1% (w/v) agarose in equilibration buffer.

12. The gels are run in a suitable vertical electrophoresis apparatus (*see* **Note 16**). We use the Ettan DALT 12 vertical system (GE Healthcare), which allows up to 12 large-format (26 × 20 cm) second-dimension SDS-PAGE gels to be electrophoresed simultaneously. The gels are run at 5 W/gel at 28°C for 45 min and then at 1 W/gel overnight at 15°C until the bromophenol blue tracking dye reaches the bottom of the gels. This takes approx 18 h for a full set of 12 gels.

13. The gels can be subjected to any suitable procedure to detect the separated proteins (*see* **Note 17**). Typical 2D separations of cardiovascular proteins using this technique are shown in **Fig. 1** (human ventricular myocardial proteins) and **Fig. 2** (human saphenous vein smooth muscle).

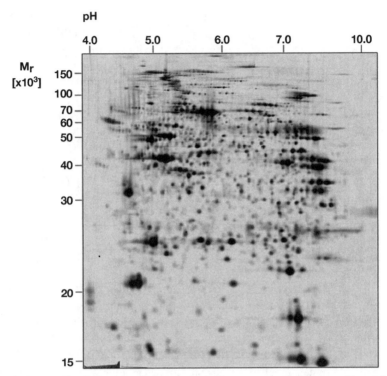

Fig. 1. A two-dimensional gel electrophoresis separation of 80 µg of heart (ventricle) proteins using an 18 pH 3.0–10.0 NL immobilized pH gradient (IPG) DryStrip in the first dimension and a 21 cm 12% SDS-PAGE gel in the second dimension. Proteins were detected by silver staining. The nonlinear pH range of the first-dimension IPG strip is indicated along the top of the gel, acidic pH to the left. The M_r (relative molecular mass) scale on the left can be used to estimate the molecular weights of the separated proteins.

4. Notes

1. We routinely use IPG gels with an 18 cm pH gradient separation distance, but it is possible to use gels of other sizes (e.g., 7-, 11-, 13-, and 24-cm pH gradient separation distance) *(3)*. Small-format gels (e.g., 7-cm strips) are ideal for rapid screening purposes or where the amount of sample is limited; for example, where the sample has been prepared by laser capture microdissection *(5)* (**Fig. 3**). Extended separation distances (e.g., 24-cm IPG strips) provide maximum resolution of complex protein patterns.
2. A wide-range, linear IPG 3.0–10.0 L pH gradient is often useful for the initial analysis of a new type of sample. However, for many samples this can result in loss of resolution in the region pH 4.0 to 7.0, in which the p*I* values of many proteins occur. This problem can be overcome to some extent with the use of a nonlinear IPG 3.0–10.0 NL pH gradient, in which the pH 4.0 to 7.0 region contains a much flatter gradient than in the more acidic and alkaline regions. This allows good sepa-

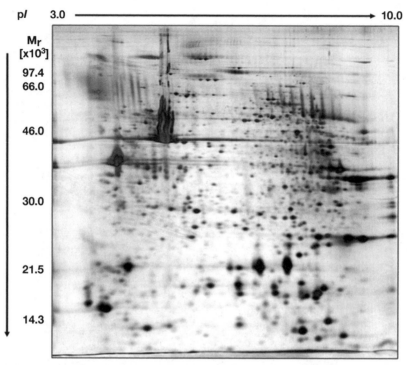

pl 3.0 ──────────────────────────────────► 10.0

Mr
[x10³]

97.4
66.0

46.0

30.0

21.5

14.3

Fig. 2. A two-dimensional gel electrophoresis separation of 400 µg of intact contractile human saphenous vein smooth muscle proteins using an 18 pH 3.0–10.0 NL
immobilized pH gradient (IPG) DryStrip in the first dimension and a 21 cm 12% SDS-
PAGE gel in the second dimension. Proteins were detected by silver staining. The nonlinear pH range of the first-dimension IPG strip is indicated along the top of the gel,
acidic pH to the left. The M_r (relative molecular mass) scale on the left can be used to
estimate the molecular weights of the separated proteins. (From **ref. *29*** with permission from Wiley-VCH.)

ration in the pH 4.0 to 7.0 region while still resolving the majority of the more
basic species. However, the use of a pH 4.0 to 7.0 IPG IEF gel can result in even
better protein separation of the acidic to neutral proteins, whereas a separate pH
6.0–9.0 or 6.0–11.0 IPG IEF gel can be used to separate the more basic proteins
(6). With complex samples such as whole tissue extracts, 2DE on a single wide-
range pH gradient reveals only a small percentage of the whole proteome because
of insufficient spatial resolution and the difficulty of visualizing low copy number proteins in the presence of the more abundant species. One approach to overcoming the problem is to use multiple, overlapping narrow-range IPGs spanning
1.0 to 1.5 pH units; an approach that has become known as "zoom gels" *(7,8)*,
"composite gels," or "subproteomics" *(9)*. Strongly alkaline proteins such as ribosomal and nuclear proteins with closely related p*I*s between 10.5 and 11.8 can be
separated using narrow-range pH 10.0–12.0 or pH 9.0–12.0 IPGs *(10)*.

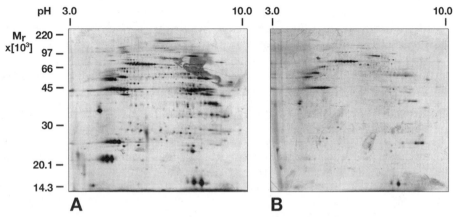

pH 3.0   10.0   3.0   10.0

M_r x[10^3]   220 — 97 — 66 — 45 — 30 — 20.1 — 14.3 —

A   **B**

Fig. 3. Two-dimensional gel electrophoresis separations of laser capture micro-dissected human cardiac tissue ([**A**] cardiac myocytes; [**B**] blood vessels) using 18 pH 3.0–10.0 NL immobilized pH gradient (IPG) DryStrip in the first dimension and 21 cm 12% SDS-PAGE gels in the second dimension. Proteins were detected by silver staining. The nonlinear pH range of the first-dimension IPG strip is indicated along the top of the gel, acidic pH to the left. The M_r (relative molecular mass) scale on the left can be used to estimate the molecular weights of the separated proteins. (From **ref. 5** with permission from Wiley-VCH.)

3. The IPG Immobiline DryStrip reswelling tray (GE Healthcare) is a grooved plastic tray with a lid designed for the rehydration of IPG DryStrip gels of any length from 7 to 24 cm in the presence of the solubilized protein sample.
4. We routinely use the Multiphor II horizontal flat-bed electrophoresis unit (GE Healthcare). Any horizontal flat-bed IEF apparatus can be used for IPG IEF, but the Immobiline DryStrip kit (*see* **Note 5**) is designed to fit the Multiphor II. Another alternative is to use the Ettan IPGphor II IEF System (GE Healthcare), an integrated system dedicated to first-dimension IEF using IPG DryStrip with built-in temperature control unit and power supply.
5. The Immobiline DryStrip kit facilitates the sample application and running of IPG IEF gels in the first-dimension of 2DE. The strip tray consists of a thin glass plate with a polyester frame. The frame acts as an electrode holder and the metal bars affixed to the frame conduct voltage to the electrodes. The electrodes, which are made of polysulphone, are moveable to accommodate gel strips of varying inter-electrode distance and have a platinum wire that rests against the electrode strip. It is also fitted with a bar, also made of polysulphone, which supports the sample cups (styrene-acrylnitrile). These cups can be used to apply sample volumes up to 100 μL as an alternative to the in-gel rehydration technique of sample application described here. The equivalent fitting for the IPGphor II IEF System is the ceramic Ettan IPGphor Manifold (GE Healthcare), which is a high-throughput accessory that also makes it possible to use sample cups.

6. It is essential that the power supply can deliver less than 1 mA at 3500 V, because these conditions are achieved during IEF of IPG gels. Power packs from some manufacturers are designed to cut out if a low current condition at high voltage is detected. The EPS 3501 XL power supply (GE Healthcare) meets this requirement. The built-in power supply of the Ettan IPGphor II System can supply up to 10,000 V. The use of such a high voltage for a short period at the end of the IEF run can help to give more highly resolved protein bands, resulting in "tighter" spots on 2D gel protein profiles.

7. Urea should be stored dry at 4°C to reduce the rate of breakdown of urea with the formation of isocyanate ions, which can react with protein amino groups to form stable carbamylated derivatives of altered charge. This effect results in the formation of additional artefactual protein spots known as "charge trains" on 2D protein profiles.

8. Lysis buffer should be prepared freshly. Small portions of lysis buffer can be stored at −80°C, but once thawed should not be frozen again.

9. We generally use the zwitterionic detergent CHAPS because this can give improved sample solubilization compared with non-ionic detergents such as Triton X-100 and Nonidet NP-40. Thiourea is a more powerful chaotropic agent, and when used in combination with urea (typically 7 *M* urea, 2 *M* thiourea) leads to improved solubilization efficiency for some types of sample *(11)*. Hydrophobic and membrane proteins usually are solubilized poorly under conditions compatible with the IEF dimension of 2DE (ionic detergents such as SDS cannot be used). Improved solubilization of this class of proteins for 2DE can be achieved using alternative linear sulphobetaine detergents such as SB3-10 or ASB-14 *(12)*.

10. For analytical purposes (e.g., silver staining), between 60 and 80 µg total protein from complex mixtures such as whole cell and tissue lysates should be applied. It is possible to obtain successful protein identification by MS on at least the more abundant protein spots using such a loading, but it is preferable for micropreparative purposes for the sample to contain between 400 µg and 1 mg total protein.

11. The total volume for rehydration must be adjusted depending on the separation length of the IPG strip used; 175 µL for 7 cm, 275 µL for 13 cm, and 600 µL for 24 cm IPG strips.

12. The temperature at which IEF with IPG is performed has been shown to exert a marked effect on spot positions and pattern quality of 2D separations *(13)*. Temperature control is, therefore, essential in order to allow meaningful comparison of 2DE patterns. Focusing at 20°C was found to result in superior 2D separations with respect to sample entry, resolution, and background staining compared with separations carried out at 10°C or 15°C *(13)*.

13. Exposure of the gel strips to the air should be as brief as possible to prevent the formation of a thin layer of urea crystals on the gel surface.

14. As a guide, for IPG IEF gel strips with an 18 cm pH gradient separation distance we use 70,000 Vh for micropreparative purposes.

15. The second SDS-PAGE dimension can also be carried out using a horizontal flatbed electrophoresis apparatus. This method is described in **ref. 14**.

16. Suitable vertical SDS-PAGE systems designed to run batches of large-format 2D gels are the Ettan DALT 6 (6 gels per run) and Ettan DALT 12 (12 gels per run) units (GE Healthcare), or the PROTEAN Plus Dodeca Cell (12 gels per run; Bio-Rad Laboratories, Watford, UK).

17. Organic dyes such as Coomassie blue R-250 and G-250 are compatible with MS, but are limited by their relative insensitivity *(15)*. Silver staining allows the detection of low nanogram amounts of protein. However, standard silver-staining protocols almost invariably use glutaraldehyde and formaldehyde, which alkylate α- and ε-amino groups of proteins, thereby interfering with their subsequent identification by MS. To overcome this problem, silver-staining protocols compatible with MS in which glutaraldehyde is omitted have been developed *(16,17)*, but these suffer from a decrease in sensitivity of staining and a tendency to a higher background. This problem can be overcome using postelectrophoretic fluorescent-staining techniques *(15)*. The best of these at present appears to be SYPRO Ruby, which has a sensitivity approaching that of standard silver staining and is fully compatible with protein identification by MS *(18,19)*. Recently, there has been increasing interest in the use of a pre-electrophoretic fluorescent staining method based on the labeling of protein samples with *N*-hydroxy succinimidyl ester derivatives of fluorescent cyanine (Cy) dyes and known as two-dimensional difference gel electrophoresis *(20,21)* (*see* Chapters 3 and 28). This approach has the advantage that a pair of protein samples can be labeled separately with Cy3 and Cy5 derivatives. The two samples can be mixed and then separated together on the same 2D gel. The resulting 2D gel is then scanned to acquire the Cy3 and Cy5 images separately. Improved quantitative accuracy of comparison of multiple pairs of samples can be achieved using a pooled internal standard labeled with a third dye, Cy2 *(22,23)*. Recently saturation labeling with cysteine-reactive Cy fluorescent dyes has been described *(24)*. This technique provides increased sensitivity for expression profiling of scarce samples such as laser-microdissected clinical specimens *(25,26)*. In addition to the aforementioned dyes, a new range of fluorescent dyes has recently become popular. It is now possible to stain 2D gels for specific proteins, such as those that are in a phosphorylated/hyperphosphorylated or glycosylated state. Pro-Q Diamond dye is a new fluorescent phosphosensor technology suitable for the detection of phosphoserine-, phosphothreonine-, and phosphotyrosine-containing proteins directly in IEF gels, SDS polyacrylamide gels, and 2D gels *(27)*. Pro-Q Emerald 300 fluorescent stain can be used for the detection of glycoproteins in polyacrylamide gels with as little as 2–4 ng of lipopolysaccharide being detectable, in contrast to 250–1000 ng required for detection with conventional silver staining *(28)*. In both cases, 2D gels can be post-stained with SYPRO Ruby dye, allowing sequential two-color detection of either phosphorylated and unphosphorylated proteins or glycosylated and nonglycosylated proteins.

References

1. McGregor, E. and Dunn, M. J. (2003) Proteomics of heart disease. *Hum. Mol. Genet.* **12,** R135–R144.

2. Aebersold, R. and Mann, M. (2003) Mass spectrometry-based proteomics. *Nature* **422,** 198–207.
3. Görg, A., Weiss, W., and Dunn, M. J. (2004) Current two-dimensional electrophoresis technology for proteomics. *Proteomics* **4,** 3665–3685.
4. Laemmli, U. K. (1970) Cleavage of structural proteins during the assembly of the head of bacteriophage T4. *Nature* **227,** 680–685.
5. De Souza, A., McGregor, E., Dunn, M. J., and Rose, M. L. (2004) Preparation of human heart for laser microdissection and proteomics. *Proteomics* **4,** 578–586.
6. Pennington, K., McGregor, E., Beasley, C. L., Everall, I., Cotter, D., and Dunn, M. J. (2004) Optimization of the first dimension for separation by two-dimensional gel electrophoresis of basic proteins from human brain tissue. *Proteomics* **4,** 27–30.
7. Wildgruber, R., Harder, A., Obermaier, C., et al. (2000) Towards higher resolution: 2D-electrophoresis of Saccharomyces cerevisiae proteins using overlapping narrow IPG's. *Electrophoresis* **21,** 2610–2616.
8. Westbrook, J. A., Yan, J. X., Wait, R., Welson, S. Y., and Dunn, M. J. (2001) Zooming-in on the proteome: very narrow-range immobilised pH gradients reveal more protein species and isoforms. *Electrophoresis* **22,** 2865–2871.
9. Cordwell, S. J., Nouwens, A. S., Verrils, N. M., Basseal, D. J., and Walsh, B. J. (2000) Sub-proteomics based upon protein cellular location and relative solubilities in conjunction with composite two-dimensional gels. *Electrophoresis* **21,** 1094–1103.
10. Görg, A., Obermaier, C., Boguth, G., Csordas, A., Diaz, J. J., and Madjar, J. J. (1997) Very alkaline immobilized pH gradients for two-dimensional electrophoresis of ribosomal and nuclear proteins. *Electrophoresis* **18,** 328–337.
11. Rabilloud, T. (1998) Use of thiourea to increase the solubility of membrane proteins in two-dimensional electrophoresis. *Electrophoresis* **18,** 758–760.
12. Santoni, V., Molloy, M., and Rabilloud, T. (2000) Membrane proteins and proteomics: un amour impossible? *Electrophoresis* **21,** 1054–1070.
13. Görg, A., Postel, W., Friedrich, C., Kuick, R., Strahler, J. R., and Hanash, S. M. (1991) Temperature-dependent spot positional variability in two-dimensional polypeptide gel patterns. *Electrophoresis* **12,** 653–658.
14. Görg, A., Obermaier, C., Boguth, G., et al. (2000) The current state of two-dimensional electrophoresis with immobilized pH gradients. *Electrophoresis* **21,** 1037–1053.
15. Patton, W. F. (2000) A thousand points of light: the application of fluorescence detection technologies to two-dimensional gel electrophoresis and proteomics. *Electrophoresis* **21,** 1123–1144.
16. Shevchenko, A., Wilm, M., Vorm, O., and Mann, M. (1996) Mass spectrometric sequencing of proteins from silver-stained polyacrylamide gels. *Anal. Chem.* **68,** 850–858.
17. Yan, J. X., Wait, R., Berkelman, T., et al. (2000) A modified silver staining protocol for visualization of proteins compatible with matrix-assisted laser desorption/ ionization and electrospray ionization-mass spectrometry. *Electrophoresis* **21,** 3666–3672.

18. Yan, J. X., Harry, R. A., Spibey, C., and Dunn, M. J. (2000) Postelectrophoretic staining of proteins separated by two-dimensional gel electrophoresis using SYPRO dyes. *Electrophoresis* **21,** 3657–3665.
19. Berggren, K. N., Schulenberg, B., Lopez, M. F., et al. (2002) An improved formulation of SYPRO Ruby protein gel stain: comparison with the original formulation and with a ruthenium II tris (bathophenanthroline disulfonate) formulation. *Proteomics* **2,** 486–498.
20. Unlu, M., Morgan, M. E., and Minden, J. S. (1997) Difference gel electrophoresis: a single gel method for detecting changes in protein extracts. *Electrophoresis* **18,** 2071–2077.
21. Lilley, K. S. and Friedman, D. E. (2004) All about DIGE: quantification technology for differential-display 2D-gel proteomics. *Expert Rev. Proteomics* **1,** 401–409.
22. Alban, A., David, S. O., Bjorksten, L., et al. (2003) A novel experimental design for comparative two-dimensional gel analysis: two-dimensional difference gel electrophoresis incorporating a pooled internal standard. *Proteomics* **3,** 36–44.
23. Yan, J. X., Devensih, A. T., Wait, R., Stone, T., Lewis, S., and Fowler, S. (2002) Fluorescence two-dimensional difference gel electrophoresis and mass spectrometry based proteomic analysis of *Escherichia coli*. *Proteomics* **2,** 1682–1698.
24. Shaw, J., Rowlinson, R., Nickson, J., et al. (2003) Evaluation of saturation labelling two-dimensional gel electrophoresis fluorescent dyes. *Proteomics* **3,** 1181–1195.
25. Kono, T., Seike, M., Mori, Y., Fujii, K., Yamada, T., and Hirobashi, S. (2003) Application of sensitive fluorescent dyes in linkage of laser microdissection and two-dimensional gel electrophoresis as a cancer proteomic study tool. *Proteomics* **3,** 1758–1766.
26. Sitek, B., Luttges, J., Marcus, K., et al. (2005) Application of fluorescence difference gel electrophoresis saturation labelling for the analysis of microdissected precursor lesions of pancreatic ductal adenocarcinoma. *Proteomics* **5,** 2665–2679.
27. Steinberg, T. H., Agnew, B. J., Gee, W. Y., et al. (2003) Quantitative analysis of protein phosphorylation status and protein kinase activity on microarrays using a novel fluorescent phosphorylation sensor dye. *Proteomics* **3,** 1244–1255.
28. Hart, C., Schulenberg, B., Steinberg, T. H., Leung, W. Y., and Patton, W. F. (2003) Detection of glycoproteins in polyacrylamide gels and on electroblots using Pro-Q Emerald 488 dye, a fluorescent periodate Schiff-base stain. *Electrophoresis* **24,** 588–598.
29. McGregor, E., Kempster, L., Wait, R., et al. (2001) Identification and mapping of human saphenous vein medial smooth muscle proteins by two-dimensional polyacrylamide gel electrophoresis. *Proteomics* **1,** 1405–1414.

2

Analyzing the Cardiac Muscle Proteome by Liquid Chromatography–Mass Spectrometry-Based Expression Proteomics

Anthony O. Gramolini, Thomas Kislinger,
Peter Liu, David H. MacLennan, and Andrew Emili

Summary

Cardiomyopathies are diseases of the heart resulting in impaired cardiac muscle function, which can lead to heart dilation or overt heart failure. These diseases represent a major cause of global morbidity and death. Innovative preventive and therapeutic measures are urgently needed for early detection, categorization, and treatment of patients at risk of cardiomyopathy. These developments will require a more complete understanding of the molecular effects of impaired cardiac function, even prior to overt disease. The use of gel-free expression proteomics in the detailed analysis of cardiac tissues should yield significant insight into the pathophysiology of these diseases.

Key Words: Cardiac muscle; multidimensional protein identification technology (MudPIT); mass spectrometry.

1. Introduction

Cardiomyopathy is a chronic condition of impaired heart function that arises as a result of genetic predisposition and environmental interactions. Because the prognostic outcomes following diagnosis are poor, earlier detection and diagnosis of cardiomyopathy represent a pressing clinical challenge. Although significant progress has been made in identifying genetic, physiological, and environmental factors that predispose individuals to cardiomyopathy, the etiology of this disease has exhibited an unanticipated level of complexity. Additional research into the molecular basis of clinically common forms of cardiomyopathy is needed urgently to speed development of rational prophylactic and therapeutic strategies.

From: *Methods in Molecular Biology, vol. 357: Cardiovascular Proteomics: Methods and Protocols*
Edited by: F. Vivanco © Humana Press Inc., Totowa, NJ

Heart muscle expresses several thousand distinct proteins *(1)*, several hundred of which are likely tissue-specific and critical for proper heart muscle function, performance, and capacity. Although a number of genes/proteins predisposing to cardiomyopathy have been identified (e.g., dystrophin, ABCA1) based on known or suggested physiological function, identification of the full set of gene products associated with this "complex trait" has proven to be a challenge. A nonbiased, comprehensive description of the proteome, or complement of expressed protein products, in healthy and diseased cardiac tissue could provide breakthrough understanding of the pathogenesis of the disease, leading to advanced diagnostic and therapeutic targets.

The proteome is defined as the entire set of proteins that is expressed (produced) in a cell, tissue, or organ at a given time and physiological state *(2)*. The proteome is a dynamic entity dictated by collective rates of gene transcription and pre- and posttranslational controls that serve collectively to regulate protein abundance, subcellular enrichment and turnover in relation to developmental and physiological cues, environmental constraints, and disease perturbations. Tandem mass spectrometry (MS/MS) the study of the structure of gas phase ions as a means to determine the identity of biomolecules—has emerged as the method of choice for large-scale experimental investigation of the proteome *(2)*. Proteins can be "sequenced" after enzymatic digestion with a site-specific protease, typically trypsin. After selecting and fragmenting peptides, the daughter ion spectra are analyzed, usually with the help of a computer-based database search algorithm, to deduce the amino acid sequence of the peptide and, hence, the identity of the corresponding parental protein *(3)*.

The complexity of the mammalian tissue proteome represents a considerable experimental challenge *(4–10)*. Effective pre-fractionation methods are, therefore, required to increase proteome coverage in order to detect low abundance signaling proteins. Historically, two-dimensional (2D)-gel electrophoresis has provided a useful method for high-resolution separation of complex protein samples, including cardiac samples *(1)*. Nonetheless, this technique is biased against the detection of membrane proteins, low-abundance proteins, and proteins with extremes in isoelectric point (pI) and molecular weight (MW). The identification and quantification of gel-separated proteins is also limited by the need to analyze many individual gel spots. To circumvent these problems, several groups have developed protein profiling strategies based on coupling high-efficiency liquid chromatography (LC)-based separation procedures with automated mass spectrometers, allowing for very large-scale "shotgun" sequencing of complex mixtures *(5–8,10,11)*. The archetypal approach, termed "Mud-PIT" (for multidimensional protein identification technology) *(11)* was pioneered in the laboratory of John Yates, III.

Together with the recent completion of the human and mouse genome sequencing projects *(12–14)*, this proteomic methodology is well-suited to systematic global protein profiling of mammalian tissue and, therefore, offers a powerful means of investigating the effects of disease and therapeutics on mouse tissue. We have begun to apply these methods to examine the biochemical and physiological changes that accompany cardiomyopathy in a comprehensive and unbiased manner. The protocols described in this chapter were adapted and optimized in our laboratory for the analysis of skeletal muscle cell lines (C2C12 cells) *(15)*, microsomal fractions from two different mouse heart muscles (PLN-KO and PLN-I40A *[16]*), and are currently being expanded to include the detailed analysis of multiple organelle fractions of cardiac tissues obtained from mouse models of dilated cardiomyopathy. A schematic overview of our procedures is outlined in **Fig. 1**.

2. Materials

All solid chemicals were from Sigma, whereas HPLC-grade acetonitrile (ACN), methanol, and water were purchased from Fisher Scientific, and heptafluorobutyric acid was obtained from BioLynx (Brockville, Ontario, Canada). Endoproteinase Lys-C was obtained from Roche Diagnostics (Laval, Quebec, Canada).

2.1. Cardiac Muscle Extraction

1. Buffer for cardiac lysis: 250 mM sucrose, 50 mM Tris-HCl, pH 7.4, 5 mM MgCl$_2$, 1 mM dithiothreitol (DTT), and 1 mM phenylmethylsulfonyl fluoride (PMSF). Store all solutions at 4°C and add DTT and PMSF fresh with each use.
2. Solution for sucrose cushion 1: 0.9 M sucrose, 50 mM Tris-HCl, pH 7.4, 5 mM MgCl$_2$, 1 mM DTT, and 1 mM PMSF.
3. Solution for sucrose cushion 2: 2 M sucrose, 50 mM Tris-HCl, pH 7.4, 5 mM MgCl$_2$, 1 mM DTT, and 1 mM PMSF.
4. Nuclear extraction buffer I: 20 mM HEPES, pH 7.8, 1.5 mM MgCl$_2$, 450 mM NaCl, 0.2 mM EDTA, and 25% glycerol.
5. Nuclear extraction buffer II: same as nuclear extraction buffer I, with addition of 1% Triton-X 100.
6. Solution for sucrose cushion: 2 M sucrose, 50 mM Tris-HCl, pH 7.4, 5 mM MgCl$_2$, 1 mM DTT, and 1 mM PMSF.
7. Mitochondrial extraction buffer I: 10 mM HEPES, pH 7.8.
8. Mitochondrial extraction buffer II: same as nuclear extraction buffer I, with addition of 1.5% Triton-X 100.
9. Beckman ultraclear centrifuge tubes (14 × 95 mm; Cat. no. 344060).

2.2. Precipitation and Digestion of Cardiac Samples and Solid-Phase Extraction

1. 150 µg Protein in aqueous or detergent solution.

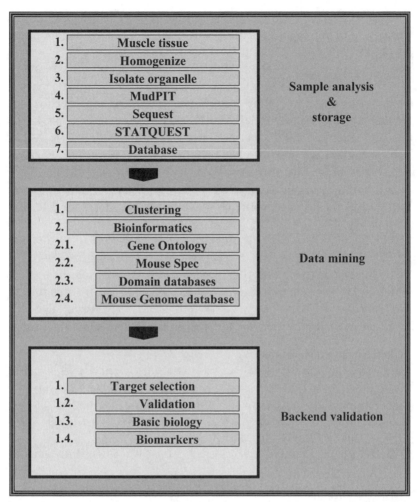

Fig. 1. Overview of an integrated heart proteomic profiling methodology. Cardiac muscle tissue is homogenized and subcellular fractions are isolated by differential ultracentrifugation in sucrose gradients. Protein extracts from each organelle are analyzed extensively by multiple independent MudPIT analyses. Generated tandem mass spectra are searched against a protein sequence database by the use of the SEQUEST and STATQUEST algorithms and subsequently filtered to minimize false-positive identifications. High-confidence protein identifications are parsed into an in-house database, and diverse data clustering and mining strategies are used to find interesting patterns of protein expression for biological validation and detailed analysis.

2. Ice-cold acetone.
3. 8 M Urea, 50 mM Tris-HCl, pH 8.5, 1 mM CaCl$_2$.
4. 50 mM Ammonium bicarbonate.

5. Endoproteinase Lys-C (Roche Diagnostics).
6. Poroszyme trypsin beads (Applied Biosystems, Streetsville, Ontario, Canada).

2.3. MudPIT Analyses

1. SPEC-Plus PT C18 cartridges (Ansys Diagnostics, Lake Forest, CA).
2. 100-µm capillary microcolumn (Polymicro Technologies, Phoenix, AZ).
3. Zorbax Eclipse XDB-C_{18} resin (Agilent Technologies, Mississauga, Ontario, Canada).
4. 5 µm Partisphere strong cation exchange resin (Whatman).
5. Solutions of: Buffer A, 5% ACN, 0.5% acetic acid, and 0.02% heptafluorobutyric acid (HFBA); Buffer B, 100% ACN; Buffer C, 250 mM ammonium acetate in buffer A; and Buffer D, 500 mM ammonium acetate in buffer A.

2.4. Bioinformatics

1. Cluster 3.0 software (java applet available from http://rana.lbl.gov/).
2. Sequest database search software (available from Thermo Finnigan).
3. STATQUEST (developed in-house; *see* **ref. 11**).
4. Swissprot annotation (http://www.expasy.org/sprot/).
5. Gene Ontology (GO) database (http://www.geneontology.org).
6. MouseSpec (http://tap.med.utoronto.ca/~posman/mousespec/).
7. GOminer (http://discover.nci.nih.gov/gominer/).
8. TreeView (http://rana.lbl.gov/downloads/TreeView/).

3. Methods

3.1. Ventricular Fractionation

1. Healthy adult mice are euthanized by administration of CO_2. The heart is removed, rinsed, and dissected to remove the atria. Ventricular tissues are washed three times in ice-cold phosphate-buffered saline (PBS) and minced finely using a razor blade or scissors. Minced samples are subsequently homogenized carefully using a loose-fitting dounce homogenizer with at least 15 strokes on ice, using ice-cold lysis buffer. All subsequent steps are performed at 4°C. The lysate is centrifuged in a benchtop centrifuge at 800g for 15 min; the supernatant serves as source of cytosol, mitochondria, and microsomes. The pellet, which contains the nuclei, is resuspended in 8 mL lysis buffer and layered onto 0.9 M sucrose, 50 mM Tris-HCl, pH 7.4 , 5 mM $MgCl_2$, 1 mM DTT, and 1 mM PMSF, and centrifuged again at 800g for 15 min. The pellet is suspended in 8 mL of 0.9 M sucrose cushion buffer and then carefully applied onto 4 mL of 2 M sucrose cushion buffer in a 13-mL ultracentrifuge tube, and pelleted at 150,000g for 60 min (Beckman SW40.1 rotor). The nuclear pellet is collected, washed once in PBS, suspended in nuclear extraction buffer I, left on ice for 15 min, and centrifuged at 8000g for 20 min. The supernatant is referred to as nuclear extract I. The pellet from this procedure is resuspended in nuclear extrac-

tion buffer II, incubated on ice for 30 min, followed by 8000*g* for 20 min. The resulting supernatant is collected and referred to as nuclear extract II. Following the ultracentrifugation, we also collect the proteins accumulated at the interface of the 250 m*M* sucrose and 0.9 *M* sucrose solutions; these proteins are highly enriched in contractile proteins. Proteins are washed twice in PBS, isolated by centrifugation at 14,000*g* for 10 min, and resuspended in mitochondrial buffer II (*see* **Note 1**).

2. Mitochondria are isolated from the crude cytoplasmic fraction by benchtop centrifugation at 8000*g* for 20 min. The supernatant is collected and used for microsomal fractions (*see* **Subheading 3.1.3.**). The pellet is incubated in 10 m*M* HEPES for 30 min at 4°C followed by brief sonication pulses at maximum setting. Samples are centrifuged at 8000*g* for 20 min and the supernatant collected (mitochondria extract I). The pellet is incubated with mitochondrial extraction buffer II for 30 min at 4°C, centrifuged at 8000*g* for 20 min, and the supernatant collected and referred to as mitochondrial extract II.

3. Finally, the microsomal fractions are isolated from the supernatant following the first 8000*g* spin in **step 2**. Samples are spun at 100,000*g* for 1 h at 4°C (Beckman SW40.1 rotor). The pellet is extracted using mitochondrial extraction buffer II, left on ice for 30 min, and centrifuged at 8000*g* for 20 min at 4°C. The supernatant is saved as the "cytosolic" fraction.

4. A schematic overview of the fractionation methodology is shown in **Fig. 2**. In addition, we perform conventional biochemical techniques, including immunoblots and enzymatic assays, to examine our fractionation methods. Results of these experiments are shown in **Fig. 3**. In **Fig. 3C**, note that we observe substantial contamination of mitochondrial ATP synthase when fractionations are performed using a polytron, compared with a dounce homogenizer, and that the addition of the sucrose cushions further lowers the amount of mitochondrial contamination in other fractions.

3.2. Digestion of Cell Extracts for MudPIT Analysis

1. One hundred and fifty micrograms of total protein from each fraction are precipitated overnight with 5 vol of ice-cold acetone followed by centrifugation at 21,000*g* for 20 min.

2. The protein pellet is solubilized in 8 *M* urea, 50 m*M* Tris-HCl, pH 8.5, at 37°C for 2 h and reduced by the addition of 1 m*M* DTT for 1 h at room temperature followed by carboxyamidomethylation with 5 m*M* iodoacetamide for 1 h at 37°C.

3. The samples are then diluted to 4 *M* urea with 100 m*M* ammonium bicarbonate, pH 8.5, and digested with a 1:150 molar ratio of endoproteinase Lys-C at 37°C overnight.

4. The following day, mixtures are further diluted to 2 *M* urea with 50 m*M* ammonium bicarbonate, pH 8.5, and a final concentration of 1 m*M* CaCl$_2$, and rotated overnight with Poroszyme trypsin beads at 30°C.

5. The resulting peptide mixtures are solid phase-extracted with SPEC-Plus PT C18 cartridges according to the manufacturer's instructions and stored at −80°C until further use.

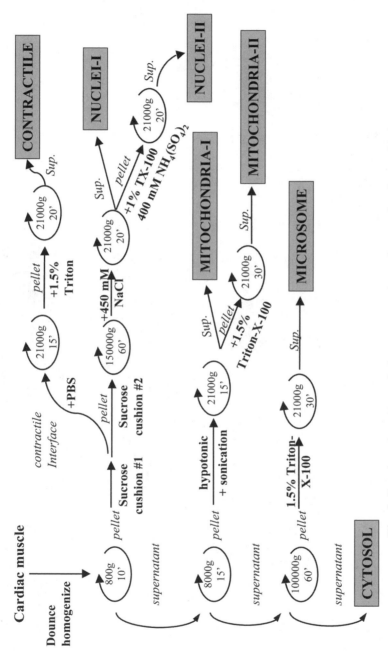

Fig. 2. Fractionation protocol. A summary schematic overview is provided.

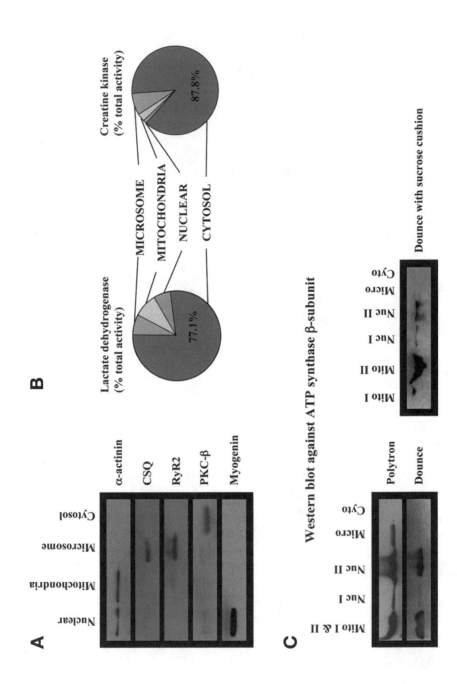

A

α-actinin
CSQ
RyR2
PKC-β
Myogenin

Nuclear
Mitochondria
Microsome
Cytosol

B

Lactate dehydrogenase
(% total activity)

77.1%

Creatine kinase
(% total activity)

87.8%

MICROSOME
MITOCHONDRIA
NUCLEAR
CYTOSOL

C

Western blot against ATP synthase β-subunit

Polytron
Dounce

Cyto
Micro
Nuc II
Nuc I
Mito I & II

Dounce with sucrose cushion

Cyto
Micro
Nuc II
Nuc I
Mito II
Mito I

22

3.3. MudPIT Analysis

1. A fully automated 12-step, 20-h MudPIT chromatographic procedure is used as described previously *(11)*. A MudPIT consists of 12 independent chromatographic steps each containing a salt bump at the beginning, which aims at moving a subset of peptides from the first dimension of the chromatography column (strong ion exchange) onto the second dimension (reverse phase) of the chromatography column. Here the peptides are separated by a conventional water/ACN gradient and directly sprayed into the mass spectrometer. The 12 steps differ by increasing concentrations in the initial salt bump used to move them onto the reverse-phase material.
2. An HPLC quaternary pump is interfaced with an LCQ DECA XP ion trap tandem mass spectrometer (ThermoFinnigan, San Jose, CA).
3. A 100-µm inner diameter-fused silica capillary microcolumn is pulled to a fine tip using a P-2000 laser puller (Sutter Instruments, Novato, CA) and packed first with 10 cm of 5-µm Zorbax Eclipse XDB-C18 resin and then with 6 cm of 5-µm Partisphere strong cation exchange resin.
4. Peptide samples are loaded manually onto a fresh column using a pressure vessel. The four buffer solutions used for the chromatography are described in **Subheading 2.3., item 5**. The first 80 min step consists of a gradient from 0 to 80% buffer B for 70 min and a hold at 80% buffer B for 10 min. The next 11 steps are 110 min each with the following profile: 5 min of 100% buffer A, 2 min of x% buffer C/D, 3 min of 100% buffer A, a 10-min gradient from 0 to 10% buffer B, and a 90-min gradient from 10 to 45% buffer B. The 2-min buffer C percentages (x) in **steps 2–12** are as follows: 10% C, 20% C, 30% C, 40% C, 50% C, 60% C, 70% C, 80 %C, 90% C, 100% C, and 100% D.

3.4. Sequest, STATQUEST, and Database Management Systems

Uninterpreted fragmentation (daughter) product ion mass spectra are sequence-mapped against a minimally redundant set of human and mouse protein sequences obtained from the SWISS-PROT and TrEMBL databases using the SEQUEST software algorithm *(3)* running on a multiprocessor computer cluster. Sequest search results are further validated using an in-house generated, probability-based computer program, termed STATQUEST *(5)*. This program automatically assigns a p-value threshold cut-off corresponding to a defined percentage likelihood of corrected peptide identification. In general, a greater than 95% likelihood of

---

Fig. 3. *(Opposite page)* Analysis of fraction purity. **(A)** Western blotting of cardiac ventricular subcellular fractions against selected marker proteins. **(B)** Normalized enzyme activity of creatine kinase and lactate dehydrogenase in subcellular fractions. **(C)** Western blotting against the mitochondrial membrane protein F_1-ATPase β-subunit in fractions isolated under different conditions: (1) use of a polytron homogenize to homogenize cardiac muscle; (2) use of a dounce homogenizer to homogenize the tissues; or (3) use of a dounce homogenizer together with an additional sucrose cushion.

correct identification is used to minimize false-positive identifications and the resulting data are parsed into an in-house SQL type database management system. The use of database management systems is highly advised for large-scale proteomics datasets, because it allows for streamlined, user-defined data queries, which greatly enhance the analysis of genome-wide proteomics projects. Furthermore, the database allows for the storage of a multitude of parameters, such as number of uniquely identified peptides, number of recorded spectra, isoelectric point (p*I*), and molecular weight of each of the identified proteins, which can be of tremendous value for further in-depth data interpretation (*see* **Note 2**).

3.5. Bioinformatics

Computational analysis of large-scale proteomics projects has become the major bottleneck of genome-wide data analysis pipelines. Therefore, a multitude of bioinformatics tools have been developed to speed up the tedious process of data processing and interpretation. In the following section, several essential tools for the analysis of expression proteomics profiles that are in current use in our laboratory will be discussed.

3.5.1. Data Clustering

The sheer size of proteomic datasets make the discovery of meaningful candidates and patterns a very difficult task. Therefore, clustering is often the starting point for grouping a set of expressed proteins based on similarities in their expression patterns. In most cases cluster analysis allows for the discovery of hidden information and regulatory patterns that have yet to be discovered. Several powerful commercial and publicly available software packages are available that are capable of performing most users tasks. A popular publicly available tool is Cluster 3.0, which is based on the original cluster tool developed by the Eisen group *(17)*. First, expression datasets are converted to tab delimited text file containing their experimental conditions (columns), identified proteins (rows) and in the case of proteomics data, some quantitative value of protein expression levels. This file is then opened with the Cluster 3.0 software tool. Next, one or more of the several mathematical models (termed similarity or distance metrics) can be applied to calculate the degree of correlation between the profiles, allowing for subsequent clustering to be performed. A second software tool, termed TreeView, allows users to display the output of the Cluster 3.0 software graphically. Individual clusters can be selected and the protein names/IDs extracted for follow-up analysis. **Figure 4** shows a typical clustergram generated for proteins detected in a series of repeat heart subcellular fractionation profiles. As a semi-quantitative estimate of protein abundance, our laboratory typically uses spectral counts (as described in **Subheading 3.6.**). To optimize

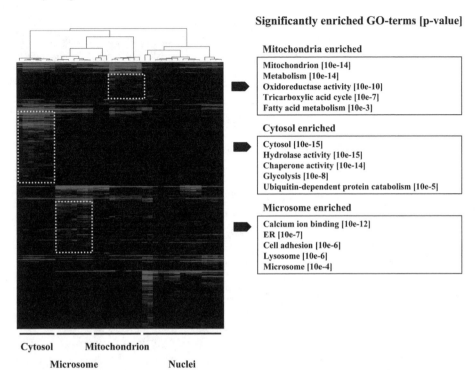

Significantly enriched GO-terms [p-value]

Mitochondria enriched

Mitochondrion [10e-14]
Metabolism [10e-14]
Oxidoreductase activity [10e-10]
Tricarboxylic acid cycle [10e-7]
Fatty acid metabolism [10e-3]

Cytosol enriched

Cytosol [10e-15]
Hydrolase activity [10e-15]
Chaperone activity [10e-14]
Glycolysis [10e-8]
Ubiquitin-dependent protein catabolism [10e-5]

Microsome enriched

Calcium ion binding [10e-12]
ER [10e-7]
Cell adhesion [10e-6]
Lysosome [10e-6]
Microsome [10e-4]

Cytosol   Mitochondrion

Microsome   Nuclei

Fig. 4. Data visualization. Four individual organelle fractions (cytosol, microsomes, mitochondria, and nuclei) from wild-type heart tissue were analyzed independently by multiple MudPIT analyses. The entire set of proteins identified was clustered using spectral counts as a quantitative estimate of relative protein abundance in each fraction. The profiles are displayed using a "heat map" format. Organelle-specific clusters displaying statistically significantly enriched membership for select functional annotation categories are highlighted by dashed boxes, together with the corresponding GO terms.

the performance of the cluster tool, blanks (i.e., where a protein was not detected in a given fraction) are usually replaced by a non-zero low value (e.g., 0.01) *(18)*.

3.5.2. Protein Annotation

1. Protein annotation refers to the known or predicted biological/molecular properties of a protein as extracted from the literature. A large number of highly useful Web pages are available to extract important protein annotations. In particular, the ExPASy molecular biology server (http://ca.expasy.org/) is a highly useful Web tool. This Web page allows biologists to extract extensive information concerning a specific protein (e.g., function, literature links, subcellular localization, etc.). Furthermore, ExPASy serves as a Web portal connecting to diverse, additional

knowledge databases, such as Mouse Genome Database (http://www.informatics. jax.org/), domain databases such as (InterPro or Pfam), and Gene Ontology. In addition, the Web portal offers a large number of useful proteomic tools and their corresponding Web pages, which should allow every user to extract a large amount of information on almost every protein.

2. The Gene Ontology Database (GO; www.geneontology.org) represents another extremely useful database. The GO consortium is aimed at providing the molecular function, biological process, and cellular component of proteins for every major organism in a defined user-friendly vocabulary *(19)*. The GO database consists of three main branches, termed biological process, molecular function, and cellular component. Each main category branches consecutively into a more detailed and complex network of GO terms describing specific functions or properties of particular proteins. By mapping proteins identified in an expression-based proteomics project onto the Gene Ontology database, biologists can accumulate important information about the proteins identified in a particular sample quickly. One of the drawbacks of the GO database is that not every identified protein can be mapped to a defined GO term. Only about 75% of the proteins identified throughout a proteomics project can be linked to one or more GO terms in the GO database.

3. Several tools are available for the biological community to map proteins to GO terms. We developed a Perl-based program termed GOClust. Tab delimited text files of proteins containing a SwissProt/TrEMBL accession number are used in this case. The final output is a series of tables of grouped proteins that share a common annotation to one or more pre-selected GO terms. Alternatively, the Java base program GoMiner (http://discover.nci.nih.gov/gominer/) can be used. Users input a list of SwissProt/TrEMBL accession numbers and the program returns a list of matching GO terms in an attractive pull-down format. An interesting feature of the GoMiner software tool is that two lists—for example, the total list of proteins identified and a subcluster of proteins found to be upregulated in a disease state—can be compared with each other. The program compares the GO terms matched to both lists and provides the user with significantly enriched GO terms in the subcluster, as compared to the total input list. This feature can be extremely useful for finding biological processes or molecular functions of proteins responsible for the development of a disease phenotype, if no hypothesis is readily available. In collaboration with Dr. Tim Hughes (University of Toronto), we have developed a similar software tool, termed MouseSpec. The program accepts protein accession numbers from either the SwissProt/TrEMBL or the IPI databases as an input. The output is a list of statistically enriched GO terms (together with their p value) as compared to a locally stored GO database. A Bonferroni correction factor can be used to correct for multi-hypothesis testing; the p value threshold deemed significant for an individual test is divided by the number of tests conducted, thereby accounting for spurious significance owing to multiple testing over all the categories in the GO database. A cut-off value of 10^{-3} is used as a final selection criterion to highlight promising, biologically interesting clusters *(20)*.

3.6. Quantitative Nature and Post-Analyses

1. Confidently identifying as many proteins as possible in a particular sample is an important task. Nevertheless, if comparison of samples, for example, wild-type vs disease, is desired, an estimation of relative protein abundance is also essential. Determination of relative protein abundance on a truly global scale is a very challenging task and several approaches have been published in past few years. The most commonly used approach for quantitative proteomics is the use of isotope labeling, particularly, the isotope coded affinity tag (ICAT) pioneered in the laboratory of Dr. Ruedi Aebersold at the Institute for Systems Biology *(21)*. In this method, two samples are labeled independently with either a light or a heavy isotope containing reagent. Samples are combined, digested enzymatically, and analyzed by LC-MS. Because isotopes possess chemically and physically identical properties, the two differently labeled peptides (light and heavy) will co-elute from the chromatographic column. Nevertheless, because they differ by a defined mass unit, the two peptides will be separated by the mass spectrometer. Integration of the area underneath both peaks permits relative quantification of the proteins identified. The ICAT methodology has been applied successfully to several biologically oriented projects *(22–24)*. However, isotope-labeling reagents, such as the ICAT reagent, are expensive, limiting their application to large-scale projects. Moreover, the quantitative integration of every co-eluting peak requires extensive computation and might only result in the accurate quantification of the higher abundant proteins *(21)*.

2. An alternative, label-free methodology based on the cumulative number of recorded spectra mapping to an identified protein has been published by the Yates laboratory *(25)*. Although not as accurate as isotope labeling, this technology has bypassed the need for expensive isotope labels and time-consuming back-end analysis tools. From our experience, this methodology generally provides a good first indication of changes in relative protein abundance, especially if a larger number of spectra are recorded for a specific protein and if the changes between two conditions (healthy and disease) are considerable. As a result, this is the method of choice for our cardiac-tissue-profiling projects.

3.7. Verification

As with all large-scale projects, there is the inherent risk of false-positive data being included in the data sets. To minimize this risk, we set our detection stringency limits quite high (i.e., greater than 95% confidence interval for protein determinations). However, in critical experiments, we employ conventional methods of validation of the bioinformatics strategies. This includes, where available, Western blot analyses, assays of enzymatic activities where appropriate, and RT-PCR to measure mRNA levels in cases where antibodies or other assays are not available.

3.8. Conclusion

Advanced high-throughput tandem MS-based shotgun protein profiling techniques and allied computational approaches can be applied to the examination of the effects of cardiomyopathy on global patterns of protein expression and accumulation in heart tissue in a set of well-defined mouse models of cardiac disease. This will provide a unique opportunity for a more complete understanding of the molecular logic that governs cardiac muscle physiology and will provide insights into the biochemical and physiological basis for the development of heart disease. If we understand these progressive processes, we will be able to design therapeutic interventions that will block progression to cardiac disease. We also envision that the information will provide us with the potential for identifying biomarkers of heart disease and even of specific forms of heart disease. If the results are duplicated in large-scale human cohorts, they permit the development of effective clinical methodologies for early intervention and the prevention of progression to heart failure in human patients.

4. Notes

1. The major difficulty in applying high-resolution, global protein analysis to muscle tissues is the presence of high concentrations of sarcomeric, mitochondrial, and cytoskeletal proteins that are not present in other tissue types. We have addressed this issue by including a gentle dounce homogenization, which minimizes the rupture of mitochondria during our fractionation, thereby diminishing the contamination of other fractions with mitochondrial proteins. In addition, to remove a substantial amount of contractile and cytoskeletal proteins, we include two sucrose cushions in our fractionation protocol. These cushions allow for a cleaner nuclei preparation with little contractile protein contamination; unfortunately, the total yield of nuclear protein is very low (*see* **Fig. 3**).

2. The extreme complexity of mammalian cardiac tissue is problematic even for high-resolution separation methodologies such as the MudPIT technique. Heart tissue contains an overwhelming number of expressed proteins (in the range of at least several thousand). Enzymatic digestion further increases this complexity to tens of thousands of peptides. Identification of proteins in MudPIT-based studies is based on the isolation and fragmentation of individual peptides eluting from the HPLC column into the mass spectrometer. However, not all peptide ions can be successfully analyzed by the mass spectrometer owing to limitations in the speed of scanning or overall duty-cycle *(25)*. This confounding issue is complicated further by the wide range in protein concentrations typically found in mammalian tissue. As a result, peptides from lower abundance proteins often elute from the chromatographic columns without ever being detected by the mass spectrometer, because they can be masked by a few very high-abundance peptide peaks generated by higher-abundance proteins (*see* **Fig. 5A**).

   Several partial solutions have been devised to surmount this problem. For instance, simplification of the protein mixture, e.g., by subcellular pre-fractionation, provides

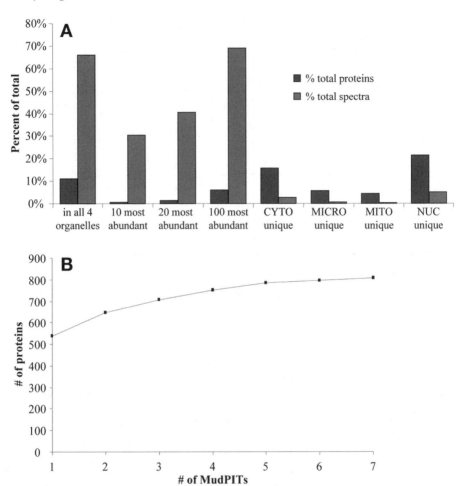

Fig. 5. Analysis of MudPIT protein detection. (**A**) Protein abundance and dynamic range. Throughout the analysis of four organelle fractions in wild-type heart, a total of 1652 proteins with a total of 79,197 spectra were detected. Displayed are the fraction (%) of proteins and spectra for specific subsets of the total identifications. These results show that although high-abundance proteins make up only a small fraction of the proteins identified, they are nevertheless preferentially detected with a large number of corresponding spectra, potentially masking proteins of lower abundance. (**B**) Random sampling and detection saturation. The total number of high-confidence proteins detected in the heart cytosol after a certain number of individual MudPIT analyses is presented. Repeated analysis results in an apparent saturation in the total cumulative number of proteins that can be identified.

a simple, yet powerful, approach for improving the odds of detection of lower-abundance proteins. Nevertheless, organellar extracts still contain a highly complex mixture of proteins, resulting in the missed identification of many expressed proteins. Organelle extracts can be further simplified using conventional fractionation methodologies aimed at protein level fractionation, such as ion-exchange or size-exclusion chromatography. However, because most proteomic profiling projects are aimed at the comparison of particular samples (e.g., healthy vs disease state), a simple comparison can result in misleading interpretations. Therefore, repeat analyses of the same sample (i.e., running multiple LC-MS analyses), can improve the overall detection coverage (total number of protein identifications) and markedly increase detection of lower-abundance proteins (*see* **Fig. 5B**).

Acknowledgments

Work in our laboratory is supported by grants from the Natural Science and Engineering Research Council of Canada (to AE), the Ontario Genomics Institute and Genome Canada (to AE and DHM), by Heart and Stroke Foundation of Ontario Grant T-5042 and CIHR Grants MT-12545 and MOP-49493 and the Neuromuscular Research Partnership Program (to DHM), and by the Muscular Dystrophy Association (to AOG). AOG is supported by a fellowship from the Heart and Stroke Foundation of Canada; TK was supported by a fellowship from the Josef Schormuller Gedachtnisstiftung.

References

1. Dos Remedios, C. G., Liew, C. C., Allen, P. D., et al. (2003) Genomics, proteomics and bioinformatics of human heart failure. *J. Muscle Res. Cell Motil.* **24,** 251–260.
2. Aebersold, R. and Mann, M. (2003) Mass spectrometry-based proteomics. *Nature* **422,** 198–207.
3. Eng, J. K., McCormack, A. L., and Yates, J. R., 3rd. (1994) An approach to correlate tandem mass spectral data of peptides with amino acid sequences in a protein database. *J. Am. Soc. Mass Spectrom.* **5,** 976–989.
4. Cagney, G. and Emili, A. (2002) De novo peptide sequencing and quantitative profiling of complex protein mixtures using mass-coded abundance tagging. *Nat. Biotechnol.* **20,** 163–170.
5. Kislinger, T. and Emili, A. (2003) Going global: protein expression profiling using shotgun mass spectrometry. *Curr. Opin. Mol. Ther.* **5,** 285–293.
6. Andersen, J. S., Lam, Y. W., Leung, A. K., et al. (2005) Nucleolar proteome dynamics. *Nature* **433,** 77–83.
7. Andersen, J. S., Wilkinson, C. J., Mayor, T., et al. (2003) Proteomic characterization of the human centrosome by protein correlation profiling. *Nature* **426,** 570–574.
8. Mootha, V. K., Bunkenborg, J., Olsen, J. V., et al. (2003) Integrated analysis of protein composition, tissue diversity, and gene regulation in mouse mitochondria. *Cell* **115,** 629–640.
9. Durr, E., Yu, J., Krasinska, K. M., et al. (2004) Direct proteomic mapping of the lung microvascular endothelial cell surface in vivo and in cell culture. *Nat. Biotechnol.* **22,** 985–992.

10. Schirmer, E. C., Florens, L., Guan, T., et al. (2003) Nuclear membrane proteins with potential disease links found by subtractive proteomics. *Science* **301,** 1380–1382.

11. Washburn, M. P., Wolters, D., and Yates, J. R., 3rd. (2001) Large-scale analysis of the yeast proteome by multidimensional protein identification technology. *Nat. Biotechnol.* **19,** 242–247.

12. Waterston, R. H., Lindblad-Toh, K., Birney, E., et al. (2002) Initial sequencing and comparative analysis of the mouse genome. *Nature* **420,** 520–562.

13. Venter, J. C., Adams, M. D., Myers, E. W., et al. (2001) The sequence of the human genome. *Science* **291,** 1304–1351.

14. Lander, E. S., Linton, L. M., Birren, B., et al. (2001) Initial sequencing and analysis of the human genome. *Nature* **409,** 860–921.

15. Kislinger, T., Gramolini, A. O., Pan, Y., et al. (2005) Proteome dynamics during C2C12 myoblast differentiation. *Mol. Cell Proteomics* **4,** 887–901.

16. Pan, Y., Kislinger, T., Gramolini, A. O., et al. (2004) Identification of biochemical adaptations in hyper- or hypocontractile hearts from phospholamban mutant mice by expression proteomics. *Proc. Natl. Acad. Sci. USA* **101,** 2241–2246.

17. Eisen, M. B., Spellman, P. T., Brown, P. O., et al. (1998) Cluster analysis and display of genome-wide expression patterns. *Proc. Natl. Acad. Sci. USA* **95,** 14863–14868.

18. Cox, B., Kislinger, T., and Emili, A. (2005) Integrating gene and protein expression data: pattern analysis and profile mining. *Methods* **35,** 303–314.

19. Ashburner, M., Ball, C. A., Blake, J. A., et al. (2000) Gene ontology: tool for the unification of biology. The Gene Ontology Consortium. *Nat. Genet.* **25,** 25–29.

20. Robinson, M. D., Grigull, J., Mohammad, N., et al. (2002) FunSpec: a web-based cluster interpreter for yeast. *BMC Bioinformatics* **3,** 35.

21. Gygi, S. P., Rist, B., Gerber, S. A., et al. (1999) Quantitative analysis of complex protein mixtures using isotope-coded affinity tags. *Nat. Biotechnol.* **17,** 994–999.

22. Brand, M., Ranish, J. A., Kummer, N. T., et al. (2004) Dynamic changes in transcription factor complexes during erythroid differentiation revealed by quantitative proteomics. *Nat. Struct. Mol. Biol.* **11,** 73–80.

23. Han, D. K., Eng, J., Zhou, H., et al. (2001) Quantitative profiling of differentiation-induced microsomal proteins using isotope-coded affinity tags and mass spectrometry. *Nat. Biotechnol.* **19,** 946–951.

24. Shiio, Y., Donohoe, S., Yi, E. C., et al. (2002) Quantitative proteomic analysis of Myc oncoprotein function. *EMBO J.* **21,** 5088–5096.

25. Liu, H., Sadygov, R. G., and Yates, J. R., 3rd. (2004) A model for random sampling and estimation of relative protein abundance in shotgun proteomics. *Anal. Chem.* **76,** 4193–4201.

3

Two-Dimensional Differential Gel Electrophoresis of Rat Heart Proteins in Ischemia and Ischemia-Reperfusion

Jun Sakai, Hironori Ishikawa, Hideshi Satoh, Setsuko Yamamoto, Shinichi Kojima, and Masaharu Kanaoka

Summary

Ischemia-reperfusion injury occurs in acute myocardial infarction, cardiopulmonary bypass surgery, and heart transplantation. However the precise mechanisms still remain unclear. In order to identify proteins that are involved in ischemia-reperfusion injury, we compared precipitated 100,000*g* fractions of normal, ischemic, and ischemic-reperfused rat hearts using two-dimensional (2D) difference gel electrophoresis (2D-DIGE). 2D-DIGE is reliable method to define quantitative protein differences, especially when subtle protein changes are under investigation. In this study, six spots that changed more than twofold and two additional spots related to these spots were detected. Five of the spots were identified by matrix-assisted laser desorption/ionization time of flight mass spectrometry as protein disulfide isomerase, one as 60 kDa heat-shock protein, and two as elongation factor Tu.

Key Words: Ischemia; ischemia-reperfusion; 2D-DIGE; PDA3; HSP60; EF-Tu.

1. Introduction

Ischemia-reperfusion injury occurs in patients undergoing revascularization by percutaneous transluminal coronary angioplasty or by thrombolysis after acute myocardial infarction, in patients undergoing cardiac surgery, and in patients with angina pectoris. Understanding the cellular mechanism of ischemia-reperfusion injury is very important for heart attack and stroke therapy. Although there have already been several reports describing the pathogenesis of ischemia-reperfusion injury (1–3), the precise mechanism still remains unclear. In this experiment, we tried to identify disease-relevant proteins and elucidate the

From: *Methods in Molecular Biology, vol. 357: Cardiovascular Proteomics: Methods and Protocols*
Edited by: F. Vivanco © Humana Press Inc., Totowa, NJ

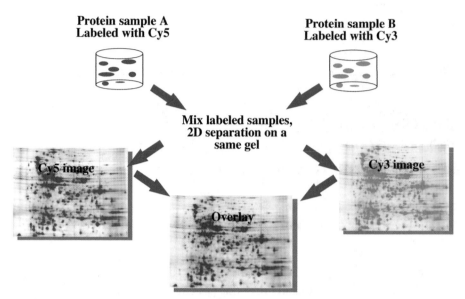

Fig. 1. An outline of two-dimensional differential gel electrophoresis analysis. Two protein mixtures derived from two different states are labeled with either Cy3 or Cy5 dye. Then both samples are combined and subjected to 2D SDS-PAGE. Each pair of Cy3- and Cy5-labeled proteins will migrate to the same position on a 2D gel. The relative quantitation is performed by using the Amersham 2920-2D Master Imager.

pathophysiological mechanisms of ischemia-reperfusion injury by using two-dimensional differential gel electrophoresis (2D-DIGE).

Proteomic analysis by two-dimensional electrophoresis (2DE) is an approach often used to detect differences in the protein expression levels among samples. However, because of variations between gels, no two gel images can be superimposed directly, and warping is required to overlay and compare them. 2D-DIGE, which was first reported by Unlu et al. in 1997 *(4)*, succeeded in overcoming these limitations by enabling different samples to be run on the same gel (**Fig. 1**). In 2D-DIGE, each sample is covalently labeled with one of a mass- and charge-matched set of fluorophores (Cy2, Cy3, and Cy5) before mixing the samples together and analyzing them *(5)*. There is a minimal dye, which reacts with the epsilon amino group of lysine side chains, and a saturation dye, which reacts with the thiol group of cysteine side chains. In the minimal dye method, less than 5% of proteins are labeled, and the labeled proteins can shift away from the unlabeled protein during electrophoresis, particularly in the case of small-molecular-weight proteins. For this reason, the gels are often poststained with a dye such as SYPRO Ruby, for example, for protein identification. In the saturation dye method, all cysteine residues within a protein are labeled. Therefore this method is much more sensitive than the minimal dye method and does not require poststaining *(6)*.

The analysis based on 2DE has another limitation, in that only proteins that are expressed at fairly high levels can be detected in the cell or sample. One approach to finding less abundant proteins is to isolate subcellular fractions (*see* Chapter 7), in which the sample is enriched by extracting a certain fraction of the cell or sample. In this study, we used precipitated 100,000g fractions containing mitochondrial proteins that may be profoundly affected by ischemia and reperfusion.

2. Materials

2.1. Perfused Rat Heart Model

1. Animals: Male Wistar/ST rats: 8–10 wk of age (Nippon SLC, Shizuoka, Japan).
2. Krebs-Henseleit buffer: 119 mM NaCl, 24.9 mM NaHCO$_3$, 4.7 mM KCl, 1.2 mM KH$_2$PO$_4$, 1.3 mM CaCl$_2$, 1.2 mM MgSO$_4$, 11 mM glucose (*see* **Note 1**).
3. A gas mixture of 95% O$_2$ and 5% CO$_2$.
4. Liquid nitrogen.

2.2. Subcelluar Fractionation

1. Protease inhibitor cocktail tablet (Roche, Mannheim, Germany).
2. Homogenization buffer: 10 mM Tris-HCl, pH 7.4, 0.32 M sucrose, 1 mM EGTA, 2 mM ethylene diamine triacetic acid, protease inhibitor cocktail tablet, 50 mM NaF, 1 mM Na$_3$VO$_4$ (*see* **Note 2**), 0.4 nM microcystin LR (*see* **Note 2**). Stable at −20°C for up to 3 mo.
3. Lysis buffer: 10 mM Tris-HCl, pH 8.0, 7 M urea, 2 M thiourea, 4% CHAPS, protease inhibitor cocktail tablet, 1 mM NaF, 1 mM Na$_3$VO$_4$, 1 mM sodium pyrophosphate. Stable at −20°C for up to 3 mo (*see* **Note 3**).
4. Protein assay kit (Bio-Rad Laboratories, Hercules, CA).

2.3. Protein Cyanine Dye Labeling

1. 1-(5-carboxypentyl)-1'-propylindocarbocyanine halide *N*-hydroxysuccinimidyl ester (Cy3) (Amersham Biosciences, Piscataway, NJ) stock solution: 1 mM Cy3 in *N,N*-dimethylformamide (DMF) (Amersham) (*see* **Notes 4** and **5**). Stored at −20°C for up to 2 mo or until the expiration date on the container, whichever is sooner.
2. 1-(5-carboxypentyl)-1'-methylindodicarbocyanine halide *N*-hydroxysuccinimidyl ester (Cy5) (Amersham) stock solution: 1 mM Cy5 in DMF (Amersham) (*see* **Notes 4** and **5**). Stored at −20°C for up to 2 mo or until the expiration date on the container, whichever is sooner.
3. pH indicator strip (Merck, Darmstadt, Germany).
4. Labeling buffer: 8 M urea, 4% CHAPS, 50 mM Tris-HCl, pH 8.0.
5. Stop solution: 10 mM lysine (Nacalai tesque, Kyoto, Japan). Stored at −20°C.

2.4. Two-Dimensional Electrophoresis

1. IPG buffer pH 3.0–10.0 NL (Amersham).
2. 2D sample buffer (2X): 8 *M* urea, 4% CHAPS, 130 m*M* dithiothreitol (DTT), 2% immobilized pH gradient (IPG) buffer, pH 3.0–10.0 NL. Stored at −20°C.
3. Rehydration buffer: 8 *M* urea, 4% CHAPS, 13 m*M* DTT, 1% IPG buffer, pH 3.0–10.0 NL. Stored at −20°C.
4. Immobiline DryStrip (Amersham): 18 cm long, pH 4.0–7.0.
5. Dimethyl silicone oil (Shin-Etsu Chemical, Tokyo, Japan).
6. Bind-Silane working solution: 8 mL ethanol, 200 µL glacial acetic acid, 10 µL bind-silane (Amersham), 1.8 mL water.
7. Monomer stock solution: 30% acrylamide, 0.8% *N,N'*-methylenebisacrylamide. Store at 4°C in the dark.
8. 4X resolving gel buffer: 1.5 *M* Tris-HCl, pH 8.8. Store at 4°C.
9. Water-saturated isobutanol: shake equal volumes of water and isobutanol in a glass bottle and allow to separate. Use the top layer. Store at room temperature.
10. Equilibration buffer: 100 m*M* Tris-HCl, pH 6.8, 8 *M* urea, 30% glycerol, 1% sodium dodecyl sulfate (SDS), 32.5 m*M* DTT.
11. Agarose overlay solution: 0.5% agarose containing a small amount of bromophenol blue.
12. Running buffer: 25 m*M* Tris-HCl, 192 m*M* glysine, and 0.1% SDS, pH 8.3.

2.5. In-Gel Digestion

1. SyproRuby solution (Molecular Probes, Eugene, OR).
2. Wash solution 1: 50% acetonitrile (ACN), 50 m*M* ammonium bicarbonate.
3. Reduction solution: 10 m*M* DTT, 50 m*M* ammonium bicarbonate.
4. Alkylation solution: 50 m*M* iodoacetamide, 50 m*M* ammonium bicarbonate. The solution must be made in a shaded tube because of an iodoacetamide is light-sensitive.
5. Wash solution 2: 100% ACN.
6. Wash solution 3: 50 m*M* ammonium bicarbonate.
7. Digestion solution: 12.5 ng/µL trypsin (Promega, Madison, WI; *see* **Note 6**), 25 m*M* ammonium bicarbonate. The pH should be between 7.5 and 8.5.
8. Extraction solution 1: 25 m*M* ammonium bicarbonate.
9. Extraction solution 2: 100% ACN.

2.6. MALDI-TOF Analysis

1. Wetting solution: 0.1% TFA, 50% ACN.
2. Equilibration solution: 0.1% TFA/water.
3. Wash solution: 0.1% TFA/water.
4. Elution solution: a saturated solution of α-cyano-4-hydroxycinnamic acid (αCHCA) (Bruker-Daltonics, Bremen, Germany) in 0.1% TFA, 50% ACN.
5. ZipTipC$_{18}$ (Millipore, Bedford, MA).

Table 1
2D DIGE Experimental Design of Rat Heart
in Ischemia and Ischemia-Reperfusion

Gel	Cy3	Cy5
1	50 µg of control sample	50 µg of ischemia sample
2	50 µg of ischemia sample	50 µg of ischemia-reperfusion sample
3	50 µg of ischemia-reperfusion sample	50 µg of control sample

3. Methods

In this experiment, we applied the minimal dye method using Cy3 and Cy5 because when we started this study, the only minimal dyes commercially available were Cy3 and Cy5. **Table 1** shows our experimental design. However, the labeling methods described here could be used just as well with three dyes instead of two. For further options and experimental details of three dyes, we refer you to the Amersham Biosciences Ettan DIGE user manual (http://www4. amershambiosciences.com).

3.1. Perfused Rat Heart Model

1. Rats are killed by cervical dislocation, the thorax is opened, and the heart is quickly removed and placed in Krebs-Henseleit buffer *(7)*.
2. The heart is then cannulated via the aorta and perfused with Krebs-Henseleit buffer, continuously bubbled with a 95% O_2 and 5% CO_2 gas mixture, at a constant flow rate of 10 mL/min using the Langendorff perfusion technique (*see* **Note 7**). The entire system is maintained at 37°C.
3. For the normal control group, the hearts are perfused for 30 min as described above. For the ischemia group, after a 30-min stabilization period, the hearts are subjected to 40 min of ischemia by shutting off the perfusion flow. The hearts are subjected to 40 min of ischemia and 20 min of reperfusion for the ischemia-reperfusion group.

   Left and right ventricles are removed and rapidly freeze-clamped with liquid nitrogen-cooled Wollenberger tongs, then pulverized to a fine powder by using a percussion mortar, which has also been cooled with liquid nitrogen.

3.2. Subcellular Fractionation

1. Aliquots of frozen tissue powders (0.3–0.5 g) are homogenized on ice in 5 v of homogenization buffer with a Physcotron homogenizer (Microtec, Chiba, Japan) at 50% of maximum speed three times for 30 s each.
2. Homogenize again with a Teflon homogenizer (Ikemoto Scientific, Tokyo, Japan) five strokes at 1000*g*. Homogenates are mixed with an equal volume of homogenization buffer (final 10% homogenate) and centrifuged at 1000*g* for 10 min to precipitate nucleus, myofibrils, and unbroken cells.

3. The 1000*g* supernatants are carefully transferred to another tubes without disturbing the pellets and further ultracentrifuged at 100,000*g* for 60 min to separate cytosolic (supernatant) and particulate (precipitate) fractions.
4. The 100,000*g* supernatants (cytosolic fraction) are carefully removed without disturbing the pellets. The 100,000*g* precipitates (particulate fraction) are dissolved in 0.3 mL of lysis buffer.
5. The protein concentration in each fraction can be determined using a protein assay kit *(8,9)*.

3.3. Protein Cyanine Dye Labeling

1. Add 3 vol of acetone to the particulate fractions. Vortex them well, keep them on ice for 1 h, and centrifuge them at 12,000*g* at 4°C for 10 min. Remove the supernatants and wash the precipitates with ice-cold 80% acetone. Then centrifuge them again (*see* **Note 8**). The precipitates are dissolved with labeling buffer (*see* **Note 9**).
2. Check to see that the pH values of sample solutions are between pH 8.0–9.0 using pH indicator strips (*see* **Note 10**).
3. To prepare the working dye solutions, dilute the Cy3 and Cy5 stock solutions with DMF at the ratio 2:5. In other words, 1 μL of the dye working solution contains 400 pmol of Cy dye.
4. Add 1 μL of the dye working solution to 50 μg sample protein for the labeling reaction. Mix the dye and protein sample thoroughly by vortexing.
5. Incubate the labeled samples for 30 min on ice in the dark.
6. To stop the labeling reactions, add 1 μL of stop solution, mix thoroughly by vortexing, and leave for 10 min on ice in the dark (*see* **Note 11**).

3.4. Two-Dimensional Electrophoresis

1. Add an equal volume of 2D sample buffer (2X) to the labeled sample and leave on ice in the dark.
2. Combine equal amounts of protein from the two samples labeled with different dyes (*see* **Table 1**).
3. Adjust the total volume of the labeled samples to 350 μL using the rehydration buffer (*see* **Note 12**).
4. Deliver 350 μL of the labeled sample solution into each slot of the rehydration tray.
5. Remove the protective cover of Immobiline DryStrips and place them gel side down in each slot.
6. Overlay each Immobiline DryStrip with 2 mL of dimethyl silicone oil to prevent evaporation and urea crystallization. The rehydration step is completed overnight at room temperature.
7. Prepare the Multiphor II Immobiline DryStrip Kit.
8. Transfer the rehydrated Immobiline DryStrips to grooves of the aligner, ensuring that the gel side is up (*see* **Note 13**).
9. Place the moistened electrode strips across the cathodic and anodic ends of the Immobiline DryStrips and set each electrode on an electrode strip.

10. For the isoelectric focusing (IEF), set the initial voltage to 500 V, then increase it up to 3500 V linearly within 1 h. Keep it at 3500 V until total focusing time reaches 80 kVh. This IEF step is performed at 20°C. After IEF, the strips can be stored at –80°C.

11. Pipet approx 2 mL of the Bind-Silane working solution onto the one glass plate and distribute equally over the plate. Leave to air-dry for 1–1.5 h.

12. Place the marker approx halfway along both sides of glass plate.

13. To prepare ten 1.0-mm thick, 12% gels, mix 383 mL of monomer stock solution, 250 mL of 4X resolving gel buffer, 2 mL of 10% SDS, 352 mL of water, 5 mL of 10% ammonium persulfate, and 330 µL of TEMED. Pour the gel solution and overlay with water-saturated isobutanol onto each gel. Allow the gels to polymerize for at least 3 h.

14. Before separation in second dimension, each strip is equilibrated with 15 mL of equilibration buffer by gently shaking for 15 min.

15. Place the strips in between the two glass plates containing the SDS-PAGE gels. Then pipet the agarose overlay solution slowly up to the top of these glass plates. Add the running buffer to the chamber.

16. The electrophoresis is performed using constant voltage of 100 V at 10°C until the bromophenol blue dye front reaches the bottom of the gel.

17. After 2D polyacrylamide gel electrophoresis, three gels are imaged directly using the Amersham 2920-2D Master Imager. This imager should be warmed up for 30 min before scanning.

18. The Cy3-labeled gel images are scanned at Ex 540 nm and Em 590 nm, whereas the Cy5-labeled gel images are scanned at Ex 620 nm and Em 680 nm. Scan times are normalized to give approximately the same level of signal for each cyanine dye-labeled sample (maximum pixel intensity about 55,000).

19. The gel images are saved as 16 bit TIFF images. TIFF images are analyzed using DeCyder-DIA version 3.0 software. In this study, spots that changed more than 1.4-fold or less than 1.4-fold were selected for picking (*see* **Fig. 2**).

3.5. In-Gel Digestion

1. Visualize the resulting gels by SYPRO Ruby staining. The SYPRO Ruby staining-gel images are scanned at Ex 400 nm and Em 633 nm using the Amersham 2920-2D Master Imager (Amersham Biosciences).

2. Gel plugs containing the selected protein spots are excised using the automated Amersham Ettan spot picker (Amersham Biosciences).

3. The gel plugs are washed with 1 mL of the wash solution 1 and dried in a SpeedVac evaporator.

4. The gel plugs are incubated in 200 µL of the reduction solution at 56°C for 60 min.

5. The reduction solution is discarded, and after cooling to room temperature, the gel plugs are incubated in 200 µL of the alkylation solution at room temperature in the dark for 30 min.

6. The alkylation solution is discarded and the gel plugs are washed twice with 200 µL of the wash solutions 2 and 3 alternately.

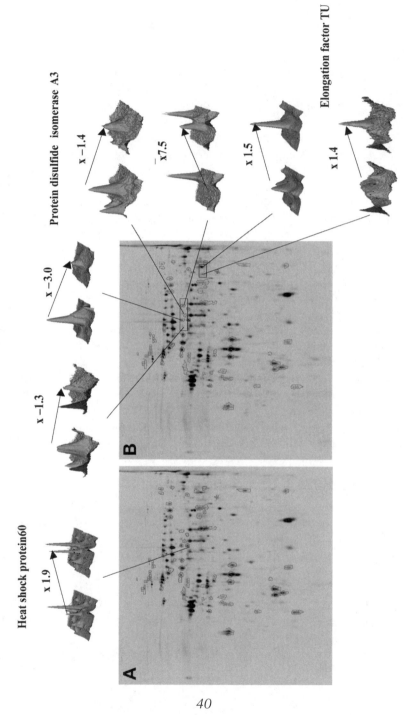

Heat shock protein60

x 1.9

Protein disulfide isomerase A3

x –1.4

x7.5

x 1.5

Elongation factor TU

x 1.4

x –3.0

x –1.3

A

B

40

7. The gel plugs are dried in a SpeedVac evaporator and incubated in 20 μL of the digestion solution (*see* **Note 13**) at 37°C overnight.
8. After digestion, products are recovered by sequential extraction with 50 μL of the extraction solutions 1 and 2 alternately. The extracts are then dried in a SpeedVac evaporator and resuspended in 10 μL of 0.1% TFA solution.

3.6. MALDI-TOF Analysis

1. Digested samples are adsorbed by the C18 resin of ZipTip, washed with 0.1% TFA, and eluted directly onto a MALDI-TOF sample plate using 2–3 μL of elution solution.
2. After the sample drops are completely dried, the MALDI-TOF sample plate is inserted to the ion source of the Reflex III MALDI-TOF MS (Bruker Daltonics).
3. The spectra data for the samples are acquired using a reflectron mode.
4. For protein identification, several search programs such as MASCOT (Matrix Science, London) can be used to search proteins in the NCBInr database.

4. Notes

1. The buffer should be bubbled with a 95% O_2 and 5% CO_2 gas mixture for at least 30 min before used for the experiment, to equilibrate the oxygen content and pH (apparently 7.4 after the bubbling). Calcium precipitation might occur when the bubbling is insufficient, but this can be re-solubilized after continuous bubbling. The buffer should be bubbled continuously until the end of the experiment.
2. Sodium orthovanadate should be activated for maximal inhibition of tyrosine phosphatases *(11)*. Prepare a 200 m*M* solution of sodium orthovanadate, and adjust the pH to 10.0. At pH 10.0 the solution will be yellow. Then boil the solution until it turns colorless, and cool to room temperature. Re-adjust the pH to 10.0. If the solution is still yellow color, boil the solution again. After the solution remains colorless, store the activated sodium orthovanadate at −20°C. Microcystin LR is highly hepatotoxic and should be handled with care.
3. The lysis and sample preparation buffers should not contain any primary amines before labeling, because primary amines will compete with the protein for CyDye DIGE Fluor minimal dyes.

Fig. 2. *(Opposite page)* Typical two-dimensional differential gel electrophoresis gel images of the 100,000*g* precipitates (particulate fractions) of rat hearts and images from the three-dimensional (3D) viewer of the Decyder software. Proteins were labeled with Cy3 or Cy5 and separated on pH 4.0–7.0 immobilized pH gradient strips in the first dimension and by a 12% isocratic SDS-PAGE gel in the second dimension. **(A)** Cy3 image of control sample; **(B)** Cy5 image of ischemia sample. We detected 1450 protein spots on this gel. Then we picked up the spots that changed more than twofold in expression level during ischemia and/or reperfusion and several additional spots related to these spots, and identified them as described *(12)*. 3D images show the different amounts of Cy3- and Cy5-labeled proteins in the identified spots.

4. The DMF must be high-quality anhydrous (specification: $\leq$0.005% H_2O, $\geq$99.8% pure) and every effort should be taken to ensure it is not contaminated with water. DMF, once opened, will start to degrade generating amine compounds. Amines will react with the CyDye DIGE Fluor minimal dyes, reducing the concentration of dye available for protein labeling.

5. Confirm that the dye solution is an intense color. During transport, the dye powder may spread around the inside surface of the tube (including the lid). If the dye is not an intense color, then pipet the solution around the tube (and lid) to ensure resuspension of dye. Vortex and spin down.

6. Other site-specific proteases such as Achromobactor proteinase I also can be used. The optional composition of the digestion solution will vary for different application.

7. The procedure to perfuse the cannulated heart after it is removed from the body should be performed as quickly as possible to avoid ischemic damage to the heart.

8. Acetone preparation can remove salts from samples. This is important for labeling and 2DE. Too much salts can interfere with Cy-labeling and electrophoresis.

9. The protein concentration of sample solutions should be between 5–10 mg/mL.

10. The pH of the sample solutions used with minimal dyes should be between pH 8.0–9.0. If the pH of samples is outside this range, the pH of samples will need to be adjusted before labeling.

11. The labeled samples can be processed immediately or stored for at least 3 mo at −70°C in the dark.

12. The total volume of the sample is required for each Immobiline DryStrip. 350 μL is applied to an 18-cm Immobiline DryStrip.

13. Do not touch the gel side.

14. After the gel plug absorbs the digestion solution, a few μL of the digestion solution should remain. If no solution remains, more digestion buffer must be added to saturate the plug.

References

1. Miyamae, M., Camacho, S. A., Weiner, M. W., and Figueredo, V. M. (1996) Alcohol consumption reduces ischemia-reperfusion injury by species-specific signaling in guinea pigs and rats. *Am. J. Physiol.* **271,** H1245–H1253.

2. Das, D. K. and Maulik, N. (1994) Antioxidant effectiveness in ischemia-reperfusion tissue injury. *Methods Enzymol.* **233,** 601–610.

3. Chien, K. R., Reeves, J. P., Buja, L. M., Bonte, F., Parkey, R. W., and Willerson, J. T. (1981) Phospholipid alterations in canine ischemic myocardium. Temporal and topographical correlations with Tc-99m-PPi accumulation and an in vitro sarcolemmal Ca2+ permeability defect. *Circ. Res.* **48,** 711–717.

4. Unlu, M., Morgan, M. E., and Minden, J. S. (1997) Difference gel electrophoresis: a single gel method for detecting changes in protein extracts. *Electrophoresis* **18,** 2071–2077.

5. Alban, A., David, S. O., Bjorkesten, L., et al. (2003) A novel experimental design for comparative two-dimensional gel analysis: two-dimensional difference gel electrophoresis incorporating a pooled internal standard. *Proteomics* **3,** 36–44.

6. Shaw, J., Rowlinson, R., Nickson, J., et al. (2003) Evaluation of saturation labeling two-dimensional difference gel electrophoresis fluorescent dyes. *Proteomics* **3,** 1181–1195.

7. Yamamoto, S., Matsui, K., Kitano, M., and Ohashi, N. (2000) SM-20550, a new Na+/H+ exchange inhibitor and its cardioprotective effect in ischemic/reperfused isolated rat hearts by preventing Ca^{2+}-overload. *J. Cardiovasc. Pharmacol.* **35,** 855–862.

8. Mizukami, Y., Hirata, T., and Yoshida, K. (1997) Nuclear translocation of PKC zeta during ischemia and its inhibition by wortmannin, an inhibitor of phosphatidylinositol 3-kinase. *FEBS Lett.* **401,** 247–251.

9. Yoshida, K., Hirata, T., Akita, Y., et al. (1996) Translocation of protein kinase C-alpha, delta and epsilon isoforms in ischemic rat heart. *Biochim. Biophys. Acta* **1317,** 36–44.

10. Albert, C. J. and Ford, D. A. (1999) Protein kinase C translocation and PKC-dependent protein phosphorylation during myocardial ischemia. *Am. J. Physiol.* **276**(2 Pt 2), H642–650.

11. Gordon, J. A. (1991) Use of vanadate as protein-phosphotyrosine phosphatase inhibitor. *Methods Enzymol.* **201,** 477–482.

12. Sakai, J., Ishikawa, H., Kojima, S., Satoh, H., Yamamoto, S., and Kanaoka, M. (2003) Proteomic analysis of rat heart in ischemia and ischemia-reperfusion using fluorescence two-dimensional difference gel electrophoresis. *Proteomics* **3,** 1318–1324.

4

Analysis of Antihypertensive Drugs in the Heart of Animal Models

A Proteomic Approach

Alberto Lázaro, Julio Gállego-Delgado, Julio I. Osende, Jesús Egido, and Fernando Vivanco

Summary

Arterial hypertension is the most frequent chronic disease and it is an important cause of morbidity and mortality in the developed world. Arterial hypertension is associated with such adverse effects as accelerated arteriosclerosis and pathological left ventricular hypertrophy, among others. The molecular mechanisms affecting left ventricular hypertrophy remain mostly unknown. The advent of proteome profiling has facilitated the elucidation of disease-associated proteins, paving the way for molecular diagnostics and the identification of novel therapeutic targets. We explored the proteomic profile of pathological left ventricular hypertrophy in comparison with normal heart in a model of rats and investigated the proteomic changes in response to different antihypertensive regimens in order to elucidate their cardioprotective effects. Here we describe in depth the protocol for this type of study.

Key Words: Hypertension; left ventricular hypertrophy; antihypertensive therapy; differential protein expression profile; cardiac index; spontaneously hypertensive rats; SHR.

1. Introduction

Cardiovascular disease (CVD) is an important cause of morbidity and mortality in the developed world. Arterial hypertension is one of the most important modifiable risk factors for the development of CVD, affecting mainly heart, kidney, brain, and eyes *(1)*. It has been estimated that arterial hypertension affects around 40% of the population in Europe and 30% of the population in North America *(2)*.

From: *Methods in Molecular Biology, vol. 357: Cardiovascular Proteomics: Methods and Protocols*
Edited by: F. Vivanco © Humana Press Inc., Totowa, NJ

The effects of untreated or poorly controlled hypertension include both vascular dysfunction (accelerated atherosclerosis and arteriosclerosis) and the development of heart diseases with pathological left ventricular hypertrophy (LVH), a major predictor of systolic heart failure and death.

LVH is considered to be a reversible state in most instances, if blood pressure (BP) is kept under tight control *(3)*. In fact, lowering BP with antihypertensive drugs reduces risk of a variety of CVD, including cardiovascular death and total mortality *(4)*. It is frequently assumed that after regression of LVH, myocardial physiology is also normalized, but little is known about the molecular mechanisms or treatment-specific effects operating in normalized myocardium.

There is a wide range of pharmacological agents currently available for the treatment of hypertension, including thiazide diuretics, α-blockers, β-blockers, calcium channel blockers, angiotensin converting enzyme inhibitors (ACEi), and angiotensin receptor blockers. Therefore a differential effect on cellular protein expression in normalized LVH could be expected.

Proteomic analysis has been shown to provide a potentially great amount of information about protein expression in different settings *(5)*. In this sense, this technique is appropriate to study differential protein profile in established LVH and in regressed LVH with different treatment modalities. Thus we have compared the heart protein expression pattern of normotensive Wistar Kyoto rats (WKY) and spontaneously hypertensive rats (SHR) with LVH and studied the effect of different antihypertensive regimens on the cardiac protein expression profile.

The proteomic analysis of LVH and regressed LVH by treatments could offer the potential to find new pathways involved in hypertensive heart disease.

2. Materials

2.1. Animal Model and Antihypertensive Therapy

1. Animals: SHR and WKY (Criffa, Barcelona, Spain; *see* **Note 1**).
2. ACEi: quinapril (as powdered hydrochloride salt; Pfizer, Barcelona, Spain). Store at 4°C.
3. Angiotensin II type 1 (AT-1) receptor antagonist: losartan (as powdered potassium salt; Merck Sharp & Dohme, Madrid, Spain). Store at 4°C.
4. α-1 Blocker receptor: doxazosin (Pfizer). Store at 4°C.
5. Tail-cuff sphygmomanometer (NARCO Biosystems, Austin, TX).
6. Niprem 1.5 software package (CIBERTEC S.A., Madrid, Spain).

2.2. Heart Retrieval and Storage

1. Sterile surgical material.
2. Pentobarbital sodium (flask containing 1 g; Braun Medical SA, Barcelona, Spain). To prepare, dissolve the powder in 20 mL of sodium saline serum. This product is stable for 1 mo at room temperature.

3. 0.9% Sodium saline serum (Braun Medical SA). Store at 4°C.
4. Liquid nitrogen. Use cryo-gloves when working with this product because it can cause skin burns.

2.3. Sample Preparation

1. Liquid nitrogen.
2. Metal mortar.
3. 100 m*M tris* (2-carboxyethyl)-phosphine hydrochloride (TCEP-HCl; Pierce, Cultek, Madrid, Spain). To prepare 500 µL, dissolve 0.01433 g in double-deionized (dd) water (*see* **Note 2**). Store in 20-µL aliquots at −20°C.
4. Lysis tissue buffer stock: To prepare 30 mL, dissolve 4.6 g of thiourea, 12.6 g of urea and 1.2 g of CHAPS (Fluka, Sigma, St. Louis, MO) in dd water. Store in 2-mL aliquots at −20°C. Just prior to use, add 20 µL of TCEP-HCl (100 m*M*) and 5 µL of Bio-Lyte Ampholyte (Bio-Rad, Hercules, CA) per mL of stock solution. The final concentrations are: 7 *M* urea, 2 *M* thiourea, 4% (w/v) CHAPS, 2 mM TCEP-HCl, and 0.5% Bio-Lyte Ampholyte.
5. Polycarbonate centrifuge tubes, 13 × 51 mm (Beckman Instruments, CA, cat. no. 349622).
6. Neutral test 1.5-mL tubes with cap-type Eppendorf® (AFORA, Barcelona, Spain, cat. no. 00298-00).
7. Sonifier B-12 cell disrupter (Branson Sonic Power Co., Danbury, CT).

2.4. Two-Dimensional Electrophoresis

2.4.1. First Dimension

1. Protean isoelectric focusing (IEF) cell and accessories (Bio-Rad).
2. ReadyStrip™ immobilized pH gradient (IPG) strips (170 mm, pH 4.0–7.0, linear gradient) (Bio-Rad). Store at −20°C (*see* **Note 3**).
3. Rehydration buffer stock: To make 25 mL, dissolve 12 g of urea, 0.5 g of CHAPS, and trace amounts of bromophenol blue in dd water. Store in 2-mL aliquots at −20°C. Just prior to use, add 10 µL of TCEP-HCl and 2 µL of Bio-Lyte Ampholyte per mL of rehydration buffer stock. The final concentrations are: 8 *M* urea, 2% (w/v) CHAPS, trace of bromophenol blue, 1% (w/v) TCEP-HCl, and 0.2% (v/v) Bio-Lyte Ampholyte.
4. Mineral oil, biotechnology-grade (Bio-Rad). Store at room temperature.
5. Electrode wicks (Bio-Rad).

2.4.2. IPG Strips Equilibration

1. Equilibration buffer: To prepare 200 mL, dissolve 72.07 g of urea, 4 g of sodium dodecyl sulfate (SDS) (Bio-Rad), and a few grains of bromophenol blue, and add 6.7 mL and 69 mL of Tris-HCl, pH 8.8, and glycerol, respectively, in dd water. Store in 50-mL aliquots at −20°C. The final concentrations are: 50 mM Tris-HCl, pH 8.8, 6 *M* urea, 30% (v/v) glycerol, 2% (w/v) SDS, and trace of bromophenol blue (a few grains).

2. Dithiothreitol (DTT), as solid powder (Bio-Rad). Store at 4°C.
3. Iodoacetamide, as solid powder (Bio-Rad). Store at 4°C.

2.4.3. Second Dimension

1. Acrylamide/bisacrylamide stock buffer: 30% (w/v) acrylamide and 0.8% (w/v) N,N'-methylenebisacrylamide, in dd water. To prepare 400 mL, dissolve 120 g of acrylamide and 3.2 g of N,N'-methylenebisacrylamide. These products are extremely toxic and it is recommended that a mask be worn during their use. Store at 4°C away from light (*see* **Note 4**).
2. 1.5 *M* Tris-HCl, pH 8.8. To prepare 500 mL, dissolve 90.75 g of Tris in dd water and adjust to pH 8.8 with HCl. Store at 4°C.
3. 10% (w/v) SDS. To prepare 25 mL, dissolve 2.5 g of SDS in dd water, and store at room temperature.
4. 10% (w/v) ammonium persulfate (APS). To prepare 1 mL, dissolve 0.1 g of APS in dd water, and store at 4°C in aliquots.
5. TEMED (Merck, Darmstadt, Germany). Store at 4°C. Wear a mask when using this product.
6. Isopropanol. Store at room temperature.
7. Running buffer (10X): 25 m*M* Tris, 192 m*M* glycine, and 0.1% (w/v) SDS. To prepare 1 L, dissolve 30.2 g of Tris, 144.2 g of glycine, and 10 g of SDS in dd water. Store at room temperature.
8. Forceps (Bio-Rad).
9. Protean II xi 2-D cell (Bio-Rad).

2.5. Staining and Scanning

1. Gel staining box (Nalgene, Rochester, NY) (*see* **Note 5**).
2. Fixer buffer: 5% (v/v) acetic acid and 30% (v/v) ethanol in dd water. To prepare 200 mL, add 10 mL of acetic acid, 60 mL of ethanol, and 130 mL of dd water. Prepare prior to use, and store at room temperature. When using all the staining chemicals, please follow **Note 1**.
3. 5% (w/v) sodium thiosulfate stock. To prepare 50 mL, dissolve 2.5 g of sodium thiosulfate in dd water. This product is stable during 6 d at room temperature.
4. Sensitize buffer: 0.8 m*M* sodium thiosulfate solution. To prepare 200 mL, dilute 800 µL of 5% sodium thiosulfate stock in 199.2 mL of dd water.
5. Silver nitrate solution: 12 m*M* silver nitrate (Riedel-deHaën, Sigma). Prepare prior to use at room temperature. For example, to prepare 200 mL, dissolve 0.408 g in dd water.
6. Developer buffer stock: make 30 min before use. To prepare 200 mL, dissolve 6 g of potassium carbonate and 50 µL of 5% sodium thiosulfate stock. Just prior to use, add 50 µL of formaldehyde (37% [w/v]) (Merck) and mix vigorously. The final concentrations are: 3% (w/v) potassium carbonate, 0.025% (v/v) formaldehyde, and 0.00125% (v/v) sodium thiosulfate.
7. Stop buffer: 2% (v/v) acetic acid and 4% (w/v) Tris-HCl in dd water. To prepare 200 mL, mix 8 g of Tris-HCl and 4 mL of acetic acid. Prepare fresh prior to use.

8. Scanner GS-800 (Bio-Rad) or other densitometer compatible with two-dimensional gel electrophoresis (2DE) analysis.

2.6. Image Analysis

1. PD Quest 2-D v 7.1.1 gel analysis software (Bio-Rad) or other similar/equivalent software compatible with 2DE analysis.

3. Methods

Experimental hypertensive model is performed with male SHR. As normotensive control, age-matched WKY rats were used ($n = 7$–8 animals per group; *see* **Notes 6** and **7**).

IEF is an electrophoretic method that separates proteins according to their isoelectric point (p*I*). The following protocol/instructions assume the use of Protean IEF cell (*see* **Subheading 3.3.1.**). Preparing the IPG strips with SDS-equilibration buffer is necessary for running the second dimension (*see* **Subheading 3.3.2.** and **Subheading 3.3.3., step 6**). The proteins are coated with SDS so cysteines are reduced and alkylated.

Second-dimension SDS-polyacrylamide gel electrophoresis (SDS-PAGE) is an electrophoretic method for separating proteins according to their molecular weights (MW). These instructions assume the use of Protean II xi 2-D cell (Bio-Rad). The second dimension was performed on 20 cm × 22 cm × 1 mm, 12.5% acrylamide/bisacrylamide SDS-PAGE (*see* **Subheading 3.3.3.**). For visualizing protein (*see* **Subheading 3.4.**), the silver-staining protocol used is based on the protocol of Blum et al. *(6)*, with the modifications made by Rabilloud *(7)*, but other protocols are possible.

3.1. Animal Model and Heart Retrieval

1. SHR at 12 wk old are randomized to treatment groups in different cages:
   a. Untreated group: without treatment.
   b. Losartan group: SHR receiving losartan (30 mg/kg/d).
   c. Quinapril group: SHR receiving quinapril (16 mg/kg/d).
   d. Combine treatment group (doxazosin plus quinapril): SHR treated with doxazosin and quinapril (15 mg/kg/d and 1.6 mg/kg/d, respectively).
2. Prepare all drugs by dilution in tap water (*see* **Note 8**).
3. All animals groups have free water and food access, in a light-, temperature-, and humidity-controlled environment.
4. To assess the effect of hypertension and the antihypertensive therapy, measure systolic blood pressure (SBP) by a tail-cuff sphygmomanometer weekly, as follows:
   a. Immobilize the rat in a special plastic cage and place the cuff sphygmomanometer in the tail (*see* **Note 9**).
   b. Measure the SBP using the Niprem 1.5 software package according to the manufacturer's protocol. An example of SBP measurements is shown in **Fig. 1**.

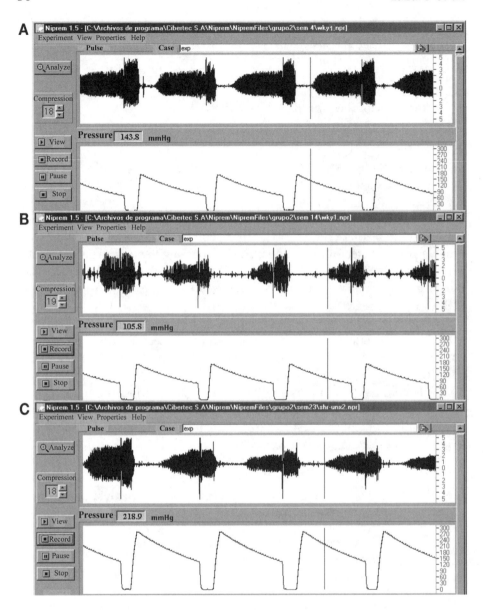

Fig. 1. Representative images of blood pressure (BP) measurements using the Niprem 1.5 software package. **(A)** Normotensive Wistar Kyoto rat under the initial stress shows a systolic blood pressure (SBP) of 143.8 mmHg. **(B)** The same animal after calming down shows a SBP of 105.8 mmHg. **(C)** Spontaneously hypertensive rat after calming down shows a SBP of 218.9 mmHg.

   c. For each animal, make three independent SBP measurements and average them.

5. At 48 wk of age, anesthetize the animals in the experimental operating room by intraperitoneal injection with pentobarbital sodium (5 mg/100 g body weight).
6. Weigh the animal, sacrifice it, and remove the heart.
7. Use cold sodium saline serum to wash the heart and weigh it to calculate the cardiac index (heart:body weight ratio $\times 10^3$; *see* **Note 10**).
8. Quickly separate the heart with scalpel in atria and ventricles and freeze the left ventricles in liquid nitrogen. Store in cryovials in liquid nitrogen.

3.2. Sample Preparation for 2DE

1. Thaw the lysis buffer stock on ice before starting the procedure (1 mL for each sample is necessary; *see* **Note 11**).
2. Add TCEP-HCl and Bio-Lyte Ampholyte for complete the lysis buffer (*see* **Subheading 2.3., item 4**).
3. Make 1-mL aliquots in 1.5-mL Eppendorf tubes, and store on ice for later use.
4. Cut a piece of left ventricle sample and transfer into a sterile metal mortar under liquid nitrogen (*see* **Note 12**).
5. Pulverize the sample.
6. Homogenize the resulting powder in 1 mL of lysis buffer (use previous aliquots; *see* **Note 13**).
7. Mix well with pipet, vortex, and place on ice for 10 min.
8. Transfer the sample into a polycarbonate centrifuge tube on ice.
9. Sonicate the tube until foam appears. Repeat this step three times (*see* **Note 14**).
10. Centrifuge the tube at 100,000g for 1 h at 4°C.
11. Carefully remove the supernatant and transfer it to another 1.5-mL Eppendorf tube.
12. Measure the protein concentration by the Bradford dye-binding procedure (*8*).
13. Store the samples at −80°C until use.

3.3. 2DE SDS-PAGE

3.3.1. First Dimension

1. Thaw the samples and mix with rehydration buffer and TCEP-HCl on ice.
2. Pipet the appropriate volume for 200 μg of solubilized cardiac proteins into a 1.5-mL Eppendorf tube at room temperature.
3. Add the appropriate volume of rehydration buffer to a final volume of 300 μL (for 170-mm long IPG strips).
4. Add 3 μL of TCEP-HCl and 0.6 μL of Byo-Lyte Ampholytes, and mix by vortex.
5. Apply the require number of samples in the channels of a focusing tray. Pipet the samples along the entire length of each channel.
6. Take the IPG strip using forceps and remove the protective plastic cover.
7. Put the IPG strip with the acrylamide facing down. Lay the end of the IPG strip in the corner of the channel and slowly lower it onto the sample (*see* **Note 15**).

8. Cover the strip with mineral oil to minimize evaporation and prevent the IPG strip from drying out.
9. IEF is carried out according to the manufacturer's protocol. The steps included in the selected program are:
   a. Rehydrate the strips at 50 V for 12 h.
   b. 500 V for 1 h using a rapid voltage ramping method.
   c. 1000 V for 1 h by linear voltage ramping method.
   d. 8000 V for 30 min by gradient voltage.
   e. 8000 V until 50,000 Vh using a rapid voltage.
   A Peltier temperature control platform maintains gels at 20°C throughout IEF *(9)*.
10. Use small pieces of wicks to remove interfering salts, inserting one wick between the strip and each electrode. Usually, it is preferable to do this once the proteins have already been focused. To be sure, we apply the wicks when the process reaches the step at 8000 V using a rapid voltage and once the voltage 8000 V is constant. Pause the power machine and use forceps to lift the ends of the IPG strips and insert the wicks over the electrodes (*see* **Note 16**).
11. Resume IEF.
12. When IEF is finished, IPG strips are stored at −80°C, in channels of a plastic tray (*see* **Note 17**).

3.3.2. IPG Strips Equilibration

1. Thaw IPG strips in a channel of plastic tray (*see* **Subheading 3.3.3., step 6**).
2. Prepare DTT 1% (w/v) equilibration buffer. For example, to prepare 20 mL, dissolve 0.2 g of DTT in 20 mL of equilibration buffer (*see* **Note 18**).
3. Prepare iodoacetamide 2.5% (w/v) equilibration buffer. For example, to prepare 20 mL, dissolve 0.5 g of iodoacetamide in 20 mL of equilibration buffer. Prepare at room temperature in the dark just prior to use (*see* **Note 18**).
4. When IPG strips are thawed, incubate with agitation in DTT equilibration buffer for 20 min (5 mL per each IPG).
5. Decant the first equilibration solution and add dd water for washing the IPG strips for 2 min without shaking.
6. Pass IPG strips to other clean channels of plastic tray.
7. Add iodoacetamide-equilibration buffer (5 mL per each strip), and incubate with agitation for 20 min.
8. After the last step, IPG strips can be loaded in the second dimension.

3.3.3. Second Dimension

1. Assemble the polymerization gel sandwich consisting of two glass plates (inner and outer) and two spacers (1 mm) in between them, according to the manufacturer's protocol (*see* **Note 19**).
2. Prepare the stock solution of polymerization by combining all reagents (per two gels): 33.36 mL of acrylamide/bisacrylamide stock buffer, 20 mL of 1.5 M Tris-HCl, pH 8.8, 800 µL of SDS 10%, and 25.44 mL of dd water.

3. Immediately prior to polymerization, add to the stock solution 400 µL of APS and 26.4 µL of TEMED. Mix gently and apply the mix to the gel sandwich using a glass pipet until the front is 0.5 cm from the top.
4. Add isopropanol in the top and slightly hit the glasses to homogenize the upper border (*see* **Note 20**).
5. The gel should polymerize in about 30–40 min.
6. As gel polymerization takes place, equilibrate the IPG strips as described in **Subheading 3.3.2.** To make the running buffer, dilute 200 mL of the running buffer stock (10X) with 1800 mL of dd water in a calibrated cylinder. Cover with Para-Film and invert to mix.
7. When the gels are polymerized, decant the isopropanol and rinse the top with dd water.
8. Add the running buffer over polymerized gel.
9. After equilibration, drain the IPG strips. Place it on a filter paper, placing the acrylamide face up for a few seconds, to remove excess equilibration buffer.
10. Place the IPG strip on top of the SDS-PAGE gel with the plastic backing against the plate (handle with forceps). With a plastic ruler or spatula, press the IPG strip until it is positioned directly on top of the SDS gel, making complete contact (*see* **Note 21**).
11. Assemble the electrophoresis unit (Protean II xi 2-D cell) and add running buffer (*see* **Note 22**).
12. Run the gels at 30 mA for 30 min followed by 55 mA for around 5 h, at 4°C (*see* **Note 23**). Stop electrophoresis when the blue front reaches the bottom.
13. When finished, remove gels and place them in a gel-staining boxes with fixing buffer.

3.4. Staining and Scanning

1. Soak the gel(s) overnight in the gel-staining box(s) with gel fixation solution (*see* **Note 24**).
2. Shake the box(s) in a platform rocker (STR6, Bibby, UK) for 30 min and repeat this step two more times, changing the fix solution 2×30 min (*see* **Note 25**).
3. Rinse with water for 4×10 min with shaking. Change the water each time.
4. To sensitize the gels, add the sensitivity buffer when the last water rinse finished. Shake slowly on the bench for 1 min (*see* **Note 26**).
5. Rinse 2×1 min in dd water.
6. Impregnate the gels for 50–60 min in silver nitrate solution with continuous shaking.
7. Rinse in dd water for 20 s.
8. Develop image for less than 20–25 min (*see* **Note 27**) in developing solution with continuous shaking and checking the appearance of the spots.
9. When development is finished, stop it with stop buffer for 20–30 min, with gentle agitation.
10. Rinse the gels twice for 10 s in the box, and take digital images of the gels in the scanner, according to the manufacturer's protocol. Save it in .TIFF format to computer analysis (*see* **Note 28**). An example of the gels obtained is shown in **Fig. 2**.
11. Finally, store the gel(s) in the box(s) in dd water at 4°C, to preserve the protein spots for further mass spectrometry (MS) analysis.

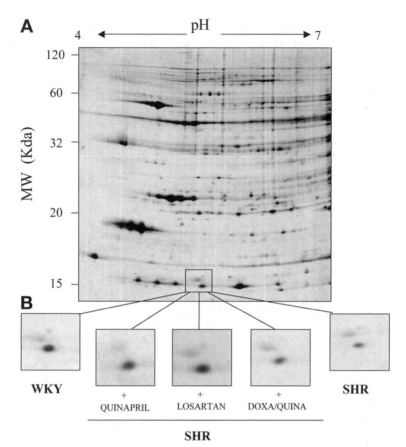

Fig. 2. (A) Representative image of two-dimensional gel electrophoresis silver-stained gel of left ventricle proteins (isoelectric focusing: 4.0–7.0 pH range, second dimension, 12.5% acrylamide) from normotensive Wistar Kyoto rat (WKY). **(B)** Enlarged images of three spots differently expressed in spontaneously hypertensive rats in comparison with normotensive rats (WKY). The effect of the different antihypertensive therapies (quinapril, losartan, and double-therapy doxazosin plus quinapril) is also shown.

3.5. Computer Analysis

1. Import digitalized images to PD Quest.
2. Analyze the spots from four different gels of different animals per group. The analysis is made according to the manufacturer's instructions (user guide).
3. To compensate for any variation in protein loading and development level of silver stain, normalize the intensity levels of each spot (protein) in the gels, using the normalization option total density in valid spots or, if saturated spots exist, the option total density in gel image.

4. The program calculates the mean values and coefficients of variation of the differentiated spots (as expression or presence/absence).

5. Analyze the differential altered proteins between WKY and SHR rats, and study the normalized, partially recovered, and deregulated proteins with antihypertensive regimens. An example of this analysis is shown in **Fig. 3**.

6. Once analysis has finished, spots of interest (with significant variations) must be digested for MS identification.

4. Notes

1. Wear gloves and use good laboratory safety precautions when using all animals/reagents and during all processes described.

2. All buffers/solutions for 2DE should be prepared in double-deionized water. This standard is referred to as "dd water" in this chapter.

3. It is important to be sure that IPG strips are not expired. This could spoil the final gel resolution.

4. To avoid for contaminants particles in the polymerization of gels for 2DE, filter this solution through a filter paper.

5. The staining boxes must be extremely clean to avoid high background and false spots.

6. The SHR model was first developed in 1963 by Okamoto and Aoki *(10)*. It is generally used for studies in hypertension and cardiovascular research. BP in the SHR increases gradually during the developmental phase (7–15 wk) before reaching a stable plateau (~200 mmHg), which is sustained.

7. This is a chronic model of hypertension but other chronic or acute models have been described *(11,12)*, for example: by infusion of angiotensin II or endothelin, with nitro-L-arginine methyl ester, and so on. This protocol is also adequate for these other models.

8. All these drugs are solvable in water. In other cases, drugs must be given *ad libitum* or by gastric gavage.

9. Observe that animals need a few minutes to calm down, so first measurements could be higher than expected owing to stress (*see* **Fig. 1**).

10. This parameter is very important to evaluate the antihypertensive therapy-induced cardiac hypertrophy regression.

11. The opened aliquots never should be refrozen.

12. This step must be done quickly because the sample tends to be thawed very fast. Sterilize the mortar at 200°C for 2 h and then keep it at −80°C for at less 45 min until use.

13. Protease inhibitor can be used to prevent protein loss if necessary.

14. This machine makes a high and unpleasant noise, thus headphones are recommended.

15. Avoid trapping air bubbles under the IPG strip. If this happens, lift it up slowly and lower it again.

16. Alternatively, Whatman paper can be used if electrode wicks are not available.

A

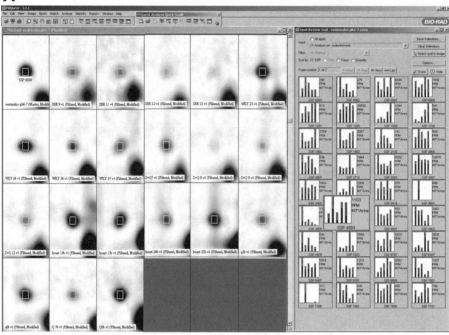

B

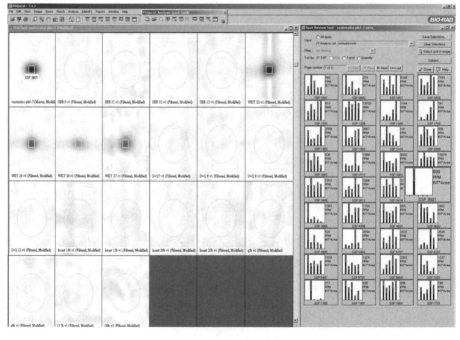

17. It is very important to place the acrylamide face up in the channel, to avoid damaging the acrylamide when the strips are thawed.
18. When mixing DTT and iodoacetamide with equilibration buffer, it is necessary to centrifuge these dissolutions, to eliminate the foam formation.
19. Wash the glass plates with normal water, dd water, and finally with ethanol. Air-dry the glasses.
20. The front must remain straight because when the IPG strip is placed on top of SDS gel, the contact must be complete. If the front is not perfect, the result of the second dimension will be wrong.
21. Be careful not to damage the gel with the spatula and do not trap air bubbles between the IPG strip and the gel surface.
22. When add the running buffer, avoid dropping it directly on the IPG strips because it could lose full contact with the SDS-PAGE.
23. Be careful, amperage higher than 60 mA can break the glasses.
24. The volumes of different reagents in the staining must completely cover the gels. We recommend using 200 mL per each gel.
25. Be careful with the shaking because the gels can break easily.
26. In this step, the time is very important because excess sensitization process results in a high level of background staining.
27. It is important to first shake gels in a shaker and afterwards use hand agitation to better control the appearance of the spots.
28. It is important to scan the gels at the same size. This will facilitate the forward computer analysis.

Acknowledgments

This work has been partially supported by FIS (PI02/1047) (PI 02/3093), Spanish Cardiovascular Network RECAVA (03/01), SAF-2004-06109, European Network (QLG1-CT-2003-01215), and the International Cardura Award (Pfizer, New York, NY).

Fig. 3. *(Opposite page)* Representative images of PDQuest v 7.1.1 software output. **(A)** In the left part, different intensities of the spot **4504** are showed. SHR, spontaneously hypertensive rats; WKY, Wistar Kyoto rats; D+Q, SHR treated with doxazosin plus quinapril (15 and 1.6 mg/kg/d, respectively); losart, SHR treated with losartan (30 mg/kg/d); q, SHR treated with quinapril (16 mg/kg/d). Histograms (in the right part), shown an average intensity of the spot for each group (n = 4 gels from different animals per each group). In this case, the average quantity decreased in SHR in comparison with WKY animals. Losartan and quinapril, but not double therapy, partially recovered it. **(B)** Protein spot **3601** is a representative example that is present only in WKY group. At right, the corresponding histogram can be observed.

References

1. Ruilope, L. M., Campo, C., Rodriguez-Artalejo, F., Lahera, V., Garcia-Robles, R., and Rodicio, J. L. (1996) Blood pressure and renal function: therapeutic implications. *J. Hypertens.* **14,** 1259–1263.
2. Wolf-Maier, K., Cooper, R. S., Banegas, J. R., et al. (2003) Hypertension prevalence and blood pressure levels in 6 European countries, Canada, and the United States. *JAMA* **289,** 2363–2369.
3. Mancini, G. B., Dahlof, B., and Diez, J. (2004) Surrogate markers for cardiovascular disease: structural markers. *Circulation* **109,** IV22–30.
4. Franco, V., Oparil, S., and Carretero, O. A. (2004) Hypertensive therapy: Part II. *Circulation* **109,** 3081–3088.
5. Aebersold, R. and Mann, M. (2003) Mass spectrometry-based proteomics. *Nature* **422,** 198–207.
6. Blum, H., Beier, H., and Gross, H. J. (1987) Improved silver staining of plant proteins, RNA and DNA in polyacrylamide gels. *Electrophoresis* **8,** 93–99.
7. Rabilloud, T., Brodard, V., Peltre, G., Righetti, P. G., and Ettori, C. (1992) Modified silver staining for immobilized pH gradients. *Electrophoresis* **13,** 264–266.
8. Bradford, M. M. (1976) A rapid and sensitive method for the quantitation of microgram quantities of protein utilizing the principle of protein-dye binding. *Anal. Biochem.* **72,** 248–254.
9. Gorg, A., Postel, W., Friedrich, C., Kuick, R., Strahler, J. R., and Hanash, S. M. (1991) Temperature-dependent spot positional variability in two-dimensional polypeptide patterns. *Electrophoresis* **12,** 653–658.
10. Okamoto, K. and Aoki, K. (1963) Development of a strain of spontaneously hypertensive rats. *Jpn. Circ. J.* **27,** 282–293.
11. Takemoto, M., Egashira, K., Usui, M., et al. (1997) Important role of tissue angiotensin-converting enzyme activity in the pathogenesis of coronary vascular and myocardial structural changes induced by long-term blockade of nitric oxide synthesis in rats. *J. Clin. Invest.* **99,** 278–287.
12. Wu, R., Laplante, M. A., and De Champlain, J. (2004) Prevention of angiotensin II-induced hypertension, cardiovascular hypertrophy and oxidative stress by acetylsalicylic acid in rats. *J. Hypertens.* **22,** 793–801.

5

A Solubility Optimization Protocol for Two-Dimensional Gel Electrophoresis of Cardiac Tissue

Brian A. Stanley and Jennifer E. Van Eyk

Summary

We outline a strategy for the optimization of buffer conditions for the solubilization, extraction, and isoelectric focusing (IEF) of proteins from cardiac tissue for two-dimensional gel electrophoresis (2DE). This strategy, which involves altering both the extraction and IEF buffers, allows one to ensure representation of the proteome that is as complete as possible. Initial buffer choices are given, as well as basic protocols for modifications. Although these conditions have been effectively demonstrated for human myocardium, in principle this procedure can be used for the initial screen of any new sample of tissue or cultured cells.

Key Words: Heart muscle; 2DE; optimization; IEF; solubilization.

1. Introduction

The analysis of proteins by two-dimensional gel electrophoresis (2DE) involves multiple separations steps. These steps include the extraction of proteins from the tissue or cell source, the first-dimension separation via isoelectric focusing (IEF), and the second-dimension separation by SDS-PAGE *(1)*. In order to maximize protein abundance and the number of proteins resolved, the composite proteins must maintain their solubility through each of these steps. This is especially true if one is interested in a whole cellular/tissue extraction in which one attempts to solubilize as much of the proteome as possible in a single extraction or a membrane protein-enriched subproteome fraction, which is notoriously difficult to solubilize *(2)*. Incomplete extraction can lead to false representations of the proteome and consequently misleading results. This is especially true in the differential solubilization of specific isoforms or modified protein forms.

From: *Methods in Molecular Biology, vol. 357: Cardiovascular Proteomics: Methods and Protocols*
Edited by: F. Vivanco © Humana Press Inc., Totowa, NJ

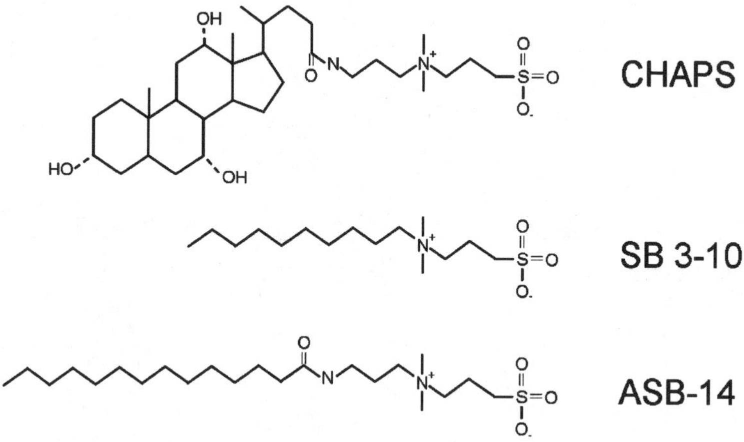

Fig. 1. Diagram of the structure of the different zwitterionic detergents.

Consequently, adequate solubility is critical in order to get a "true" representation of the cellular proteome. However, because proteins are exposed to different microenvironments with different buffer restrictions between 2DE steps, buffer replacement between steps may be required for optimal 2DE.

The solubilzation of proteins in an aqueous buffer requires proteins to unfold and be stabilized while their component side chains are exposed to highly polar water molecules *(2,3)*. This state is often energetically unfavorable because pockets of hydrophobic residues will tend to exclude water molecules, forcing the protein to fold. Although this is most problematic for highly hydrophobic proteins, it can cause serious problems for any protein with a hydrophobic core. By including a chaotropic agent (e.g., urea, thiourea) in the buffer, this energy barrier is reduced, allowing the protein to unfold. Including a detergent at this stage increases the stability of the protein in the unfolded state and allows effective solubility. Consequently, the amount/type of chaotropic agent and detergent present in a buffer can have a great effect on the ability of proteins to solubilize.

The best known agent for protein solubilization is sodium dodecyl sulfate (SDS) because it is a very effective chaotropic agent, as well as a highly ionic detergent *(4,5)*. This makes it ideal for agent for protein solubilization during the second-dimension separation. However, its ionic nature makes it incompatible with IEF, which has led to the use of high concentrations of urea (>7 *M*) combined with zwitterionic detergents (e.g., CHAPS) as the primary protein-solubilizing agents for both protein extraction from tissue as well as IEF. Unfortunately, CHAPS is not the most effective zwitterionic detergent for the solubilization of many proteins as seen in human cardiac tissue *(6)*. Alternative zwitterionic detergents with long-chain sulfobetaine groups have recently become commercially available (e.g., ASB-14, SB-13; *see* **Fig. 1**) that have demonstrated an

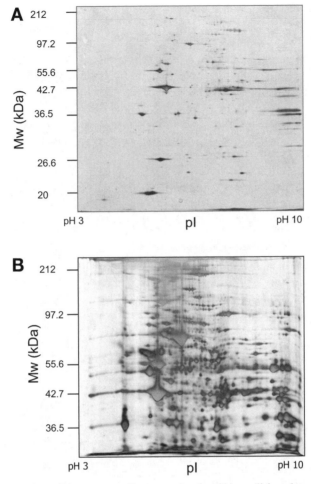

Fig. 2. Separation of human cardiac proteins by **(A)** traditional two-dimensional electrophoresis conditions (using extraction and isoelectric focusing buffer 1 from methods and a 12% acrylamide second dimension) and **(B)** an optimized combination of conditions (using extraction buffer 3 [*see* **Subheading 2.1.**] and focusing buffer 1 [*see* **Subheading 2.2**] run on an 8% acrylamide second dimension).

increased effectiveness at solubilizing membrane and/or highly hydrophobic proteins in different pH ranges in human cardiac tissue *(6)*. However, because each tissue/species contains proteins with different primary structure, different detergents may be appropriate for a given tissue/species. Correspondingly, one must often experiment with different detergents and chaotropic agents in both the extraction and IEF buffer to determine the optimum combination of buffers used for 2DE. We demonstrate here a methodology for optimizing 2DE patterns and an initial combination of buffers to utilize (*see* **Fig. 2**).

2. Materials

2.1. Protein Extraction from Tissue (see Note 1)

1. Buffer 1: 20 m*M* Tris-HCl, pH 6.8 (molecular biology-grade), 7 *M* urea (electro-phoresis-grade), 2 *M* thiourea (electrophoresis-grade), 4% CHAPS.
2. Buffer 2: 20 m*M* Tris-HCl, pH 6.8, 7 *M* urea, 2 *M* thiourea (same grade as extraction buffer 1), 4% ASB-14 (Calbiochem, EMD Biosciences, San Diego, CA).
3. Buffer 3: 20 m*M* Tris-HCl, pH 6.8, 5 *M* urea, 2 *M* thiourea, 2% CHAPS (same grades as extraction buffer 1), 2% 3-(Decyldimethyl-ammonio) propane sulfonate inner salt (SB3-10) (Sigma, St. Louis, MO).
4. Buffer 4: 20 m*M* Tris-HCl, pH 6.8, 2% SDS (electrophoresis-grade).
5. Duall brand beveled glass tissue grinder (Kontes, Vineland, NJ).
6. 2D quantification kit (GE Healthcare).
7. Con-torque high-torque low-rpm drill press (Erbach Corp, Ann Arbor, MI).

2.2. Isoelectric Focusing (see Note 2)

1. Buffer 1: 7 *M* urea, 2 *M* thiourea, 4% CHAPS (same grades as extraction buffer 1), 1% dithiothreitol (DTT) (molecular biology-grade), 0.5% pH 3.0–10.0 Biolyte-Ampholytes (Bio-Rad, Hercules, CA).
2. Buffer 2: 7 *M* urea, 2 *M* thiourea, 1% DTT, 4% ASB-14, 0.5% ampholytes.
3. Buffer 3: 5 *M* urea, 2 *M* thiourea, 1% DTT, 2% CHAPS, 2% SB3-10, 0.5% ampholytes (same grades as extraction buffer 2 and IEF buffer 1).
4. 18 cm pH 3.0–10.0 immobilized pH gradient (IPG) strips (Bio-Rad) (*see* **Note 3**).
5. IEF focusing tray (18 cm).
6. Protean IEF cell (Bio-Rad).

3. Methods

3.1. Protein Extraction From Tissue

1. Weigh tissue.
2. Homogenize tissue in 10 times weight-to-volume of a extraction buffer 1 using the tissue grinder attached to the con-torque press at 4°C (*see* **Note 4**).
3. Transfer the homogenate to a centrifuge tube.
4. Centrifuge at 18,000*g* for 15 min at 4°C.
5. Remove supernatant and transfer to a new centrifuge tube.
6. Centrifuge at 18,000*g* for 5 min at 4°C.
7. Transfer new supernatant to 600 µL Eppendorff tubes in 50-µL aliquots.
8. Determine protein concentration using a 2D quantification kit (GE Healthcare).
9. Freeze tubes in liquid nitrogen and store at −80°C.
10. Repeat extraction using each of the common extraction buffers listed in **Subheading 2.1.**

3.2. Isoelectric Focusing

1. Mix 200 µg of protein from one of the extracts with IEF buffer 1 to obtain a final volume of 350 µL (*see* **Note 5**).

Table 1
Grid of Possible Conditions

	IEF buffer		
Extraction buffer	Buffer 1	Buffer 2	Buffer 3
Buffer 1	No. spots	No. spots	No. spots
Buffer 2	No. spots	No. spots	No. spots
Buffer 3	No. spots	No. spots	No. spots
Buffer 4	No. spots	No. spots	No. spots

2. Using distilled, deionized water, wet paper wicks and lay them over the electrodes.
3. Pipet the sample into the middle of the IPG tray such that no air bubbles are present and liquid is spread out along length of tray.
4. Lay IPG strip over the sample gel side down such that the ends are touching the electrodes and no air bubbles are present under the strip.
5. Overlay with mineral oil and focus strips as described in manufacturer's protocol depending on pH gradient used.
6. In order to ensure reproducibility between runs, only one type of extraction and focusing buffer should be used for a given IEF run.

3.3. Determining Optimum Conditions

1. Following IEF, run the second dimension as described in manufacturer's protocol and stain gels using any method (*see* **Note 6**).
2. Using a gel analysis software program determine the number of spots per gel.
3. Record this number in **Table 1**.
4. Repeat procedure using each combination of common extraction and IEF buffers listed in **Subheadings 2.1.** and **2.2.** (*see* **Note 7**).
5. Repeat using 100 and 400 µg protein loads (*see* **Note 8**).
6. The conditions that demonstrate the maximum solubility will be apparent based on the coordinate in the table that has the maximum number of spots.
7. If the results are not satisfactory, additional reducing agents such as approx 1.5% (v/v) 2-hydroxyethyl disulfide (or DeStreak reagent; GE Healthcare) may be added to the IEF buffers. Alternatively, replacing the zwitterionic detergents in the material section with others such as IGEPAL (Sigma), or Triton X-100 in either extraction (*see* **Subheading 2.1.**) or IEF buffer (*see* **Subheading 2.2.**) may be beneficial. In addition, including 20% glycerol in either the extraction or IEF buffer may assist solubility.

3.4. Conclusion

Tissue samples taken from different species and different organs contain different populations of proteins. As such, the buffer used for maximum protein solubility during extraction will differ depending on the tissue examined. However, this buffer may not be appropriate or effective at IEF for 2DE. We have outlined a protocol for the effective screening of different buffers in order to determine maximum protein solubility for 2DE. For this protocol we have listed common buffers that are useful; however, these buffers should be considered solely as a "starting point" and many other possible buffers are available for this purpose.

4. Notes

1. SB3-10 cannot be solubilized in buffers with greater than 5 M urea. All percentages given are in (w/v).
2. The ampholytes used here should be the same as that used in the IPG strips. All percentages given are in (w/v) except ampholytes, which are (v/v).
3. Any pH gradient can be used, however, pH 3.0–10.0 strips allow for a general overview of a large pH range.
4. The glass tissue grinder can be connected to the con-torque press by attaching a small piece of rubber tubing over the grinder.
5. Any load of interest can be used at this stage, but using a 200 µg load allows the properties of the IEF buffer to dominate. However, it is important to keep the concentration of SDS below 0.05% for IEF because the charge will inhibit focusing (maximum ~40 µL of extraction buffer 4).
6. Depending on the location of the majority of protein spots, different concentrations of acrylamide may be used in the second dimension. A couple of good initial concentrations to use are 8% and 12% acrylamide. This will separate both low- (12% acrylamide) and high- (8% acrylamide) molecular-weight proteins.
7. Any combination of buffers can be used for this optimization; however, the buffers listed in the materials section have been shown previously to be effective at solubilizing human myocardial proteins. Different tissue types may require different concentrations of detergents; however, the detergents listed here are a useful place to start.
8. Different protein concentrations should be used because this will highlight one of the key problems with SDS as a solubilizing agent. In many cases, the high protein load will only be possible with extracts containing zwitterionic detergents, owing to the 40 µL restriction on buffers containing 2% SDS. For instance, at low protein loads, the SDS may be superior; however, higher protein loads will reveal more protein species with the extracts containing zwitterionic detergents.

References

1. Gorg, A., Weiss, W., and Dunn, M. J. (2004) Current two-dimensional electrophoresis technology for proteomics. *Proteomics* **4**, 3665–3685.
2. Molloy, M. P. (2000) Two-dimensional electrophoresis of membrane proteins using immobilized pH gradients. *Anal. Biochem.* **280**, 1–10.
3. Herbert, B. (1999) Advances in protein solubilization for two-dimensional electrophoresis. *Electrophoresis* **20**, 660–663.

4. Harder, A., Wildgruber, R., Nawrocki, A., Fey, S. J., Larsen, P. M., and Gorg, A. (1999) Comparison of yeast cell protein solubilization procedures for two-dimensional electrophoresis. *Electrophoresis* **20,** 826–829.
5. Ames, G. F. and Nikaido, K. (1976) Two-dimensional electrophoresis of membrane proteins. *Biochemistry* **15,** 616–623.
6. Stanley, B. A., Neverova, I., Brown, H. A, and Van Eyk, J. E. (2003) Optimizing protein solubility for two-dimensional analysis of human myocardium. *Proteomics* **3,** 815–820.

6

A Method for the Effective Depletion
of Albumin From Cellular Extracts

Application to Human Myocardium

Brian A. Stanley and Jennifer E. Van Eyk

Summary

Proteomic analysis of large numbers of proteins is assisted if each protein species is present at approximately equal concentrations. As such, the extraction of proteins from tissue samples should be designed to maintain a limited dynamic range in the concentration of proteins present. However, in many tissue extracts a high concentration of serum albumin exists from tissue perfusion and/or an inability to effectively rinse the tissue owing to surgical limitations. The analysis of these tissues would be assisted if contaminating serum albumin could be reduced. This chapter outlines a protocol for the effective reduction of serum albumin levels from human myocardium extracts enriched for soluble cytoplasmic proteins.

Key Words: Tissue extraction; serum albumin; protein abundance.

1. Introduction

In order to perform a proteomic analysis on a tissue sample, the constituent proteins must first be extracted while maintaining their solubility for downstream separation (e.g., two-dimensional [2D] gels, liquid chromatography, etc.). However, in many cell types a considerable difference can exist between the most abundant and the least abundant proteins *(1)*. This large dynamic range of proteins can cause considerable bias in proteins that are effectively analyzed owing to technical limitations in proteomic methodologies (i.e., limited dynamic range and/or sensitivity of 2D gel stains) *(2,3)*. This is the case for cardiac tissue in which contractile proteins account for the majority of protein mass within cardiomyocytes. One method to observe these lower abundant proteins is through

From: *Methods in Molecular Biology, vol. 357: Cardiovascular Proteomics: Methods and Protocols*
Edited by: F. Vivanco © Humana Press Inc., Totowa, NJ

the use of sequential extractions to select different subproteomes for future analysis. This is the basis of multiple sequential extraction methods, including the "IN Sequence" method developed in our lab (*see* Chapter 8). In many sequential extraction protocols (including multiple commercially available kits), a common initial step is to obtain the most soluble proteins by homogenizing the tissue in a neutral buffer (e.g., HEPES, pH 7.4). In cardiac tissue this is a useful extract because it minimizes the quantity of myofilament proteins present in this initial extract. Unfortunately, the most soluble protein often found in tissue extracts is serum albumin, owing to serum contamination of the tissue. This contamination is greater in highly perfused tissues such as myocardium, as well as in human tissue in which large amounts of saline washes during collection may not be feasible owing to surgical constraints. Regardless of the extraction method used, this contaminating serum albumin is problematic because it creates a wide dynamic range of proteins present in these cytoplasmic-enriched extracts that should be removed. We demonstrate here a modified version of an albumin purification procedure first developed by Hao *(4)*. This method can be used on cytoplasmic-enriched extracts containing high levels of albumin in order to reduce the dynamic range of these extracts, as well as to reduce the sample-to-sample variation as a result of variable albumin contamination (*see* **Fig. 1**). This methodology utilizes a selective salt/ethanol precipitation to precipitate most sample pro-teins while maintaining the albumin in solution.

2. Materials

2.1. Albumin Depletion

1. A cytoplasmic-enriched protein extract from human myocardium (1 mL) (*see* Chapter 8).
2. 25 m*M* HEPES, pH 7.4, 1% sodium dodecyl sulfate (SDS).
3. 25 m*M* HEPES, pH 7.4, 0.1% SDS, 4 *M* urea buffer.
4. 95% Ethanol stock (1 mL) (Sigma, St. Louis, MO).
5. 100 µL of 5 *M* NaCl stock.
6. Labquake end-over-end shaker (Barnstead/Themolyne, Dubuque, IA).
7. Acetone kept at −20°C.

3. Methods

3.1. Albumin Depletion

1. Perform the initial homogenization on a tissue sample of approx 100 mg such that there is a cytoplasmic extract with serum albumin contamination of approx 800 µL (*see* **Note 1**).
2. Transfer a volume of the extract to be depleted (e.g., 150 µL) to a centrifuge capable tube (e.g., 1.5-mL Eppendorff).

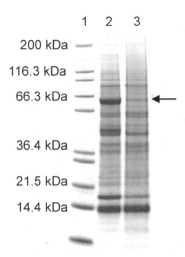

Fig. 1. 10 μg of cytoplasmic protein-enriched extract from human myocardium before albumin depletion (lane 2) and after albumin depletion (lane 3) separated by one-dimensional SDS-PAGE gel (4–12% Nupage gel, MES buffer; Invitrogen, Carlsbad, CA). Lane 1 contains molecular-weight markers (Mark XII, Invitrogen). The position of albumin is shown by the arrow.

3. Increase the cytoplasmic-enriched extract to a final concentration of 0.15 M NaCl (using a 5 M NaCl stock solution; approx 4.7 μL of stock solution/150 μL extract) (*see* **Note 2**).
4. Transfer to an end-over-end shaker and rotate for 1 h at 4°C (*see* **Note 3**).
5. Using a 95% ethanol stock solution, bring the salt/extract mixture to a final concentration of 47% ethanol (152 μL of 95% ethanol/155 μL of salt/extract mixture) (*see* **Note 2**).
6. Transfer to an end-over-end shaker and rotate for 1 h at 4°C.
7. Centrifuge mixture at 12,000g for 15 min at 4°C.
8. Remove supernatant and store at −80°C (*see* **Note 4**).
9. Suspend the pellet in the original volume (i.e., 150 μL) of 25 mM HEPES, pH 7.4, 1% SDS (*see* **Note 5**).

3.2. Acetone Precipitation for 2D Electrophoresis

1. Precipitate the solution obtained in **Subheading 3.1.** by adding 8X (v/v) of acetone (stored at −20°C) to the suspended pellet (e.g., 1200 μL of acetone for 150 μL extract) and let sit overnight at −20°C (*see* **Notes 3** and **6**).
2. Centrifuge at 18,000g for 15 min at 4°C.
3. Remove supernatant and discard.
4. Re-suspend the pellet in 100 μL of 25 mM HEPES, pH 7.4, 0.1% SDS, 4 M urea (*see* **Note 7**).

3.3. Conclusions

We demonstrate here a protocol for the effective depletion of serum albumin contamination from cytoplasmic extracts of myocardial tissue. Although human myocardium was used as the tissue source in this protocol, this procedure is valuable as it can be used for any extract containing large amounts of albumin. However, it must be stressed that owing to sequence variation in albumin from different species, the specificity of precipitation will vary from species to species.

4. Notes

1. This protocol can be applied to any protein extract that contains soluble cytoplasmic extracts. However, the protocol as listed uses the first extract from the "IN sequence" protocol (*see* Chapter 8).
2. Albumin exhibits a considerably higher solubility compared with other proteins at these particular concentrations of NaCl and ethanol (0.15 M; 47%), which allows it to stay in solution, whereas other proteins precipitate. However, some albumin does precipitate, and some other proteins stay in solution under these conditions. The concentration of either NaCl or EtOH can be increased with the result that other proteins will precipitate more completely; however, it will also increase the amount of albumin present in the pellet. Consequently, it is a trade-off between total protein yield and maintaining reduced albumin in the pellet.
3. In order to ensure reproducibility between runs, a consistent timing of this step is important.
4. The supernatant contains the majority of the albumin and other proteins that did not precipitate. Consequently, this fraction should be stored for future analysis.
5. There is a large amount of salt present, and as such, the extract should only be used for downstream methods, which are compatible (i.e., one-dimensional SDS-PAGE).
6. This step is necessary for those types of analyses that are intolerant of high concentrations of salt. This includes 2D SDS-PAGE, because the high concentration of salt can affect the ability of the proteins to focus in the first dimension.
7. This buffer is suitable for many downstream applications; however, complete solubilization of this pellet may require incubation at 55°C for 10 min combined with considerable vortexing. This can cause protein carbamylation. As such, higher levels of urea can be used instead of the high temperature.

References

1. Corthals, G. L., Wasinger, V. C., Hochstrasser, D. F., and Sanchez, J. C. (2000) The dynamic range of protein expression: a challenge for proteomic research. *Electrophoresis* **21,** 1104–1115.
2. Rabilloud, T. (1999) Silver staining of 2-D electrophoresis gels. *Methods Mol. Biol.* **112,** 297–305.
3. Nishihara, J. C. and Champion, K. M. (2002) Quantitative evaluation of proteins in one- and two- dimensional polyacrylamide gels using a fluorescent stain. *Electrophoresis* **23,** 2203–2215.
4. Hao, Y. (1979) A simple method for the preparation of human serum albumin. *Vox Sang.* **36,** 313–320.

II

HEART SUBPROTEOMES

7

Subcellular Fractionation

Mariana Eça Guimarães de Araújo and Lukas Alfons Huber

Summary

The successful combination of highly sensitive mass spectrometry and pre-fractionation techniques has provided a powerful tool to detect dynamic changes in low abundant regulatory proteins at the organelle level. Subcellular fractionation, being flexible, adjustable (both in cell and tissues), and allowing the analysis of proteins in their physiologic/intracellular context, has become the most commonly used preparative/enrichment method. This chapter introduces state-of-the-art subcellular fractionation protocols and briefly discuss their suitability, advantages, and limitations.

Key Words: Subcellular fractionation; homogenization; gradient centrifugation; organelles; nuclei; cytosol; total membranes; peripheral membrane proteins; integral membrane proteins.

1. Introduction

The present estimate of the current number of proteins within a cell now reaches 100,000. This number represents the diversity generated by hundreds of chemical modifications that occur upon protein synthesis, altering the chemical and physical properties of proteins. Furthermore, protein abundance may range between 7 and 10 orders of magnitude, with the result that high-abundance proteins usually mask the presence of low-abundance ones during analyses. Because of the limited power of analytical separation techniques presently applied to protein profiling and expression analyses, prefractionation strategies are required to reduce sample complexity (1), and allow the detection of less abundant proteins such as kinases, phosphatases, and GTPases.

This chapter focuses on subcellular fractionation as the first and essential step among enrichment techniques. Subcellular fractionation—allowing the separation of organelles based on their physical or biological properties—consists of two major steps:

From: *Methods in Molecular Biology, vol. 357: Cardiovascular Proteomics: Methods and Protocols*
Edited by: F. Vivanco © Humana Press Inc., Totowa, NJ

1. Disruption of the cellular organization (homogenization).
2. Fractionation of the homogenate to separate the different populations of organelles *(2,3)*.

   The objective of the homogenization step is to release the organelles and other cellular constituents as a free suspension of intact individual components *(4)*. The assessment of the homogenate, either by phase-contrast microscopy or latency measurement of organelle content markers, is an important control because the quality of the homogenate critically influences the success of the following steps. Unfortunately, the cytoplasmic and cytoskeletal organization varies enormously between different tissue culture cell lines, implying that homogenization conditions must be optimized for each cell line *(3)*.

   Several techniques exploiting physical and biological parameters have been used to isolate organelles and membranes upon homogenization. Among those, centrifugation is considered the most effective, being both easy to set up and ideally combined with analytical proteomics techniques *(5)*. Here we introduce several centrifugation methods, from the simple separation of cytosol and total membranes, to a more refined organelle-based separation. However, it must be emphasized that complete purification is—with few exceptions—rarely possible *(1,3)*. Different subcellular compartments share similar physical properties and co-migrate at least to some extent in conventional gradients *(5,6)*.

2. Materials

2.1. Standard Homogenization Protocol

1. Homogenization buffer (HB): 250 mM sucrose, 3 mM imidazole, pH 7.4, 1 mM EDTA, double-distilled (dd)H$_2$O, protease inhibitors.
2. Phosphate-buffered saline$^-$ (PBS$^-$): 137 mM NaCl, 2.7 mM KCl, 1.5 mM KH$_2$PO$_4$, 6.5 mM Na$_2$HPO$_4$, 1 mM CaCl$_2$, 1 mM MgCl$_2$, 0.5X protease inhibitors (optional: some cells already break during scraping).

2.2. Homogenization Protocol for Cells Requiring Hypotonic Shock

1. Buffer A (HBA): 3 mM imidazole, pH 7.4, 1 mM EDTA, ddH$_2$O, protease inhibitors.
2. Buffer B (HBB): 500 mM sucrose, 3 mM imidazole, pH 7.4, 1 mM EDTA, ddH$_2$O, protease inhibitors.
3. PBS$^-$: 137 mM NaCl, 2.7 mM KCl, 1.5 mM KH$_2$PO$_4$, 6.5 mM Na$_2$HPO$_4$, 1 mM CaCl$_2$, 1 mM MgCl$_2$, 0.5X protease inhibitors (optional: some cells already break during scraping).

2.3. Latency Measurement

2.3.1. Fluid Phase Internalization of Horseradish Peroxidase on Dishes

1. Horseradish peroxidase (HRP).

2. Phosphate-buffered saline$^+$ (PBS$^+$): 137 mM NaCl, 2.7 mM KCl, 1.5 mM KH$_2$PO$_4$, 6.5 mM Na$_2$HPO$_4$, 1 mM CaCl$_2$, 1 mM MgCl$_2$.
3. PBS$^+$/bovine serum albumin (BSA): PBS$^+$ containing 5 mg/mL BSA.
4. Internalization medium: Modified Eagles' medium (MEM); 10 mM HEPES, pH 7.4, 5 mM D-glucose.

2.3.2. Determination of HRP

1. HB plus (HB$^+$): 250 mM sucrose, 3 mM imidazole, pH 7.4, 1 mM EDTA, 0.03 µM cycloheximide, protease inhibitors.
2. O-dianisidine. This is a very toxic and light-sensitive reagent. Wear gloves at ALL times, keep the reagent light protected, and perform the reactions in the dark.
3. 0.3% H$_2$O$_2$, 0.5 M NaPO$_4$, pH 5.0, 2% Triton X-100, 1 mM KCN.

2.4. Separating Integral From Peripheral Membrane Proteins

1. 0.1 M Na$_2$CO$_3$, pH 11.0 (for addition to membrane pellet).
2. 0.2 M Na$_2$CO$_3$, pH 11.0 (for addition to solubilized membranes).

The solutions should be prepared fresh and kept on ice.

2.5. Subcellular Fractionation Using a Step Gradient

High percentage sucrose solutions take some time to dissolve and should, therefore, be prepared the day before the experiment (*see* **Note 1**).

1. HB: 250 mM sucrose, 3 mM imidazole, pH 7.4, 1 mM EDTA, ddH$_2$O, protease inhibitors.
2. 25% (w/w) sucrose solution in ddH$_2$O: 0.806 M sucrose, 3 mM imidazole, pH 7.4, 1 mM EDTA, ddH$_2$O. The refractive index of this solution should be 1.3723 at 20°C.
3. 35% (w/w) sucrose solution in ddH$_2$O: 1.177 M sucrose, 3 mM imidazole, pH 7.4, 1 mM EDTA, ddH$_2$O. The refractive index of this solution should be 1.3904 at 20°C.
4. 62% (w/w) sucrose solution in ddH$_2$O: 2.351 M sucrose, 3 mM imidazole, pH 7.4, 1 mM EDTA, ddH$_2$O. The refractive index of this solution should be 1.4463 at 20°C.

2.6. Subcellular Fractionation Using a Continuous Gradient

Please take into consideration the information provided in **Subheading 2.5.** of this materials chapter concerning the preparation of sucrose solutions.

1. HB: 250 mM sucrose, 3 mM imidazole, pH 7.4, 1 mM EDTA, ddH$_2$O, protease inhibitors.
2. 10% (w/w) sucrose solution in ddH$_2$O: 0.3 M sucrose, 3 mM imidazole, pH 7.4, 1 mM EDTA, ddH$_2$O. The refractive index of this solution should be 1.3479 at 20°C.
3. 40% (w/w) sucrose solution in ddH$_2$O: 1.375 M sucrose, 3 mM imidazole, pH 7.4, 1 mM EDTA, ddH$_2$O. The refractive index of this solution should be 1.3997 at 20°C.

Fig. 1. Gentle scraping of a cellular monolayer. Dish was placed on a ice-cold metal plate and washed with cold phosphate-buffered solution (PBS). Cells were scraped in one continuous 360° movement using a rubber policemen followed by the removal of a small white sheet of cells from the scraper by fast vertical shaking within the PBS solution.

3. Methods

3.1. Standard Homogenization Protocol

For mechanically breaking the plasma membrane, we describe a needle/ syringe-based protocol that does not require specialized equipment, is easy to handle, and gives reproducible results.

1. Passage cells so that they are approx 80–90% confluent at the start of the experiment (*see* **Note 2**).
2. The subsequent harvesting, homogenization, and fractionation steps should be performed at 4°C. Also, keep all solutions on ice and add protease inhibitors to the buffers.
3. Wash the plates three times with ice cold PBS. Add PBS⁻ and scrape the cells in one continuous movement using a rubber policeman (*see* **Fig. 1**). Cells should come off the plate in form of small white sheets. Transfer the cells to 15-mL tubes using a transfer pipet. Centrifuge at 400*g* for 5 min, at 4°C.
4. Resuspend each pellet in three times the pellet volume (3 vol) of HB using a cut pipet tip. The aim is to re-buffer the cell pellet and it is not necessary to remove all existing clumps. Centrifuge at 1100*g* for 10 min, 4°C.
5. Resuspend each pellet *gently* with HB using a cut wide tip. Verify that no cell clumps are visible and avoid formation of air bubbles while pipetting (*see* **Note 3**). Split the suspension into several 15-mL tubes (1 mL/tube maximum).

6. Attach a 22G needle to a 1-mL syringe. The syringe should be pre-washed with ice-cold HB (*see* **Note 4**) and all air bubbles removed from the needle.
7. Pass the suspension through the syringe 2–10 times. Monitor the homogenization by phase contrast microscopy. Please note that the number of strokes is highly dependent on the manual force of the operator and should, therefore, be carefully controlled. Homogenization is best achieved at a ratio of 70–80% clear nuclei to intact cells (*see* **Note 5**).
8. Add 700 µL HB to each 1 mL of homogenized cells. Centrifuge at 1600*g* for 10 min, 4°C.
9. Carefully retrieve the postnuclear supernatant (PNS) and the nuclear pellet. Leave an aliquot of both for later analysis. The PNS is now ready for fractionation. For additional analysis, nuclei can be purified from the pellet fraction.

3.2. Homogenization Protocol for Cells Requiring Hypotonic Shock

Certain cells are more difficult to homogenize either because they tend to round up when scraped and/or because they do not form strong intercellular contacts. These cells need to be swollen prior to homogenization (*see* **Note 6**).

1. Perform **steps 1** and **2** of the standard homogenization protocol.
2. Re-buffer the cells using three times the pellet volume (3 vol) of HBA in such a way that the entire pellet floats up. Centrifuge 10 min at 178*g*, 4°C and aspirate the HBA buffer.
3. Re-suspend the cells *gently* in 0.5–1 vol of HBA using a cut blue tip. Check for the absence of cellular clumps and keep the cells on ice for additional 10–20 min (swelling).
4. Add an equal volume of HBB to that of HBA used. Carefully and slowly agitate the tube until no visible smears remain. Split the suspension into several 15-mL tubes (1 mL/tube maximum).
5. Perform **steps 5–9** from **Subheading 3.1.** of the standard homogenization protocol.

3.3. Latency Measurement

The latency measurement allows the quantification of disrupted organelles during homogenization using internalized phase markers such as HRP or avidin *(4)* (*see* **Note 7**).

3.3.1. Fluid-Phase Internalization of HRP on Dishes

1. Use a metal plate to cool dishes to ice temperature using a basket filled with crushed ice. Assure that the dishes are set completely flat.
2. Wash dishes three times with PBS⁺ for 2–3 min on a shaking platform.
3. Place the dishes for 2–3 s on a 37°C water bath and replace the PBS⁺ with warmed-up internalization medium containing 3.2 mg/mL avidin or 1.8 mg/mL HRP (*see* **Note 8**).
4. Incubate dishes on a rocking platform at 37°C for the desired time (*see* **Table 1**). A waterbath can also be used.

Table 1
Required Internalization Times for Different Endocytic Compartments

Fraction	Time of internalization
Early endosomes	5 min
Carrier vesicles	10 min + 30 min chase upon microtubules depolymerization
Late endosomes	10 min + 30 min chase

Depolymerization of microtubules is achieved with 1–2 h incubation in the presence of 10 μM nocodazole and addition of this drug to subsequent buffers up to the internalization step *(4)*.

If chase is required upon internalization, perform the following steps before proceeding to **step 5** of the fluid-phase internalization protocol (*see* **Subheading 3.3.1.**):

Immediately upon internalization, replace the dishes on ice and wash them two to three times with pre-warmed PBS+/BSA.

Change the PBS+/BSA solution to internalization medium containing 2 mg/mL BSA and incubate the dishes at 37°C for the necessary time.

5. Cool the dishes again and wash them three times with PBS+/BSA (5 min per wash).
6. Wash twice with PBS+. The dishes are ready for homogenization.
7. Homogenize cells according to the standard homogenization protocol or the protocol for cells requiring hypotonic shock (*see* **Subheadings 3.1.** or **3.2.** of this chapter). Collect aliquots of the nuclear pellet and PNS.
8. The remaining PNS can then be used to purify the fraction of interest (e.g., perform a step gradient and collect interphases to analyze the integrity of late and early endosomes).

3.3.2. Determination of HRP

1. Dilute the PNS in HB+ and centrifuge 30 min at 260,000g, 4°C.
2. Collect the supernatant. Re-suspend the pellet in HB+.
3. Using clean glassware mix 12 mL of 0.5 M NaPO$_4$, pH 5.0, and 6 mL of 2% Triton X-100 with 100.8 mL of ddH$_2$O. Gently dissolve 13 mg of *O*-dianisidine into the previous solution and finally add 1.2 mL of 0.3% H$_2$O$_2$. Do not stir the solution. Be careful when weighing *O*-dianisidine; it is a carcinogenic and highly toxic substance.
4. Prepare blanks and standards diluting 1–10 ng of HRP in HB+.
5. Dilute samples, blanks, and standards 1:10 into the solution previously prepared as explained in **step 3**. Quickly mix and place the plate in the dark at room temperature. Record the time.
6. When the mixture turns brownish, stop the reaction by adding 0.01 times the reaction volume (0.01 vol) of 1.0 mM KCN. Read the absorbance at 455 nm. Results are expressed as optical density (OD) units/min. Make a standard curve using the concentrations and the absorbance values obtained for the standards, and calculate the amount of HRP present in the PNS and in each sample. The reaction is in the linear range up to an OD of about 1.5.

7. Calculate the HRP specific activity of each sample as ng of HRP/µg of protein (*see* **Table 2**). Calculate the fraction's enrichment as its HRP activity per HRP activity of the PNS. Determine the yield of each fraction as percentage of its specific activity per total activity present in the PNS.

3.4. Preparation of Cytosol and Total Membranes

Upon homogenization, a low-speed centrifugation step removes nuclei together with cell debris, unbroken cells, and some large subcellular components. The PNS contains the cytosol and the organelles in free suspension *(5,7)*. Using the prepared PNS and with a single centrifugation step, one can separate the cytosol from total membranes, thereby reducing the complexity of the sample for further analyses.

1. Prepare a PNS fraction according to the protocols described in **Subheadings 3.1.** and **3.2.**
2. Centrifuge the PNS at 100,000*g*, 1 h, 4°C.
3. Carefully remove the supernatant (cytosolic fraction) and place it into a fresh tube. Immediately use this fraction or store at −80°C upon snap-freezing in liquid nitrogen.
4. Re-suspend the pellet containing the membranes in an appropriate detergent-containing buffer (*see* **Note 9**). If not for immediate use, store the total membranes at −80°C upon snap-freezing in liquid nitrogen.

3.5. Separating Integral From Peripheral Membrane Proteins

We present here an extraction protocol using sodium carbonate at high pH *(7,8)*, that breaks organelle membranes (they form membrane sheets instead of round-up intact forms), while at the same time shaves of all the peripheral attached proteins. Using this method, one can extract peripheral membrane proteins both from the internal and the external face of the membranes.

1. Prepare a total membrane fraction as described in **Subheading 2.4.**
2. Re-suspend the membrane pellet in 0.1 *M* Na_2CO_3, pH 11.0, freshly prepared. Final volume should be 150–200 µL for ~50 µL pellet. If membranes in suspension are used, dilute them 1:3 in 0.2 *M* Na_2CO_3, pH 11.0. Please take into account that the final volume has to fit into a centrifuge tube!
3. Incubate on ice for 15 min.
4. Centrifuge the membranes at 230,000*g*, 4°C for 10 min.
5. Collect the supernatant and store on ice.
6. Re-suspend the pellet in 0.1 *M* Na_2CO_3, pH 11.0 as in this **Subheading** and repeat **steps 2–4**. Poll the supernatants (peripheral membrane proteins). If not for immediate use, store at −80°C upon snap-freezing in liquid nitrogen.
7. Re-suspend the pellet (integral/transmembrane proteins) in an appropriate detergent-containing buffer (*see* **Note 9**).

Table 2
Latency's Calculation of Organelle Content Markers[a]

Fraction	Volume (µL)	Prot. protein (µg)	Prot. conc. (µg/µL)	HRP (ng)	HRP (ng/µL)	HRP activity	Enrichment	Tot. HRP (ng)	Yield (% HRP)
Late	7	200	0.679	0.226	0.361	0.181	0.797	0.169	36.1	1.52
Endosomes	8	200	0.641	0.214	0.525	0.263	1.229	0.261	52.5	2.21
	9	200	0.635	0.212	0.708	0.354	1.672	0.355	70.8	2.98
Early	13	200	0.868	0.289	2.953	1.477	5.103	1.083	295.3	12.41
Endosomes	14	200	0.894	0.298	3.157	1.579	5.297	1.124	315.7	13.27
	15	200	0.833	0.278	3.627	1.814	6.531	1.386	362.7	15.25
PNS		240	2.103	2.103	9.913	9.913	4.714		2379.12	100.00

[a]Calculations are given for six example fractions. Early endosomes were labeled after 5 min HRP internalization. Cells were homogenized according to our standard protocol (*see* **Subheading 3.1**). The PNS obtained was further fractionated on a continuous gradient (*see* **Subheading 3.7**). Upon centrifugation, we determined the amount of HRP in each fraction and the corresponding fraction's latency.

8. If the fractions are to be loaded directly on SDS-PAGE, proteins present in the supernatant (peripheral membrane proteins), should be precipitated, for instance, by using the Wessel-Fluegge method *(9)* (*see* **Note 10**). The obtained pellet can then be re-suspended in standard Laemmli sample buffer *(10)*.

3.6. Subcellular Fractionation Using a Step Gradient

Before starting, add protease inhibitors and cycloheximide to the sucrose solutions. Cycloheximide inhibits the release of ribosomes from the endoplasmic reticulum; the high-protein content makes this organelle quite heavy and, therefore, easier to separate from other lighter organelles *(6)*. (Work strictly on ice!)

1. Adjust the sucrose concentration of the PNS to 40.6% by adding 62% sucrose (usually 1:1.2 [v/v]) (*see* **Note 12**). Control the sucrose concentration of the solution using a refractometer. Add HB or 62% sucrose if further adjustment is required.
2. Load the diluted PNS on the bottom of an ultracentrifuge tube. The choice of tube to use will depend on the volume of the PNS and the ultracentrifuge rotor of choice.
3. Overlay sequentially with 1.5 vol of 35% sucrose, 1 vol of 25% sucrose, and, finally, fill up the rest of the tube with HB. Mark the interfaces with a waterproof pen.
4. Mount the tubes on the appropriate rotor and centrifuge at $210,000g$, 4°C, for 1.5 h. It is very important to balance the tubes with an identical gradient.
5. After centrifugation, one should be able to detect a milky band of membrane particles at each interface (*see* **Fig. 2**). Collect the different interfaces and an aliquot of the sucrose cushions. It is easier to collect an interface with the top cushion still remaining.
6. Fractions should be snap-frozen in liquid nitrogen and stored at −80°C. Upon storage, thaw out the samples as rapidly as possible prior to use or analyses.
7. Using any of the standard methods to measure protein concentration, calculate yield and enrichment of each fraction relative to the initial PNS.
8. Prepare a gradient profile. In brief, load equal volumes of each of the fractions collected on an SDS-PAGE. Use antibodies against well-established organelle markers in order to determine the degree of separation achieved *(6)*.

3.7. Subcellular Fractionation Using a Continuous Gradient

Continuous gradients provide better resolution than step gradients at the expense of yield. After centrifugation to equilibrium, the organelles are distributed throughout the gradient according to their densities. The presence of subdomains within an organelle can be responsible for small differences in density, resulting in co-migration of organelles in several adjacent fractions. This effect also increases the final volume required to collect a specific organelle.

For this protocol it is advantageous to have access to both a gradient-forming centrifuge (e.g., gradient master, BioComp) and a fraction collector (e.g., Auto Density Flow, Labconco), both sensitive and costly pieces of equipment. Such instruments can largely improve handling and reproducibility. As mentioned

Fig. 2. Step gradient tube after ultracentrifugation. Organelles appear in distinct interfaces (milky bands).

for the step gradient protocol, add protease inhibitors and cycloheximide to the sucrose solutions before starting. Work strictly on ice and remember to keep a small aliquot of the PNS as a reference sample. Warm the 10% and 40% sucrose solutions to room temperature before starting.

1. Draw a mark for 50% of the volume of the tubes using the small metal block provided with the gradient master.
2. Fill half the tube with 10% solution. Using a metal syringe (also from the gradient master equipment), load the 40% sucrose underneath the 10% solution until the interphase between both is found exactly at the 50% volume mark.

3. Carefully close the tubes with the long black lids (*see* **Note 13**). Level the gradient master and perform the adequate run (*see* instruction manual for the equipment).
4. Remove the caps with care and place the gradients on ice. Once cooled, the gradient can be overloaded with the prepared PNS. Avoid disturbing the existing gradient when performing this step. Spin gradients overnight at 210,000g, 4°C.
5. Using the centering jaws, fix the centrifuged tube to the fraction collector. Turn the instrument on until the probe tip reaches the surface of the fluid.
6. Collect the fractions. Calculate the number of drops to place in each 1.5-mL precooled tube considering the total volume of the gradient prepared. The lightest fraction is labeled as fraction 1.
7. Fractions should be frozen in liquid nitrogen and stored at −80°C. Upon storage, thaw the samples as rapidly as possible prior to use.
8. Analyze the fractions as explained in **Subheading 3.6., steps 7** and **8**.

4. Notes

1. The concentration of all sucrose solutions must be controlled using a refractometer. This is a fragile instrument that should be handled with care.
2. In certain cell types, the passage number influences the homogenization result. Always use cells with a known passage number. Confluence of cells in the plate is another critical parameter (cell–cell interactions must be broken during homogeneization). Check that they have reached at least 80% confluence before starting. Finally, care should be taken that plates are set flat on the shelf of the incubator, so that a proper and even monolayer can be formed.
3. Homogenization is easier at high density of cells 20–30% (v/v); therefore, keep volume of HB low.
4. The presence of air bubbles in the needle generates unpredictable forces, leading to bad homogenization or extensive nuclear breakage.
5. Cytoplasmic aggregates are often found during homogenization owing to the presence of cytoskeletal elements that entrap organelles. These aggregates should be re-suspended as their presence might result in loss of yield owing to sedimentation *(3)*. In over-homogenized cells, on the other hand, organelles will break, releasing their content. As a consequence, the cytosolic fraction will be contaminated with typical organelle/membrane-bound material. Also, nuclear breakage releases DNA and acts as a potential source for aggregates *(5)*. Therefore, it is crucial to access the quality of the homogenization (phase-contrast microscopy or latency measurement).
6. The hypotonic shock renders the cells more fragile and, therefore, extra care should be taken when using this protocol. The presence of sucrose in the HBB is necessary to assure membrane integrity.
7. In brief, the protocol consists of an internalization step followed by thorough wash steps to remove any nonspecifically adsorbed marker. Upon homogenization, HRP amounts in the different fractions are quantified in a spectrophotometer at 455 nm

(brown product is formed upon the reaction of the peroxidase with O-dianisidine and H_2O_2).

8. HRP is a relatively stable enzyme and the assay is very sensitive. Avoid cross-contaminating samples or standards.

9. Any standard cell lysis buffer or HB with 1% Triton X-100 can be used. The choice of buffer depends mostly on the downstream application.

10. The Wessel/Fluegge precipitation method *(9)* is based on a defined methanol–chloroform–water mixture for quantitative precipitation of soluble as well as hydrophobic proteins from dilute solutions. The efficiency of this method is not affected by the presence of detergents, lipids, salt, buffers, and β-mercaptoethanol.

11. During centrifugation, organelles sediment (or float) until they reach their isopycnic position within the density gradient. Depending on their content, protein/lipid ratio, shape, and size, different organelles will have different densities. The degree of separation during gradient centrifugation will also depend on the nature of the medium. Although sucrose is the most common, there are many other alternatives, e.g., Ficoll, Percoll, Nycodenz, or Metrizamide *(11,12)*. The choice of gradient might be cell type- or tissue-dependent *(3)*. Based on this knowledge, the step gradient protocol was designed so that different organelles are enriched at different interphases between cushions of sucrose solutions. For instance, late endosomes and lysosomes will be found floating on the interface between 25% sucrose and HB; early endosomes and carrier vesicles will be enriched at the interphase between 35 and 25%; and finally, heavy membranes like Golgi and the endoplasmic reticulum membranes will be found floating at the interphase between 40.6 and 35% *(3,6,7)*.

12. Please note that endosomes are very fragile under these conditions. Ensure the solution is mixed both thoroughly and gently. Confirm the homogeneity of the solution under a lamp (no smears should be visible).

13. Be sure to remove all existing air bubbles because they will interfere with gradient formation. When closing the tubes with the black lids, it is possible that a bit of the 10% solution flows through the hole in the lid. Remove this excess of solution before starting the gradient master run.

Acknowledgments

We are thankful to Taras Stasyk for careful reading and comments. Work in the Huber laboratory is supported by the Austrian Genome Program (GEN-AU, Austrian Proteomics Platform) and by the FWF–SFB021 "Cell Proliferation and Cell Death in Tumors."

References

1. Huber, L. A. (2003) Is proteomics heading in the wrong direction? *Nat. Rev. Mol. Cell. Biol.* **4,** 74–80.

2. Stasyk, T. and Huber, L. A. (2004) Zooming in: fractionation strategies in proteomics. *Proteomics* **4,** 3704–3716.

3. Pasquali, C., Fialka, I., and Huber, L. A. (1999) Subcellular fractionation, electromigration analysis and mapping of organelles. *J. Chromatogr. B. Biomed. Sci. Appl.* **722,** 89–102.

4. Gruenberg, J. and Howell, K. E. (1988) Fusion in the endocytic pathway reconstituted in a cell-free system using immuno-isolated fractions. *Prog. Clin. Biol. Res.* **270,** 317–331.

5. Huber, L. A., Pfaller, K., and Vietor, I. (2003) Organelle proteomics: implications for subcellular fractionation in proteomics. *Circ. Res.* **92,** 962–968.

6. Fialka, I., Pasquali, C., Lottspeich, F., Ahorn, H., and Huber, L. A. (1997) Subcellular fractionation of polarized epithelial cells and identification of organelle-specific proteins by two-dimensional gel electrophoresis. *Electrophoresis* **18,** 2582–2590.

7. Pasquali, C., Fialka, I., and Huber, L. A. (1997) Preparative two-dimensional gel electrophoresis of membrane proteins. *Electrophoresis* **18,** 2573–2581.

8. Fujiki, Y., Hubbard, A. L., Fowler, S., and Lazarow, P. B. (1982) Isolation of intracellular membranes by means of sodium carbonate treatment: application to endoplasmic reticulum. *J. Cell Biol.* **93,** 97–102.

9. Wessel, D. and Flügge, U. I. (1984) A method for the quantitative recovery of protein in dilute solution in the presence of detergents and lipids. *Anal. Biochem.* **138,** 141–143.

10. Laemmli, U. K. (1970) Cleavage of structural proteins during the assembly of the head of bacteriophage T4. *Nature* **227,** 680–685.

11. Fleischer, S. and Kervina, M. (1974) Subcellular fractionation of rat liver. *Methods Enzymol.* **31(Pt A),** 6–41.

12. Fleischer, S. and Kervina, M. (1974) Long-term preservation of liver for subcellular fractionation. *Methods Enzymol.* **31(Pt A),** 3–6.

8

Subfractionation of Heart Tissue

The "In Sequence" Myofilament Protein Extraction of Myocardial Tissue

Lesley A. Kane, Irina Neverova, and Jennifer E. Van Eyk

Summary

Proteomic analysis of heart tissue is complicated by the large dynamic range of its proteins. The most abundant proteins are the myofilament proteins, which comprise the contractile apparatus. This chapter describes a protocol for fractionation of heart tissue that extracts the myofilament proteins into a separate sample fraction, allowing analysis of lower-abundance proteins. Importantly, this is performed in a manner that is compatible with two-dimensional electrophoresis and high-performance liquid chromatography, two of main technologies of proteomics. The method produces three fractions based on solubility at different pHs: (1) cytoplasmic-enriched extract (neutral pH), (2) myofilament-enriched extract (acidic pH), and (3) membrane protein-enriched pellet. Fractionation of heart tissue in this manner provides the basis for in-depth proteomic analysis.

Key Words: Myofilaments; subproteome fractionation; striated muscle.

1. Introduction

The most abundant proteins in the heart are the myofilament proteins, which facilitate and control muscle contraction. The major myofilament proteins include actin, heavy- and light-chain myosin, actinin, myosin binding protein C, desmin, tropomyosin, and the troponin complex, which consists of troponin I, troponin T, and troponin C. These proteins are essential to the function of the heart and, as such, are very important to analyze; however, their high abundance can also cause great difficulty in proteomic studies of the heart. All proteomic technologies struggle with the limitation of dynamic range (the range of protein quantities that can be detected by a specific method). Too much of any one

From: *Methods in Molecular Biology, vol. 357: Cardiovascular Proteomics: Methods and Protocols*
Edited by: F. Vivanco © Humana Press Inc., Totowa, NJ

protein (or group of proteins) will inhibit the ability to detect and analyze lower abundant proteins. One simple solution to this problem is subfractionation. For heart tissue, extracting the myofilament proteins is a convenient way to create reproducible subproteomes. However, the myofilament proteins are not easily soluble and readily aggregate. Previous extraction protocols have therefore required high salt levels and/or high detergent concentrations *(1)*. Both of these conditions are not compatible with two-dimensional gel electrophoresis (2DE) and high-performance liquid chromatography, two of the main separation technologies of proteomics. Also, these extractions were often long processes that could cause artificial posttranslational modifications within the proteome *(2)*. Our lab has developed a myofilament extraction method (termed "IN sequence" extraction) based on protein solubility at various pH levels, which is compatible with both of these technologies and preserves the original proteome during isolation *(3,4)*. The first extraction is performed at neutral pH (7.4) in the presence of a variety of kinase, phosphatase, and protease inhibitors to preserve the intact proteome. At this pH, only the most soluble proteins in the cell remain in solution and this extract is therefore termed the cytoplasmic-enriched extract. The insoluble portion of the first extract is then homogenized in an acidic solution in which more proteins are extracted (pH 2.2). Because the major myofilament proteins are soluble at this very acidic pH, this extract is the myofilament-enriched extract. There is a set of proteins that are not soluble under either of these conditions and remain as an insoluble pellet. This fraction contains mostly membrane proteins, which require high levels of detergent and/or urea to solubilize. This extraction protocol is unique in that it is not only robust and reproducible, but also reduces the likelihood of artificial modifications and is applicable to both cardiac and skeletal muscle *(2)* (**Fig. 1**). This protocol is also adaptable to both large and small tissue amounts *(5)*. The IN sequence extraction allows for subfractionation and thus the in-depth analysis by several proteomic techniques of both high- and low-abundance proteins present in heart tissue.

2. Materials

1. HEPES extraction buffer: 25 mM HEPES, pH 7.4, 50 mM NaF, 0.25 mM Na$_3$VO$_4$, 0.25 mM phenylmethylsulfonyl fluoride (PMSF), 2.5 mM EDTA, 1.25 μM leupeptin, 1.25 μM pepstatin A and 1 μM aprotinin. Prepare HEPES extraction buffer fresh immediately prior to use; PMSF inactivates quickly upon exposure to water (*see* **Notes 1** and **2**).
2. TFA extraction buffer: 1% (v/v) trifluoroacetic acid (TFA) (pH 2.2), 1 mM Tris (2-carboxyethylphosphine) hydrochloride (TCEP). Use caution; TFA is a dangerous acid and is generally kept as a 10% stock solution. Prepare TFA extraction buffer directly prior to use.

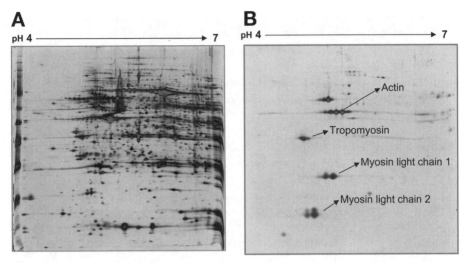

Fig. 1. 2D SDS-PAGE gels of the (**A**) cytoplasmic-enriched extract, 250 µg load and (**B**) myofilament-enriched extract, 20 µg load. (B) Some major myofilament proteins have been indicated. These images highlight the benefit of removing the myofilament proteins because even at the high load of 250 µg, the myofilament proteins are not overwhelming the lower abundant ones and on the myofilament gel even 20 µg is enough to saturate the staining on the gel.

3. Methods

3.1. HEPES Extraction

1. All steps must be performed at 4°C in a cold room or on ice.
2. Homogenize tissue in a Duall homogenizer with four times w/v HEPES extraction buffer (*see* **Note 3**).
3. Transfer homogenate to 1.5-mL microcentrifuge tubes.
4. Spin at 18,000*g* for 15 min (*see* **Note 4**).
5. Remove supernatant and keep as "HEPES fraction 1" (*see* **Note 5**).
6. Resuspend pellet in the same volume as **step 1** and spin at 18,000*g* for 15 min.
7. Remove supernatant and combine with "HEPES fraction 1." The combined supernatants are the final "HEPES extract" (cytoplasmic-enriched fraction).
8. Keep pellet for TFA extraction.

3.2. TFA Extraction

1. All steps must be performed at 4°C or on ice.
2. Resuspend the pellet from HEPES extraction in three times original tissue w/v in TFA extraction buffer.
3. Return solution to the homogenizer and grind until full solubilization (*see* **Note 6**).
4. Transfer homogenate to 1.5-mL microcentrifuge tubes.
5. Spin TFA suspension at 18,000*g* for 15 min.

6. Keep supernatant as the TFA fraction (myofilament-enriched).
7. Save pellet (membrane protein-enriched).

4. Notes

1. A 20 mM Tris buffer, pH 7.8, has been successfully used in place of the HEPES though it is thought that the better buffering capacity of HEPES at pH 7.4 makes it a more reproducible extraction. NaF, Na_3VO_4, and PMSF (kinase and phosphatase inhibitors): stock solutions of these inhibitors can be prepared ahead and stored for several months. (PMSF stock solution made in isopropanol). Leupeptin, pepstatin A, and aprotinin (protease inhibitors): stock solutions of these inhibitors should be prepared and aliquoted to avoid freeze–thaw cycles. Store at −20°C.
2. The pepstatin A, leupeptin, and aprotinin are standard protease inhibitors that can be replaced by commercial protease cocktails.
3. If the tissue sample is larger than 50 mg, the tissue must be powdered prior to homogenization. This must be done under liquid nitrogen to preserve the integrity of the proteome. The volumes can also be adjusted to perform this protocol on isolated myocytes (*see* ref. *4*).
4. The speed of this spin must be at least 16,000g, but can be faster depending on the maximum speed of the centrifuge. It must be consistent between extractions.
5. At this step it is best to leave a small amount of supernatant behind to avoid contaminating the HEPES fraction with the soft pellet.
6. The percent TFA in the buffer (and thus the pH of the buffer) can be adjusted to solubilize slightly different levels of the myofilament proteins. We found 1% TFA to be the most efficient for extraction; however, if this does not effectively solubilize the proteins in a different sample type, the percent can be modified (*6*).

References

1. Solaro, R. J., Pang, D. C., and Briggs, F. N. (1971) The purification of cardiac myofibril with Triton X-100. *Biochim. Biophys. Acta* **245,** 259–262.
2. Simpson, J. A., Iscoe, S., and Van Eyk, J. E. (2003) Myofilament proteomics: unraveling contractile dysfunction, in *Genomic and Proteomic Analysis of Cardiovascular Disease: Molecular Mechanism, Therapeutic Targets and Diagnostics* (Van Eyk, J. E. and Dunn, M., eds.), Wiley and Son, Germany, pp. 317–342.
3. Neverova, I. and Van Eyk, J. E. (2002) Application of reversed phase high performance liquid chromatography for subproteomic analysis of cardiac muscle. *Proteomics* **2,** 22–31.
4. Arrell, D. K., Neverova, I., Fraser, H., Marban, E., and Van Eyk, J. E. (2001) Proteomic analysis of pharmacologically preconditioned cardiomyocytes reveals novel phosphorylation of myosin light chain 1. *Circ. Res.* **89,** 480–487.
5. McDonough, J. L., Neverova, I., and Van Eyk, J. E. (2002) Proteomic analysis of human biopsy samples by single two-dimensional electrophoresis: coomassie, silver, mass spectrometry, and Western blotting. *Proteomics* **2,** 978–987.
6. Canton, M., Neverova, I., Menabo, R., Van Eyk, J. E., and Di Lisa, F. (2004) Evidence of myofibrillar protein oxidation induced by postischemic reperfusion in isolated rat hearts. *Am. J. Physiol. Heart Circ. Physiol.* **286,** H870–H877.

9

Optimization of Cardiac Troponin I Pull-Down by IDM Affinity Beads and SELDI

Diane E. Bovenkamp, Brian A. Stanley, and Jennifer E. Van Eyk

Summary

Cardiac troponin I (cTnI) is a key regulator of cardiac muscle contraction. Upon myo-cardial cell injury, cTnI is lost from the cardiac myocyte and can be detected in serum, in some cases with specific disease-induced modifications, making it an important diag-nostic marker for acute myocardial injury. Presently, hospital laboratories use enzyme-linked immunosorbent assays to detect cTnI, but this type of analysis lacks information about modified forms of protein (degradation or phosphorylation) that may give a more specific diagnosis from either serum or biopsies. Because cardiac and serum tissues are widely used for proteomic analysis, it is important to detect these cTnI posttranslational modifications. Therefore, we have chosen to optimize the enrichment and detection of cTnI protein by IDM Affinity Bead pull-down and surface-enhanced laser desorption/ ionization time of flight mass spectrometry (SELDI-TOF-MS or SELDI) analyses. By adjusting the chemical compositions of the buffers, we have retained antibody specificity and enriched for different forms of cTnI and its associated proteins.

Key Words: Troponin; cTnI; IDM affinity beads; surface-enhanced laser desorption ionization; SELDI; mass spectrometry; SDS-PAGE; Western blotting; antibody; cardiac; serum; purification; posttranslational modification; phosphorylation.

1. Introduction

Troponin I is an essential component of the contractile apparatus in skeletal and cardiac muscle, yet the isoforms in both tissues have unique amino acid sequences that can be detected by separate sets of antibodies (*1*). The presence of cardiac troponin I (cTnI) in the serum indicates a rupture of cardiac myo-cytes, and is therefore used as one biomarker for acute myocardial injury (AMI) (**refs.** *2* and *3*; for a review of cTnI as a specific cardiac marker, *see* **ref.** *4*).

From: *Methods in Molecular Biology, vol. 357: Cardiovascular Proteomics: Methods and Protocols*
Edited by: F. Vivanco © Humana Press Inc., Totowa, NJ

Troponin is a trimeric complex, consisting of cTnI (inhibits the binding of myosin to actin, in diastole), cTnT (anchors troponin onto the tropomyosin-actin thin filament), and cTnC (the calcium sensor, where binding calcium will release the TnI inhibition, in systole) *(5–8)*. Both cTnI and cTnT are the current gold standard for diagnosing AMI *(9,10)*; however, evidence from our lab has demonstrated a need for increased detection sensitivity for an earlier diagnosis *(11)*.

cTnI is regulated by β-adrenergic stimulators (such as protein kinase A), which make it a diphosphorylated protein, and it is modified during certain forms of heart disease. Specific proteolysis of the C- and N-terminus of cTnI and addition of phosphate groups are carried out by a large number of kinases, including protein kinase C, p21-activated kinase, and others. Therefore, cTnI protein can be present as an assortment of pathologically modified forms *(4)*.

Because these three troponin proteins naturally associate in muscle, when one protein is purified, the other forms are associated (unless the extraction buffer is designed to break them apart). In addition, owing to its insolubility and basic isoelectric point p*I* (~9.5), TnI is difficult to analyze using conventional techniques and requires stringent buffer conditions *(12)*. However, antibody-binding may be disrupted by a high stringency (and nonspecific protein binding to the matrix may be produced by low stringency).

In this protocol, we have expanded the manufacturer's basic protocols for IDM (Interaction Discovery Mapping) Affinity Bead and SELDI purification (both with a carbonyl diimidazole pre-activated surface to form a covalent link with antibodies *[13]*) (*see* **Note 1**) to enrich for cTnI, its disease-associated site-specific fragments *(3,14)*, its posttranslational modifications (such as phosphorylation *[15]* and oxidation *[16]*), and cTnI-associating proteins.

2. Materials

2.1. Protein Sources and Production of Lysate

1. Myocardium (e.g., normal or ischemic human or mouse heart tissue), or isolated TnI or troponin complex (purified Human cTnI (HyTest Ltd., Turku, Hämeenkatu, Finland, Cat. no. 8T53); or recombinant, purified cTnI protein *[17]*); or serum.
2. Extraction buffer (20–400 mM HEPES, pH 7.4, 0–0.5 M NaCl, plus inhibitor cocktail: 1 μM leupeptin, 1 μM pepstatin, 0.26 μM aprotinin, 50 mM NaF, 0.2 mM Na_3VO_4); or cTnI resuspension buffer (7 M urea, 5 mM EDTA, 15 mM mercaptoethanol, 20 mM Tris, pH 7.5).
3. Duall glass homogenizer (Kimble/Kontes, Vineland, NJ) connected to a ConTorque (Eberbach, Ann Arbor, MI) motor.
4. Bradford Protein Assay (Bio-Rad, Hercules, CA).
5. SealRite® 1.5-mL natural microcentrifuge tubes (USA Scientific Inc., Ocala, FL); or microcentrifuge tubes, 0.6-mL low retention, flat cap (Fisher, Hampton, NH).

2.2. Protein Binding to IDM Affinity Beads

1. IDM Affinity Beads (Ciphergen Biosystems, Inc., Fremont, CA).
2. 200-µL pipet tips; razor blade.
3. Water: deionized, distilled water purified on a Nanopure Diamond UV System; Barnstead International, Dubuque, IA. Used in all future sections.
4. Microcentrifuge tubes: 0.6-mL low retention, flat cap (Fisher, Hampton, NH).
5. Coupling buffer (50 mM sodium bicarbonate, pH 9.2).
6. Pulldown antibody: murine monoclonal anti-cTnI antibody 8I-7 (Spectral Diagnostics, Inc., Toronto, ON, Canada, cat. no. MA-1040).
7. End-over-end rotator: Thermolyne* LabQuake™ tube rotator (Barnstead International).
8. 0.5 M Tris HCl, pH 9.0, 0.1% Triton X-100.
9. 1 mg/mL bovine serum albumin (BSA) in 0.5 M Tris HCl, pH 9.0, 0.1% Triton X100.

2.3. Optimization of IDM Bead Washing Conditions and Elution of Bound Protein

1. Wash buffer (0.25–1 M urea, 0.05–0.5 M NaCl, 0.1% CHAPS, 50 mM Tris-HCl, pH 7.2).
2. Phosphate-buffered saline (PBS): pH 7.2, (1X = 137 mM NaCl, 2.7 mM KCl, 4.3 mM Na$_2$HPO$_4$, 1.4 mM KH$_2$PO$_4$).
3. Elution buffer: 0.1% trifluoroacetic acid (TFA)/80% acetonitrile (ACN).
4. NuPAGE® LDS Sample Buffer; 10X NuPAGE® reducing agent (Invitrogen, Carlsbad, CA).

2.4. SDS-PAGE and Western Blotting

1. 4–12% NuPAGE Novex Bis-Tris Gels, NuPAGE MES or MOPS running buffer, MagicMark™ XP Western Protein Standards (Invitrogen), Novex electrophoresis apparatus.
2. 1X NuPAGE transfer buffer (Invitrogen), 20% methanol.
3. Immobilon-P transfer membrane (Millipore, Billerica, MA), 100% methanol, 3M Electrophoresis paper, Bio-Rad mini electroblotting apparatus.
4. Western blocking reagent (Roche, Basel, Switzerland).
5. 1X TBS: 20 mM Tris-HCl, pH 7.5, 150 mM NaCl.
6. Primary antibody: murine monoclonal anti-cTnI antibody 8I-7 (Spectral Diagnostics, cat. no. MA-1040).
7. TBST: 1X TBS, 0.1% Tween-20.
8. Secondary antibody: alkaline phosphatase-conjugated AffiniPure goat anti-mouse IgG (H+L; Jackson Immunoresearch Laboratories, Inc., West Grove, PA).
9. Immun-Star™ Chemiluminescent Protein Detection Systems (Bio-Rad).
10. X-OMAT LS film (Kodak, Rochester, NY).

2.5. Silver Staining

1. Fix (50% methanol/5% acetic acid); and 50% methanol (Fisher, Hampton, NH).
2. Sensitizing solution: 1.3 mM sodium thiosulphate (Fisher).
3. Silver solution: 5.9 mM silver nitrate (Fisher). Keep at 4°C.
4. Developer: 18.8 mM sodium carbonate, 0.02% formaldehyde (Fisher).
5. Stop solution: 5% acetic acid (Fisher).

2.6. Detection of cTnI by SELDI (see Notes 2 and 3)

1. RS100 ProteinChip® Arrays, A-H Format with a 96-well bioprocessor (Ciphergen Biosystems, Inc., Fremont, CA).
2. Coupling buffer (50 mM NaHCO$_3$, pH 9.2).
3. Pulldown antibody: murine monoclonal anti-cTnI antibody 8I-7 (Spectral Diagnostics, cat. no. MA-1040).
4. Blocking solution: 3.3 mg/mL BSA in 1X PBS, pH 7.2.
5. 1X PBS, pH 7.2.
6. Binding buffer: 1X PBS, pH 7.2, 0.1% Triton X-100.
7. Wash buffer: 1 M urea, 50 mM Tris HCl, pH 7.2, 0.5 M NaCl, 0.1% CHAPS.
8. 15-mL polypropylene, conical tubes (Fisher).
9. Energy-absorbing molecules (EAM), SPA Kit (Ciphergen Biosystems, cat. no. C300-0002) 5 mg dissolved in 200 μL 1% TFA, plus 200 μL ACN (final = 12.5 μg/μL in 0.5% TFA, 50% ACN; Fisher).
10. ProteinChip Reader and Software (ProteinChip System II) and All-in-1 Peptide Standard external calibrant (Ciphergen Biosystems).

3. Method

3.1. Preparation of Protein Lysates

3.1.1. Human or Mouse Heart

1. Weigh 30–50 mg of frozen human (mouse) heart and add to 500 μL of extraction buffer in a beveled Duall glass homogenizer. Screw the glass bevel into rubber tubing on the con-torque motor and homogenize the tissue until it is solubilized.
2. Transfer the lysate into microcentrifuge tubes and centrifuge for 15 min at 18,000g at 4°C.
3. Sterilize the homogenizer in bleach solution before re-use.
4. Estimate protein concentration by Bradford Protein Assay.
5. Freeze at −80°C in aliquots until use.

3.1.2. Purified Human cTnI

1. Resuspend 0.1 mg of freeze-dried protein into 100 μL of cTnI resuspension buffer by pipetting up and down.
2. Freeze at −80°C in 2-μL aliquots until use (concentration is 1 μg/μL).

3.2. Protein Binding to IDM Affinity Beads

1. Let IDM bead suspension come to room temperature for 30 min.
2. Draw 10 µL water into a 200-µL pipet tip and scratch the volume onto the tip with a razor blade (add a larger volume of beads for a larger amount of protein). Expel the water and use this tip to transfer exactly 10 µL of beads (let them settle and transfer only beads) into a 0.6 mL siliconized tube with 500 µL of water. Make enough tubes for each sample plus controls. **Note:** For this section, "wash" means inverting the tubes by hand at least 10 times to mix.
3. Wash the beads four times 500 µL with water (to remove the manufacturer-supplied 0.1% acetic acid/dimethylsulfoxide buffer). After each addition of water wash, let the beads settle for approx 30 s (do not centrifuge; the dense zirconia skeleton allows the beads to quickly settle). For all of the following steps, leave enough liquid to cover the beads after buffer removal, so the beads do not dry.
4. Add 200 µL coupling buffer to the beads and to that add 5 µg 8I-7 antibody.
5. Incubate end-over-end at 4°C overnight (8–12 h).
6. Remove antibody supernatant and place it into a clean 0.6-mL tube. (Keep this fraction on ice for analysis by SDS-PAGE and Western blotting = lane 1 in **Fig. 1B**). Wash beads with 200 µL coupling buffer (to remove trapped, unbound antibody), then wash beads with 200 µL Tris/Triton X buffer (to end-cap reactive groups). Discard both solutions.
7. Add 150 µL of 1 mg/mL BSA in Tris/Triton X buffer to beads (to block nonspecific binding). Incubate end-over-end for 1 h at room temperature.
8. While blocking the beads, prepare the protein samples. Either add 10–100 µg of human or rat heart lysate into 1X PBS (for a total volume of 200 µL), or 0.6 µg of cTnI purified protein into either 200 µL of 1X PBS or into a mixture of 50 µL normal human serum plus 150 µL PBS.
9. Remove and discard BSA solution from beads and wash three times 500 µL with 1X PBS.
10. Add protein samples or buffer controls to the appropriately labeled tubes. Incubate tubes end-over-end for 1 h at room temperature.

3.3. Optimization of IDM Bead Washing Conditions and Elution of Bound Protein

Note: For all washes in this section, "wash" means end-over-end mixing on LabQuake for 2 min at room temperature.

1. Remove protein sample supernatant (the unbound proteins) and save it in a clean tube for analysis (lane 2 in **Fig. 1B**).
2. Wash the beads four times 300 µL with appropriate wash buffer (save all of the washes for analysis: lanes 3–6 in **Fig. 1B**). Adjust the components of the manufacturer's suggested stringent wash buffer (1 *M* urea, 50 m*M* Tris HCl, pH 7.2, 0.5 *M* NaCl, 0.1% CHAPS) to optimize binding of cTnI from your choice of tissuesource. Wash beads with:

A

	urea (M)				NaCl (M)				Tris HCl	CHAPS	Summary
	0.25	0.5	0.75	1	0.05	0.1	0.25	0.5	50 mM	0.10%	
urea	X	X	X	X	√	√	X	X	•	•	X
NaCl	√	√	√	•	√	√	X	X	•	•	√ ≤ 0.1
sequential											
1	√	√	√	√	√	√	√	√	•	√	√
2	√	√	√	√	√	√	√	√	•	•	√
3	√	•	√	√	√	√	√	√	•	•	√
4	√	√	•	•	√	√	√	√	•	•	√
5	√	√	√	•	√	•	√	√	•	•	√
6	√	√	√	•	√	√	√	•	•	•	X

B

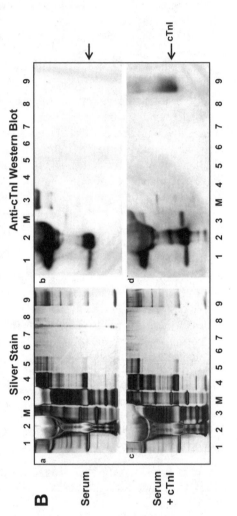

Silver Stain   Anti-cTnI Western Blot

Serum

Serum + cTnI

cTnI

a. The same concentrations of Tris, NaCl, and CHAPS with increasing concentrations of urea (0.25 or 0.5 or 0.75, 1 *M*.

b. The same concentrations of urea, Tris, and CHAPS with increasing concentrations of NaCl (0.05 or 0.1 or 0.25 or 0.5 *M*).

c. Sequentially added buffer components (in order: 50 m*M* Tris HCl, pH 7.2 + 0.1% CHAPS + 0.5 *M* urea + 1 *M* urea + 0.1 *M* NaCl + 0.5 *M* NaCl; **Fig. 1A**). Once the buffer has been optimized, confirm binding in another experiment, washing four times with the final optimized wash buffer (1 *M* urea, 50 m*M* Tris HCl, pH 7.2, 0.1 *M* NaCl, 0.1% CHAPS; **Fig. 1B**).

3. Wash two times 500 µL PBS (save the first wash for analysis: lane 7 in **Fig. 1B**).
4. Wash two times 500 µL water (save the first wash for analysis: lane 8 in **Fig. 1B**).
5. Either elute off of beads for matrix-assisted laser desorption ionization (MALDI) analysis, with 20 µL of 1% TFA/50% ACN and spot with matrix of choice (data not shown), or, for SDS-PAGE and Western blotting, add 20 µL NuPAGE LDS sample buffer, gently vortex for 5 min, and boil for 10 min at 70°C (load an aliquot in lane 9 in **Fig. 1B**).

3.4. SDS-PAGE and Western Blotting

1. Load protein samples (equivalent percent volumes) onto pre-cast 4–12% *bis-tris* NuPAGE gels and electrophorese with NuPAGE MOPS or MES running buffer (*see* **Note 4**).

Fig. 1. *(Opposite page)* (**A**) Optimization of IDM bead binding to cardiac Troponin I (cTnI). Listed in the rows of this chart are the three types of optimization (only changing urea, only changing NaCl, and sequentially adding each buffer component from less stringent to more stringent). The columns list the various solutions used in the experiments. Solutions included in one particular experiment (a row) are represented with a dot; those that are not are represented with a slash. A checkmark means that cTnI is still bound to the beads and X means cTnI is no longer binding to the beads (as determined by Western blotting). For all experiments, 0.6 µg of purified cTnI was spiked into 1X phosphate-buffered solution (PBS). NaCl seems to be more important for cTnI 8I-7 antibody binding, because changing urea alone does not change binding, but reducing NaCl to 0.1 *M* allows retention of cTnI on the beads. Summaries of the experimental results are listed in the column on the far right. (**B**) Optimized conditions for cTnI binding to 8I-7 antibody-bound IDM Affinity Beads. Parts a and b show the binding of serum alone, c and d show the binding of 0.6 µg purified cTnI spiked into normal human serum (a 23-yr-old male). Parts b and d are Western blots (with anti-cTnI 8I-7) of gels identical to the silver-stained gels shown in a and c Lane 1, unbound antibody (heavy and light chains are detected in b and d owing to direct binding of secondary antibody); lane 2, unbound protein (8I-7 binds nonspecifically to the albumin and IgG in the serum sample); lanes 3–6, washes number 1 to 4 (1 *M* urea, 0.1 *M* NaCl, 50 m*M* Tris-HCl, pH 7.2, 0.1% CHAPS); lane 7, 1X PBS wash; lane 8, water wash; lane 9, cTnI eluted off of beads by boiling in LDS sample buffer; lane M, MagicMark™ XP Western Protein Standards (80, 60, 50, 40, 30, 20 kDa). A comparison of lane 9 from b and d shows that cTnI specifically binds to IDM Affinity Beads under these conditions.

2. After separation, transfer proteins to Immobilon-P membrane (pre-wet in 100% methanol) with 1X NuPAGE transfer buffer (20% methanol) at 30 V constant for 1 h.
3. Rinse blot with water and block overnight, rocking, at 4°C in 10 mL of a 1:10 dilution of Western blocking reagent in 1X TBS.
4. Dilute primary antibody (0.5 µg/mL final) in a 1:100 dilution of Western blocking reagent in 1X TBS. Add 10 mL per membrane and incubate for 1 h at room temperature.
5. Wash four times 5 min with TBST at room temperature.
6. Dilute secondary antibody (0.3 µg/mL final) in a 1:100 dilution of Western blocking reagent in 1X TBS. Add 10 mL per membrane and incubate for 1 h at room temperature.
7. Wash four times 5 min with TBST at room temperature.
8. Detect the protein by following manufacturer's instructions for Immun-Star™ Chemiluminescent Protein Detection Systems and expose to X-OMAT LS film (*see* **Figs. 1Bb,d**).

3.5. Silver Staining (see Note 5)

1. This protocol is based on the method by Shevchenko et al. *(18)*. All steps are done at room temperature and gels are rocked in glass dishes on an orbital shaker. For each of the following steps, add approx 100 mL solution (for a mini-gel).
2. Add fix to the gels (duplicates of the gels run in **Subheading 3.4.1.**) and shake for 1 h, then shake for 15 min in 50% methanol. Re-hydrate gels for a total of 1 h with three changes of water.
3. Incubate for 1 min in sensitizing solution and two times 45 s in water, before shaking in silver solution for 15 min.
4. Wash two times 45 s in water, quickly rinse with developer, and then incubate with fresh developer until the proteins have stained to appropriate levels, depending on how much protein was loaded per lane (1–12 min).
5. Rinse once and then add fresh stop solution, shaking for 1 h. Shake in two changes of water for at least 1 h before placing the gel on a scanner (*see* **Figs. 1Ba,c**).

3.6. Detection of cTnI by SELDI (see Notes 2 and 3)

1. Apply 5 µg antibody and 2 µL 50 mM NaHCO$_3$, pH 9.2, onto one spot of an RS100 array (8 spots per chip). Use a new spot for each sample or control. Incubate at 4°C overnight in a humid chamber.
2. Put the RS100 chip(s) into a bioprocessor, to increase the loading volume capacity.
3. Remove antibody and apply 4 µL of blocking solution to the spot. Incubate, shaking, for 1 h at room temperature (to block nonspecific sites).
4. During this incubation, dilute various amounts of the heart protein lysate samples into binding buffer (total volume of 200 µL) in duplicate, including negative and positive controls.

5. Remove the blocking solution and wash for 2 min on a shaker with 200 µL of 1X PBS. Apply protein solutions and incubate, shaking, for 2 h at room temperature.
6. Remove protein solutions and wash (30 s, shaking) with 200 µL binding buffer, wash (30 s, shaking) with 200 µL 1X PBS, wash (30 s, shaking) with 200 µL wash buffer, and wash (2 min, shaking) with 200 µL 1X PBS. Disassemble the chips from the bioprocessor and quickly rinse each array separately with water for 10 s in a 15-mL conical tube.
7. Allow the surface to air-dry (5 min) and add two times 1 µL EAM/sinapinic acid solution to each spot (or EAM of choice) (allow to air-dry fully between each application). Analyze with the ProteinChip Reader (*see* **Fig. 2**).

3.7. Concluding Remarks

To enrich for cTnI, the most important factor to affect antibody–protein interaction was the amount of salt in the solution (concentrations above 0.1 M NaCl were disruptive). The binding of cTnI was not affected by the urea concentrations studied (0.25–1 M). These IDM bead pull-down and SELDI methods were optimized for human and mouse cardiac tissues. If cTnI is to be isolated from tissues of other animals, then optimization of this method may be required (as seen from low binding of rat heart lysate; data not shown). Human and mouse cTnI fragments and posttranslational modifications (such as phosphorylation) can be easily analyzed through Western blotting (with antibodies generated against different regions or posttranslational modifications of cTnI) or SELDI or MALDI-TOF MS methods (to detect whole-protein molecular-weight shifts).

4. Notes

1. Any type of affinity column or matrix can be used for this method, but these specific buffer conditions have been determined for the IDM Affinity Beads and RS100 ProteinChip (SELDI).
2. Any MALDI-TOF can be used to detect cTnI (MALDI); however, it does not use an affinity matrix (only SELDI does).
3. Proteins can be digested while still bound to the antibody-coupled beads and used for MALDI-TOF analysis (*see* **ref. *19***).
4. You can also use 10 or 12% NuPAGE® *bis-tris* gels or *tris*-glycine gels for electrophoresis.
5. Coomassie blue or any other protein stain method can be used. However, these methods can be less sensitive than silver staining and you may not see low abundance proteins/contaminants.

Acknowledgments

The authors would like to thank Dr. Jane Ding and Dr. Lee Lomas (from Ciphergen Biosystems, Inc.) and Dr. Ralf Labugger for their initial work with

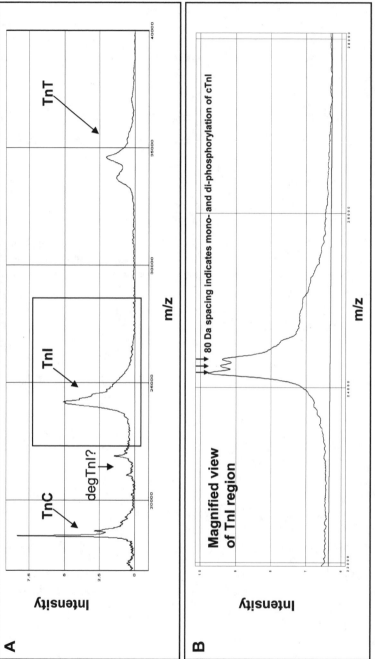

Fig. 2 (A) SELDI profile of captured troponins using the anti-cardiac Troponin I (cTnI) antibody 8I-7 from a human patient with heart failure. The different troponin species are noted, along with possible degradation products of TnI. (Extracted in 20 m*M* HEPES, pH 7.4, and 0.5 *M* NaCl). **(B)** Expanded view of the cTnI peak (similar to the region boxed in A) captured with 8I-7 antibody from ischemic mouse heart. Illustrated are the 80-Da spaced peaks, which represent the phosphorylated states (none, single, and double). (Extracted in 20 m*M* HEPES, pH 7.4.)

IDM Affinity Bead and SELDI purification of cTnI, and Dr. Ger J.M. Stienen for his normal rat heart lysate (data not shown). This work was supported by grants from the NHLBI (contract N01-HV-28180), the Donald W. Reynolds Foundation and funds from the Daniel P. Amos Family Foundation supporting the Johns Hopkins Bayview Proteomic Center.

References

1. Wu, A. H. (2004) The role of cardiac troponin in the recent redefinition of acute myocardial infarction. *Clin. Lab. Sci.* **17**, 50–52.
2. McDonough, J. L., Labugger, R., Pickett, W., et al. (2001) Cardiac troponin I is modified in the myocardium of bypass patients. *Circulation* **103**, 58–64.
3. Labugger, R., Organ, L., Collier, C., Atar, D., and Van Eyk, J. E. (2000) Extensive troponin I and T modification detected in serum from patients with acute myocardial infarction. *Circulation* **102**, 1221–1226.
4. McDonough, J. L. and Van Eyk, J. E. (2004) Developing the next generation of cardiac markers: disease-induced modifications of troponin I. *Prog. Cardiovasc. Disease* **47**, 207–216.
5. Gordon, A. M., Homsher, E., and Regnier, M. (2000) Regulation of contraction in striated muscle. *Physiol. Rev.* **80**, 853–924.
6. Solaro, R. J. and Van Eyk, J. (1996) Altered interactions among thin filament proteins modulate cardiac function. *J. Mol. Cell Cardiol.* **28**, 217–230.
7. Metzger, J. M. and Westfall, M. V. (2004) Covalent and noncovalent modification of thin filament action: The essential role of troponin in cardiac muscle regulation. *Circ. Res.* **94**, 146–158.
8. Solaro, R. J. (2003) The special structure and function of troponin I in regulation of cardiac contraction and relaxation. *Adv. Exp. Med. Biol.* **538**, 389–401.
9. Apple, F. S., Murakami, M. M., Pearce, L. A., and Herzog, C. A. (2002) Predictive value of cardiac troponin I and T for subsequent death in end-stage renal disease. *Circulation* **106**, 2941–2945.
10. Apple, F. S., Wu, A. H., and Jaffe, A. S. (2002) European Society of Cardiology and American College of Cardiology guidelines for redefinition of myocardial infarction: how to use existing assays clinically and for clinical trials. *Am. Heart J.* **144**, 981–986.
11. Colantonio, D. A., Pickett, W., Brison, R. J., Collier, C. E., and Van Eyk, J. E. (2002) Detection of cardiac troponin I early after onset of chest pain in six patients. *Clin. Chem.* **48**, 668–671.
12. Simpson, J. A., Iscoe, S., and Van Eyk, J. E. (2003) Myofilament proteomics: understanding contractile dysfunction in cardiorespiratory disease, in *Proteomic and Genomic Analysis of Cardiovascular Disease* (Van Eyk, J. E. and Dunn, M. J., eds.), Wiley-VCH, Germany, pp. 317–337.
13. Tang, N., Tornatore, P., and Weinberger, S. R. (2004) Current developments in SELDI affinity technology. *Mass. Spectrom. Rev.* **23**, 34–44.

14. Katrukha, A. G., Bereznikova, A. V., Filatov, V. L., et al. (1998) Degradation of cardiac troponin I: Implication for reliable immunodetection. *Clin. Chem.* **44,** 2433–2440.
15. Ardelt, P., Dorka, P., Jaquet, K., et al. (1998) Microanalysis and distribution of cardiac troponin I phospho species in heart areas. *Biol. Chem.* **379,** 341–347.
16. Wu, A. H., Feng, Y. J., Moore, R., et al. (1998) Characterization of cardiac troponin subunit release into serum after acute myocardial infarction and comparison of assays for troponin T and I. American Association for Clinical Chemistry Subcommittee on cTnI Standardization. *Clin. Chem.* **44(6 Pt 1),** 1198–1208.
17. Foster, D. B., Noguchi, T., VanBuren, P., Murphy, A. M., and Van Eyk, J. E. (2003) C-terminal truncation of cardiac troponin I causes divergent effects on ATPase and force: implications for the pathophysiology of myocardial stunning. *Circ. Res.* **93(10),** 917–924.
18. Shevchenko, A., Wilm, M., Vorm, O., and Mann, M. (1996) Mass spectrometric sequencing of proteins silver-stained polyacrylamide gels. *Anal. Chem.* **68(5),** 850–858.
19. Labugger, R., Simpson, J. A., Quick, M., et al. (2003) Strategy for analysis of cardiac troponins in biological samples with a combination of affinity chromatography and mass spectrometry. *Clin. Chem.* **49(6 Pt 1),** 873–879.

10

Proteomic Analysis of the Subunit Composition of Complex I (NADH:Ubiquinone Oxidoreductase) From Bovine Heart Mitochondria

Ian M. Fearnley, Joe Carroll, and John E. Walker

Summary

Complex I from the inner membranes of mammalian mitochondria is a complicated membrane-bound assembly of redox centers (flavin mononucleotide cofactor, iron sulphur centers) and at least 46 different proteins. The hydrophobic nature of its membrane-bound subunits and the complexity of subunit content present a substantial analytical challenge. The complete protein chemical analysis of complex I purified from bovine mitochondria required the resolution of subunits by one-dimensional and two-dimensional electrophoresis, reverse-phase chromatography, and combinations of these techniques. These subunits were characterized by mass spectrometry (MS)-based protein identification methods, requiring both peptide mass fingerprinting and amino acid sequencing by tandem MS. The components were identified also and characterized by measurements of subunit molecular mass. These strategies have provided a comprehensive view of the subunit content of the intact complex, its structural domains, and stable subunit modifications.

Key Words: Mitochondria; respiratory complex; complex I; NADH:ubiquinone oxidoreductase; mass spectrometry; RP-HPLC; electrophoresis; isoelectric focusing; chromatography; protein modification; protein complex.

1. Introduction

Complex I (NADH:ubiquinone oxidoreductase) is one of five protein complexes in the inner mitochondrial membrane that comprise the respiratory chain *(1,2)*. The enzyme catalyses the initial steps in the oxidative phosphorylation process, the oxidation of NADH, and the transfer of electrons to ubiquinone *(3–5)*. This electron transfer is coupled to the removal of protons from the mitochondrial matrix and so complex I contributes to the generation of the proton electrochemical potential gradient across the inner mitochondrial membrane. These processes are poorly understood and a characterization of the molecular

From: *Methods in Molecular Biology, vol. 357: Cardiovascular Proteomics: Methods and Protocols*
Edited by: F. Vivanco © Humana Press Inc., Totowa, NJ

details of the complex is underway. Complex I has an overall L-shape, with one arm in the plane of the membrane and the other extending into the mitochondrial matrix. This overall shape is common to the enzyme isolated from bacteria, fungi, and mitochondria *(6–8)*. The mammalian mitochondrial enzyme is the most complicated, in terms of subunit composition being an assembly of 46 different proteins approaching 1 MDa by mass. Seven components, the ND subunits ND1–ND6 and ND4L, are encoded by the mitochondrial genome *(9)* and comprise the most hydrophobic portions of the complex. The remaining subunits are products of nuclear genes and are imported into mitochondria. Bovine complex I is the best-characterized mitochondrial enzyme and the subunit composition together with the amino acid sequences of subunits were established by direct protein chemical analysis of purified enzyme complexes in combination with analysis of their corresponding cDNAs and genes *(10–19)*. Subsequently these data have been used as a template for the analysis of other mitochondrial enzymes, human complex I purified by affinity-purification techniques *(20)*, and neuronal complex I from rodent tissues *(21)*.

A proteomic characterization of mammalian complex I requires effective separation of all the constituents, together with a detailed examination of fractions using mass spectrometric methods of protein identification. During the original characterization of complex I and its primary structure *(3,17,22)*, a significant advantage was gained by analyzing the subcomplexes of the enzyme produced upon treatment with chaotropic detergents and resolution by ion exchange chromatography. These subcomplexes, Iα, Iβ, and Iλ, correspond to different structural domains of the complex and so they have simpler polypeptide compositions *(22,23)*. The analysis of these subcomplexes has considerably reduced the analytical challenge posed by the intact enzyme.

In the following sections, both chromatographic and electrophoretic strategies for subunit separation and the characterization of the subunit content of bovine complex I are outlined. The combination of these methods allowed the complete characterization of bovine complex I *(5,24)* (*see* **Note 1**). Four additional subunits were detected, characterized, and assigned to different domains of the enzyme *(24–26)*. More recently, modified peptides have been isolated from subunits and the posttranslational modifications of all the nuclear encoded components of complex I have been defined *(27,28)*.

2. Materials

2.1. Preparation of Complex I and Its Subcomplexes From Bovine Heart Mitochondria

Complex I was solubilized from mitochondrial membranes with N-dodecyl–β-D-maltoside (DDM) and purified by two chromatography steps, involving an

Table 1
ASB-14 Rehydration Solutions for 2D Electrophoresis

Reagents	Volume (µL)		
	2% Solution	1% Solution	0% Solution
9 *M* Urea, 2.6 *M* thiourea[a]	389.0	389.0	389.0
30% ASB-14[b]	33.4	16.7	0.0
IPG buffer[c]	2.5	2.5	2.5
Bromophenol blue[d]	5.0	5.0	5.0
1 *M* Dithiothreitol[e]	9.0	9.0	9.0
Water (deionized)	61.1	77.8	94.5

[a]Prepare 50 mL and freeze aliquots (approx 1.2 mL). Deionize with Amberlite MB-1 prior to use. Final 7 *M* urea, 2 *M* thiourea concentrations.

[b]Preparation of a stock 30% ASB-14 solution: Weigh 150 mg ASB-14 into a 2-mL plastic tube, then add water to the 0.5 mL gradation on the tube. Dissolution is aided by gentle mixing and a brief centrifugation. The final volume adjustment to 0.5 mL is achieved with the aid of a P1000 pipet. Store in the fridge and discard after approx 1 mo.

[c]Use the immobilized pH gradient (IPG) buffer with same pH range as the IPG strip being used. Final 0.5% concentration.

[d]Bromophenol blue is prepared as a saturated solution in water and stored in the fridge.

[e]Dithiothreitol is prepared as a 1 *M* solution and portions are stored frozen. Do not re-freeze.

initial ion exchange chromatography step on a column of Q-Sepharose followed by an ammonium sulphate precipitation and then gel filtration on a column of Sephacryl S-300 HR. These procedures are described in detail elsewhere *(24,29)*.

Subcomplexes Iα and Iβ are prepared from the intact enzyme by treatment with the detergent lauryldimethylamine oxide (LDAO) and ion exchange chromatography on Q-Sepharose *(22,26)*. Subcomplex Iλ corresponds to the mitochondrial matrix domain and contains a subset of the components found in subcomplex Iα. The preparation of Iλ involves treatment of the intact complex with LDAO, ultracentrifugation through a sucrose gradient, followed by gel filtration of the subcomplex *(23,25)*.

2.2. Electrophoresis

1. Immobilized pH gradient (IPG) strips and buffers were purchased from Amersham, now GE Healthcare (Amersham, UK) (*see* **Tables 1–3**).
2. Preparation of a stock solution of Coomassie blue G250 dye *(30)*:
   a. Dissolve 10 g G250 (Brilliant Blue G, Aldrich 20,139-1) in 7.5% (v/v) acetic acid (250 mL) at 70°C. Add ammonium sulphate slowly while stirring until the solution turns clear(ish). Usually up to 75 g of ammonium sulphate or approx 30% final concentration are required.
   b. Cool to room temperature and discard the supernatant.

Table 2
Equilibration Solution for IEF Gel Strips
Prior to Separation on SDS-Polyacrylamide Gels[a]

Reagent	4% SDS solution	Final concentration (approx)
9.5 M Urea[b]	6.314 mL	6 M
Glycerol	3.0 mL	30%
SDS	0.4 g	As labeled
3 M Tris, pH 8.45[c]	167 µL	50 mM
Bromophenol blue[d]	Trace	Trace

[a]Recipe for a final volume of approx 10 mL, required for equilibrating a 7-cm immobilized pH gradient strip.

[b]Deionize with Amberlite MB-1 prior to use.

[c]The Tris stock used depends on the second dimension gel system. The pH 8.45 stock is for a Tricine gel *(36)*.

[d]Bromophenol blue is prepared as a highly concentrated solution in a 300 mM Tris solution pH 8.45.

IEF, isoelectric focusing; SDS, sodium dodecyl sulfate.

Table 3
Composition of Gel Solution for a 13% Acrylamide Tricine Gel[a]

Component	Volume (mL)
30% Acrylamide, 0.8% *bis*-acrylamide	19.5
3 M Tris-HCl, pH 8.45	15.0
80% Glycerol (v/v)	7.5
30% SDS (w/v)	0.15
Water (deionized)	2.60
10% Ammonium persulphate (w/v)	0.225
N,N,N,N-Tetramethylethylenediamine (TEMED)	0.023

[a]The components required for a gel volume of 45 mL are illustrated.
SDS, sodium dodecyl sulfate.
From **ref. *36***.

    c. Dissolve the Coomassie precipitate in a solution of 40 mL water, 50 mL ethanol, and 10 mL acetic acid.
3. Staining solution:
    a. Prepare a solution of 2.6% (v/v) orthophosphoric acid, 6% (w/v) ammonium sulphate.
    b. Add stock Coomassie G250 dye solution to a final 0.1% concentration. Mix well and use immediately *(30)*.
4. 0.1 M Tris-HCl, pH 7.0, buffer.
5. 25% (v/v) methanol.

2.3. Reverse-Phase High-Pressure Liquid Chromatography

2.3.1. Protein Separations

1. Reverse-phase (RP) columns (1 mm internal diameter [id] × 100 mm) containing Aquapore RP-300, C8, 300Å, 7 μm reverse-phase silica (Brownlee, Perkin-Elmer).
2. High-pressure liquid chromatography (HPLC)-quality solvents are obtained as follows: water (HPLC gradient-grade, Fischer), acetonitrile (acetonitrile 190, far UV grade) (Romil Ltd., Waterbeach, UK) and trifluoroacetic acid (TFA) (Applied Biosystems sequencer-grade, Warrington, UK).
3. Solvent A: 0.1% (v/v) aqueous TFA.
4. Solvent B: 80% acetonitrile (ACN) in water containing approx 0.09% TFA. The concentration of TFA in Solvent B is adjusted to reduce absorbance changes owing to solvent composition during gradient separations.

2.3.2. Capillary HPLC of Peptides

1. In the capillary HPLC separation of peptides for liquid chromatography mass spectrometry (LC-MS), peptides are separated on a column of PepMap C18 column (180 μm × 100 mm; Dionex, Leeds, UK). Solvent A is 0.1% (v/v) aqueous formic acid, containing 5% ACN and Solvent B; 0.1% formic acid in 95% ACN.

2.4. Preparation of Stock Trypsin Solution for the In-Gel Digestion

1. Stock trypsin solution: A stock trypsin solution is prepared by dissolution of 1 vial Trypsin-25 μg/vial, (sequencing grade: Roche Molecular Biochemicals) into 1 mM HCl (100 μL) to a concentration of 250 ng/μL. The stock solution is stored at −20°C and is stable to several freeze/thaw cycles. The working solution (12.5 ng/μL) is prepared by a 20-fold dilution of the stock into 20 mM Tris-HCl, pH 8.0, 5 mM CaCl$_2$ and is used the same day.

3. Methods

3.1. Sample Preparation for Electrophoresis and Chromatography (see Note 2)

3.1.1. Sample Preparation for Two-Dimensional Gel Electrophoresis: Organic Solvent Precipitation

Before two-dimensional (2D) gel electrophoresis, salts, excess detergents, and lipids are removed from samples of complex I by precipitation with chloroform/methanol (31).

1. Add 2 vol of a chloroform/methanol solution 2:1 (v/v) to 1 vol of a sample of aqueous complex I in a 1.5-mL plastic disposable tube. Vortex briefly.
2. Centrifuge the tube in a benchtop centrifuge at 15,900g for 5 min. The solution forms two phases with a protein precipitate at the interface. This interface precipitate is retained.
3. Remove the upper aqueous layer by aspiration. Tilt the tube so that the precipitate slides to the side of the tube, allowing the lower chloroform phase to be removed.

4. The precipitate is washed with methanol. Add 3 vol of methanol to the tube, and centrifuge as in **step 2**.
5. Remove the supernatant by aspiration and blotting. Re-dissolve the protein precipitate immediately in a solution containing urea, thiourea, amidosulfobetaine (ASB-14), and IPG buffer, as described in **Subheading 3.4.1.** (*see* **Note 3**).

3.1.2. Sample Preparation for 2D Gel Electrophoresis: Dialysis and Sample Concentration

Salts and other low molecular weight charged contaminants that interfere with isoelectric focusing (IEP) may be removed by dialysis. When necessary, the dialysis step can be followed by a concentration step to obtain a suitable sample (10–20 mg/mL) for analysis.

1. Dialyze samples against a 20 mM Tris-HCl, pH 7.4, containing 0.05% DDM, at 4°C. A Slide-A-Lyzer (Pierce) with a 3.5-kDa cut-off is convenient for small volumes. A precipitate may form during dialysis. Harvest the dialyzed solution plus any precipitate.
2. When a concentration step is required, this can be performed using an Ultrafree-0.5 filter unit (Millipore, Bedford, MA) with a 5-kDa cut-off. The filter device is placed in a benchtop centrifuge and spun at 8100g at 4°C. The centrifugation is continued until the desired final volume is achieved.

3.1.3. Sample Preparation for RP-HPLC Separation

Excess detergents are removed from the protein sample by precipitation of the samples with cold ethanol (*see* **Note 4**).

1. Protein samples (up to 50 µL) are placed in 1.5-mL disposable tubes.
2. Add a 20X volume of cold ethanol, vortex briefly, and chill overnight or for several h at −20°C.
3. Collect the precipitate by centrifugation in a benchtop centrifuge (15,900g, 5 min).
4. Remove supernatant by aspiration, drainage, and blotting (*see* **Note 3**).
5. Redissolve the precipitate. For RP-HPLC separations, the precipitate is re-dissolved in 6 M guanidine hydrochloride containing 0.1% TFA.

3.2. N-Terminal Sequence Analysis of Proteins

Following electrophoretic separations, proteins are transferred to (poly)-vinylidene difluoride membrane (PVDF; Immobilon P, Millipore) *(32)*. The immobilized protein is analyzed by automated Edman degradation using a protein sequencer (Model 494 protein sequencer, Applied Biosystems, Warrington, UK).

1. PVDF membranes are placed directly in the reaction chamber of the protein sequencer with a cartridge seal only. No Polybrene or proprietary equivalent carrier is used.
2. Perform at least 10 cycles of Edman degradation on the samples in order that sequences arising from multiple protein components can be distinguished.

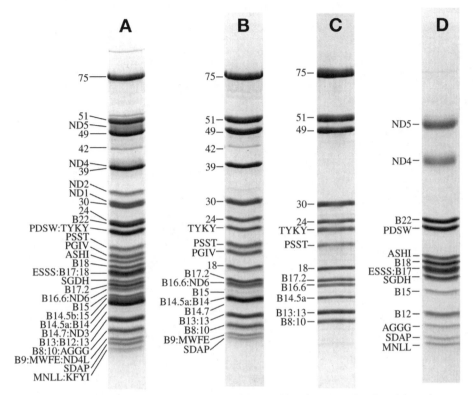

Fig. 1. Resolution of the subunit compositions of bovine complex I and its subcomplexes on one-dimensional gels. Parts **(A)**, **(B)**, **(C)**, and **(D)** show complex I and subcomplexes Iα, Iλ, and Iβ, respectively (*see* **Note 5**). The subunits were fractionated by SDS-PAGE, and bands detected by staining were analyzed by peptide mass fingerprinting and tandem mass spectrometry. The locations of subunits are indicated on the left-hand sides of gels. From **ref. 24**, with permission.

3.3. One-Dimensional Gel Electrophoresis

Subunits of complex I and subcomplexes were separated by SDS-PAGE in 12–22% acrylamide gradient gels (*33*). The gels are stained with Coomassie blue dye. The subunit separations obtained are shown in **Fig. 1** (*see* **Note 5**). The most important aspects of our gel running protocol that may differ from other laboratory protocols are referred to in **Subheading 4.2.2.** (*see* **Notes 6–8**).

3.4. 2D Electrophoresis

The protocol assumes the use of an IPGphor isoelectric focusing system (Amersham BioSciences/GE Healthcare) for the initial separations by IEF (*34*). This instrument includes an 8000 V power supply and built-in temperature control. IPG strips are used and are rehydrated in the IPGphor strip holders prior to

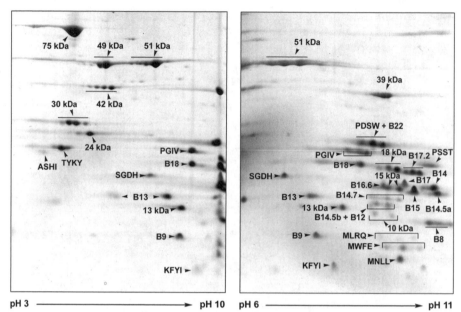

Fig. 2. Resolution of the subunits of complex I on 2D gels. First dimension: immobilized pH gradient (IPG) 3.0–10.0 (left hand panel) or IPG 6.0–11.0 (right hand panel), second dimension: tricine-SDS-13% PAGE. Proteins were identified by peptide mass fingerprinting and tandem mass spectrometry analyses. The figure is a summary of several experiments and is reproduced from ref. *(26)* with permission (*see* **Note 9**).

use. The following description of 2D gel electrophoresis (2DE) assumes some familiarity with the methods and only an overview of the 2D gel procedure is presented. However the more crucial elements of the procedure are described in detail. The 2D separations obtained are shown in **Fig. 2** (*see also* **Note 9**).

3.4.1. Solubilization of Samples and IEF

Samples for 2DE (IEF followed by SDS-PAGE) are applied to the IPG strips for IEF during the rehydration step of the protocol. The rehydration solution has a final composition of approx 7 M urea, 2 M thiourea, 18 mM dithiothreitol (DTT), 0.5% IPG buffer, trace of bromophenol blue, and 0.15% ASB-14 *(35)*.

Initially, protein samples are dissolved in a solution with an ASB-14 concentration of 1%. After dissolution, the ASB-14 concentration is reduced to 0.15% by dilution (*see* **Note 10**). Following centrifugation this supernatant (125 µL per strip) is used for simultaneous sample loading and IPG strip rehydration. Details of the solutions used in this procedure are shown in the **Table 1**. The following steps are used with a 7 cm IEF gel strip and are carried out at room temperature.

1. A chloroform/methanol precipitate is dissolved in 20 μL of a 1% ASB-14 solution. Alternatively a soluble sample must be in a volume of 10 μL or less. A 10-μL volume of a 2% ASB-14 solution is added to this sample with the final volume adjusted to 20 μL with the 0% ASB-14 solution (*see* **Table 1** for composition of ASB-14 solutions). Samples are mixed gently by aspiration with a Gilson pipet tip to aid solubilization.

2. Dilute the 20-μL solution to a final ASB-14 concentration of 0.15%, by the addition of 113.4 μL of the 0% ASB-14 solution. Following dilution and gentle vortexing, remove any insoluble material by centrifugation at room temperature in a benchtop minifuge at 15,900*g* for 10 min. Use the supernatant for rehydration of the gel strip, 125 μL per 7-cm strip (*see* **Notes 10** and **11**).

3. Place the sample solution in a central position in an IPGphor strip holder, avoiding formation of bubbles. Remove the protective cover from the IPG strip, and position the strip, gel side down, into the holder. The strip is held from one end during this process and lifted up and down gently to distribute the rehydration solution along the underside, again being careful not to introduce any bubbles. Then the strip is overlaid carefully with IPG cover fluid (light mineral oil). Approximately 600 μL of cover fluid is sufficient and this is applied in 200-μL portions, drop-wise from each end of the strip, gradually moving to the center. Finally, the cover is placed on top of the strip holder and it is transferred to the platform of the IPGphor unit.

4. Rehydration of gel strips is performed for 12 h at 20°C, applying a 20 V potential.

5. The IEF step is carried out at 20°C using the following steps: pH 3.0–10.0 gel, 250 Vh (maximum 500 V), 500 Vh (maximum 1000 V), 8000 Vh (maximum 8000 V) or pH 6.0–11.0 gel, 250 Vh (maximum 500 V), 500 Vh (maximum 1000 V), 24,000 Vh (maximum 8000 V). A current limit of 50 μA is applied per strip.

6. After focusing is complete, proceed to the second-dimension separation or store the IPG strips in 15-mL screw-capped conical plastic tubes (Sarstedt) at −20°C.

3.4.2. Alkylation of Cysteines and Equilibration of IEF Strips for SDS-PAGE

The IPGphor gel strip is equilibrated for the SDS-polyacrylamide gel separation and cysteine residues in proteins are modified by alkylation. Details of the equilibration solution used are shown in **Table 2**. It is convenient to use a cut-down 10-mL plastic pipet for the incubations, carefully sealing the ends with parafilm. Note that disposal of waste thiourea may be restricted, with disposal down the sink being prohibited. Avoid contact with solutions containing thiourea and iodoacetamide.

1. IPG strips are equilibrated using two 5-mL changes of equilibration solution.

2. The first 5-mL portion of equilibration solution (*see* **Table 2**) contains 1% (w/v) DTT and is carried out for 15 min at room temperature.

3. The second 5-mL portion of equilibration solution contains 2.5% (w/v) iodoacetamide and is also carried out for 15 min at room temperature.

4. The gel strip is finally rinsed briefly in gel running buffer prior to positioning on top of the second-dimension separating gel.

3.4.3. Second-Dimension SDS-Polyacrylamide Gel Separation

The second dimension is based on the Tricine (*N*-[2-hydroxy-1.1-*bis*(hydroxymethyl)ethyl]glycine) gel system and employs a constant resolving 13% acrylamide gel concentration *(36)*. The required volume of gel solution will vary depending on the size of gel to be poured. A typical composition for a volume of 45 mL is given in **Table 3**. The gels are generally poured the day before use and are stored at 4°C. The compositions of the electrophoresis buffers are: upper electrode buffer: 0.1 *M* Tris, 0.1 *M* Tricine, 0.1% sodium dodecyl sulfate SDS (w/v). Lower electrode buffer: 0.2 *M* Tris-HCl, pH 8.9 *(36)*. No stacking gel is employed.

1. The IEF gel strip is secured on top of the second-dimension gel with a sealing solution of 0.5–1.0% (w/v) agarose dissolved in upper tank buffer and containing a trace of bromophenol blue.
2. Electrophoresis is carried out at room temperature. The gels are electrophoresed at a low current of 10 mA until the bromophenol blue dye from the agarose sealing solution forms a clear band in the resolving gel. Then electrophoresis is continued at 150–170 V for the remainder of the run.
3. Following electrophoresis the gel is washed in 50% (v/v) aqueous methanol (2 × 15 min) and then briefly (approx 5 min) in deionized water prior to staining with colloidal Coomassie (*see* **Subheading 3.4.4.**).

3.4.4. Procedure for Staining Gels
With Colloidal Coomassie Dye G250

1. Stain the gel in the G250 stain solution overnight, with shaking, in a container with a lid.
2. Rinse the gel for 10 min in 0.1 *M* Tris-HCl, pH 7.0, to neutralize the acid.
3. Briefly rinse the gel in 25% (v/v) methanol to clean dye from the gel surface.
4. Place the gel in deionized water, record the gel image, and excise protein spots for further analysis as required.

3.5. Chromatographic Purification of Subunits by RP-HPLC

The subunits of intact complex I and subcomplexes are separated and purified by RP-HPLC on a microbore (1 mm id) column. The procedure assumes the use of an HPLC system capable of performing gradient separations at low flow rates (50 µL/min) compatible with microbore chromatography (*see* **Note 12**). The HPLC system requires a UV detector and low volume flow cell for monitoring absorbance at 225 nm and either a manual (Rheodyne) injector (100 µL loop) or an appropriate autosampler.

1. HPLC system setup: A RP-HPLC column (Aquapore RP-300, 1 mm id × 100 mm, Perkin-Elmer Life Sciences) is equilibrated with solvent A, 0.1% TFA, at a flow rate of 50 µL/min (*see* **Subheading 2.3.** and **Note 13**). It is advisable to condition the HPLC column and monitor artifact peaks by performing a gradient separation, without any sample in advance of each separation.
2. Samples of complex I or subcomplexes (approx 200 µg; *see* **Subheading 3.1.3.**) are re-solubilized in a solution (100–200 µL) containing 3–6 *M* guanidine hydrochloride and 0.1% TFA.
3. The solution is centrifuged to remove insoluble material using a benchtop centrifuge for 1.5-mL disposable tubes, 13800*g*, 5 min.
4. Load the supernatant onto the column using a sample loop or autoinjector in volumes up to 60 µL. If the sample volume exceeds 60 µL, then multiple loadings of supernatant may be necessary, while the column is washed with solvent A (100%).
5. Maintain the solvent flow at the initial conditions (100% solvent A) until the guanidine from each application of sample has passed through the column and the absorbance baseline has stabilized.
6. Apply a linear gradient of ACN to the column (0–80% solvent B in approx 60 min). Monitor the eluate by 225 nm absorbance and collect fractions corresponding to peaks for further analysis.
7. Assess the subunit content of fractions by electrospray ionisation (ESI) MS analysis, either on-line or off-line to the RP separation, and measurement of subunit molecular masses (*see* **Notes 12** and **13**). Alternatively, analyze the fractions by 1D SDS-PAGE and identify the proteins by peptide mass fingerprinting (PMF) of protein bands (*see* **Subheading 3.6.** and **Note 14**).

3.6. Protein Identification by MS

MS is a particularly valuable tool for protein identification. Protein sequences can be identified and correlated with the patterns of bands or spots from electrophoretic separations by analysis of proteolytic digests in PMF experiments. The digests are analyzed by matrix-assisted laser desorption/ionization time-of-flight mass spectrometry (MALDI-TOF-MS) producing peptide mass fingerprints, although identifications are also obtained from tandem MS analyses of the peptide mixtures. The subunits of complex I are also recognized from measurements of their molecular masses by ESI-MS *(37)*. A prerequisite for PMF and peptide sequencing by tandem MS is the production of proteolytic digests from gel-bound protein samples.

3.6.1. In-Gel Proteolysis With Trypsin

A manual procedure for obtaining tryptic digests from proteins within acrylamide gel pieces after gel purification is outlined. The method is based on the protocol described by Shevchenko et al. *(38)* except that the steps involving alkylation of cysteine residues are omitted and Tris-HCl replaces the ammonium bicarbonate buffers.

1. Excise the protein spots or bands of interest. Slice gel bands into small pieces. Place the pieces in a 0.5-mL or 1.5-mL disposable plastic tube (the tube is pre-rinsed with 50% aqueous ACN and dried).
2. Wash the gel slice in Milli-Q water (100 µL), 1 h. This and subsequent washing steps are carried out at room temperature.
3. Remove the water and wash the gel slice in 20 mM Tris-HCl, pH 8.0 (100 µL), 1 h.
4. Repeat the wash with Tris-HCl solution in **step 3**.
5. Remove the Tris-HCl solution and wash the gel slice in 50% ACN, 20 mM Tris-HCl, pH 8.0 (100 µL), 30 min.
6. Remove the 50% aqueous ACN solution.
7. Dehydrate the gel pieces by just covering with ACN (20 µL). The gel pieces turn white within 10 min. Dry the gel completely in a vacuum centrifuge for 20 min or longer until completely dry.
8. Rehydrate the dried gel pieces with a solution of Trypsin (Roche, sequencing grade), (12.5 ng/µL) in 20 mM Tris-HCl, pH 8.0, 5 mM CaCl$_2$. Approximately 7 µL trypsin solution is required.
9. After rehydration of the gel, add 20 mM Tris-HCl, pH 8.0, 5 mM CaCl$_2$ to just cover the gel pieces, so that the gel does not dry out.
10. Digest overnight at 37°C. The incubation is best performed in an oven rather than a water bath.
11. After a brief centrifugation step, remove the digestion solution, and place it in another (washed) disposable tube.
12. Extract the peptides with 5% formic acid for 1 h with occasional vortexing. Add sufficient solution just to cover the gel pieces.
13. Remove the 5% formic acid extract and perform a second extraction with a solution of 60% aqueous ACN containing 4% formic acid for 1 h with occasional vortexing.
14. The extracts of peptides are combined (usually) and a portion (0.3–0.5 µL) analyzed by MALDI-TOF-MS. The remainder is retained for further analysis by LC-MS/MS and is stored at −20°C.

3.6.2. Protein Identification by PMF

Samples of protein digests for protein identification are analyzed initially by MALDI-TOF-MS and PMF data obtained for comparison to sequence databases. A detailed description of mass spectrometer operation is not intended, but the general principles of protein identification are outlined.

1. A small portion (0.3–0.5 µL) of peptide extract of the in-gel protein digest is transferred onto a MALDI target plate.
2. Before the droplet dries, add an equal volume of matrix solution, (α-cyano-4-hydroxycinnamic acid) on top of the peptide extract droplet. Avoid touching the target plate with the tip (*see* **Note 15**).

3. Allow sample and matrix to dry and co-crystallize (better left to dry unaided).
4. The target plate is transferred into MALDI instrument and placed under vacuum.
5. Examine sample by MALDI-TOF-MS. Acquire mass spectra in the *m/z* range 700–3000 *m/z*.
6. Calibrate the MALDI-TOF spectrum. The best calibrations are obtained with internal calibration peaks. We use the calcium-related matrix ion and two bovine trypsin autolysis peaks, present in most spectra, as internal calibrants with masses of 1060.048, 2163.057, and 2273.160, respectively.
7. A list of monoisotopic masses (peaklist) is obtained from the spectra. This step is performed with the aid of proprietary software. The peaklist is compared with a database of protein sequences compiled from SWISSPROT (http://www.expasy.org) and also against "in-house" databases containing only the sequences of known subunits of oxidative phosphorylation protein complexes. A search algorithm provided within MassLynx (version 3.4) (Micromass/Waters, Manchester, UK) is one of many suitable software packages for this purpose (*see* **Note 16**).
8. Following the identification of proteins, review the data and inspect the MALDI-TOF spectrum for indications of additional components not identified in the first comparison. Then select peptide ions for confirmatory or exploratory tandem MS experiments if required.

3.6.3. Protein Identification by Tandem MS

Additional and more detailed analyses of protein digests are performed by tandem MS. The tandem MS approach produces a combination of mass and partial sequence data useful for making high-specificity comparisons with databases of known protein sequences. The following protocol uses a Q-TOF1 tandem mass spectrometer equipped with a "nano-flow" ESI interface, a capillary HPLC (CapLC) (Micromass/Waters) online to the instrument (1 µL/min flow rate) and a capillary autosampler. A PepMapC18 reverse-phase column (180 µ*M* × 100 mm; LC Packings, Amsterdam, Netherlands) is used for peptide separation. A separate RP trap cartridge (0.32 mm id × 5 mm) also containing PepMap C18 is installed in front of the analytical column (*see* **Note 17**). The LC system is configured with a 10-port valve allowing eluate from the trap cartridge to bypass the analytical column until sample loading and desalting operations are completed (*see* **Note 18**).

1. The mass spectrometer is tested, tuned, and calibrated using the collision-induced fragments of a synthetic peptide [Glu¹]fibrinopeptide B (Sigma-Aldrich; 500 fmol/µL).
2. The LC system is equilibrated with solvent A containing 0.1% formic acid in 5% aqueous ACN.
3. A preliminary LC-MS analysis is carried out without injection of sample in order to ensure that the system is free from contamination with either polymeric material or the column memories of previous separations.

4. Dilute the peptide mixture recovered from the gel to an ACN concentration of less than 10% with a solution of 2% aqueous formic acid. Usually, a dilution of at least fivefold is required.

5. A portion of the diluted sample is introduced with the aid of the autosampler. The described autosampler loads a maximal volume of only 6 µL and therefore multiple injections of sample are often required while the cartridge is eluted with Solvent A.

6. An LC-MS/MS analysis is performed. The trap cartridge is switched in series with the analytical column and peptides are eluted with a gradient of ACN in a period of approx 35 min. During this period the mass spectrometer is directed to perform an MS analysis on the eluate, scanning a mass range of 400–1400 *m/z* in a period of 1 s. Then MS/MS spectra were acquired automatically on up to three ions (1 s per ion, scanning a mass range of 50–1800 *m/z*) using data-dependent precursor ion selection. The mass spectrometer is directed to acquire data on multiply charged peptide ions distinguishing peptides from most of the chemical noise.

7. The LC-MS/MS analysis is repeated without any sample application in order to avoid column memory effects affecting subsequent LC-MS analyses. The data from this second LC-MS/MS analysis is retained and examined for peptide ions that were not recovered from the column during the initial separation.

8. Peptide fragment ion spectra are compared with sequence databases using the program MASCOT via an internet interface (*39*; http://www.matrixscience.com). The fragment ion spectra are also interpreted manually, where necessary, assembled into Peptide Sequence Tags and compared with a nonredundant protein sequence database maintained by the European Bioinformatics Institute (Hinxton, UK) using the algorithm Peptide Search (*40*) at EMBL Heidelberg (http://www.embl-heidelberg. de) or proprietary software from Micromass/Waters.

3.6.4. Measurement of Molecular Masses of Subunits

Protein molecular masses provide unique identification keys to the subunits of complex I enabling the protein compositions of fractions to be ascertained by ESI-MS (*17,37*). These measurements can also provide information on the presence or absence of posttranslational modifications of proteins (*see* **Note 19**). A list of molecular masses for the authentic bovine subunits is provided (*5,24*).

Molecular masses of subunits were measured by ESI-MS using a Sciex API III⁺ triple quadrupole mass spectrometer (MDS Sciex, Concord, ON, Canada), although simpler instruments incorporating ESI and a single mass analyzer such as a quadrupole or TOF analyzer could also be used. Samples were introduced by flow-injection analyses or in LC-ESI-MS experiments by direct transfer of the HPLC eluent to the mass spectrometer (*see* **Notes 20** and **21**). In flow-injection analysis described here, a solution of 50% aqueous ACN (approx 3 µL/ min) was delivered through a Rheodyne injector valve into the instrument with the aid of an HPLC pump. Protein samples were introduced via the Rheodyne injector.

1. Tune the mass spectrometer and calibrate the instrument over the expected mass range. A mixture of (poly)propylene glycols is used to calibrate the instrument over a range of *m/z* 59–2010.
2. Verify the instrument calibration with a stable protein of known molecular mass. Introduce approx 5 µL of a 2 µ*M* solution of horse heart myoglobin (average molecular mass 16,951.4 Da) in 1% formic acid/50% aqueous ACN into the instrument. Record the electrospray MS spectrum between 600 and 1800 *m/z*. An acceptable calibration would generate a protein mass for myoglobin within 1–1.5 Da of the expected value.
3. Prepare protein samples in acidic solutions containing ACN. Protein concentrations between 0.5–5 µ*M* are required in solutions, containing acid and organic solvent (usually ACN), and free from salts, including buffer salts and detergents (*see* **Note 22**). The sample preparation is vital for successful measurements.
4. Samples (5–10 µL) are introduced into the instrument via a Rheodyne injector valve.
5. Record electrospray spectra in MS mode (*see* **Note 23**). A mass range of 700–1700 *m/z* is suitable for most protein electrospray spectra, although some adjustment of mass range may be required.
6. Measure molecular masses by recognition of the series of multiply charged protein ions produced by ESI. This is aided by proprietary software supplied with mass spectrometers. In these cases the programs MacSpec 3.3 or BioMultiview (MDS Sciex, Canada) were used. In some instances, electrospray spectra were deconvoluted and placed onto a molecular mass scale to aid the interpretation of these data.

3.7. Analysis of Posttranslational Modifications

Posttranslational modifications in complex I subunits were indicated by measurements of protein molecular mass by ESI-MS *(3,5,17,19,28,37)*. Subsequently the identification and localization of these modifications required further analysis of these subunits and in particular their modified peptides. A variety of enzymatic digests were employed in these experiments, in order to isolate suitable modified peptides for analysis by tandem MS (*see* **Table 4**). The subsequent analysis of modified peptides by tandem mass spectrometric methods was uncomplicated once the appropriate peptides, with suitable molecular masses and properties, had been isolated. These experiments identified the 13 Nα-acetyl and one Nα-myristoyl modifications on 14 subunits of complex I and characterized the multiple methylation modifications of histidines in subunit B12 *(28)*.

4. Notes

1. Although the number of components in complex I is small compared with the thousands of proteins in an organelle such as the mitochondrion or even the tens of thousands in the cellular proteome, a comprehensive analysis of subunit composition of complex I remains a major analytical task. The protein chemical analy-

Table 4
Mass Spectrometric Characterization of Modified Peptides From Subunits of Complex I From Bovine Mitochondria

Subunit	Peptide[a]	Residues	m/z (z+)	Mass (MH+) Observed	Mass (MH+) Calculated	Sequence[b]
B22	T1	1–14	797.01 (2)	1593.02	1592.81	Ac-AFLSSGAYLTHQQK
B18	T1	1–7	417.83 (2)	834.66	834.56	Myr-GAHLAR
B17.2	T1-2	1–9	594.44 (2)	1187.88[c]	1171.69	Ac-MELLQVLKR
B17	R1	1–9	611.18 (2)	1221.36	1221.61	Ac-SGYTPEEKLR
B16.6	M1	1–10	486.40 (2)	971.80[d]	1019.52	Ac-AASKVKQDM
B15	R1	1–9	563.68 (2)	1126.36	1126.55	Ac-SFPKYEASR
B14.7	D1	1–9	603.70 (2)	1206.40	1206.66	Ac-AKTVLRQYW
B14.5a	Y1-2	1–9	561.26 (2)	1121.52	1121.57	Ac-ASATRFIQW
B14.5b	Y1	1–11	649.56 (2)	1298.12	1297.61	Ac-MMTGRQGRATF
B14	D1	1–22	743.91 (3)	2229.73	2230.23	Ac-AASGLRQAAVAASTSVKPIFSR
B13	T1-2	1–6	336.27 (2)	671.54	671.45	Ac-AGLLKK
B12[e]	T1	1–12	1321.07[f]	1321.07	1320.62[e]	Ac-AHGH*GH*EHGPSK[f]
B9	K1	1–9	523.70 (2)	1046.40	1046.60	Ac-AERVAAFLK
B8	K1	1–12	592.05 (2)	1183.10	1182.71	Ac-AAAAIRGVRGK

[a]In the "peptide" column, the letters T, R, M, D, Y, and K indicate peptides obtained by cleavages with trypsin, endoproteinase Arg C, cyanogen bromide, endoproteinase Asp-N, chymotrypsin, and endoproteinase Lys-C, respectively. Ac- and Myr- denote acetyl- and myristyl-, respectively.

[b]Sequences deduced from the data are underlined.

[c]The difference between the observed and calculated values (16 Da) corresponds to the oxidation of the N-terminal methionine residue.

[d]The C-terminal methionine has been converted to homoserine lactone by cyanogen bromide cleavage, leading to the difference (48 Da) between the observed and calculated values. The protonated molecular masses calculated for the modified peptides are 1187.68 and 971.52 Da respectively.

[e]Six differently modified versions of this peptide were characterized. This version is N-acetylated and contains two methylated histidine residues H*.

[f]Data obtained from analysis of singly charged ions using a tandem mass spectrometry instrument (Q-TOF) with MALDI ionization. From **ref. 28**.

sis of this complex is made difficult by the complicated nature of the subunit composition and in particular by the hydrophobic nature of many of its membrane-bound subunits. They pose a challenge to the existing methods of protein purification and protein analysis.

No single separation technique permits the complete definition of subunit composition. For example, subunits with masses lower than 20 kDa are not all resolved by simple 1D SDS-PAGE and the extremely basic isoelectric point of many subunits causes difficulties in 2D electrophoretic analyses. The most hydrophobic subunits are neither detected on 2D gels nor recovered from RP-HPLC separations. Therefore multiple separation methodologies are required to display and define the subunit composition of the intact bovine enzyme and subcomplexes *(24)*.

Many protein spots, protein bands, or fractions recovered from these separations still contain multiple components and careful and rather detailed analyses of these are required. As a protein identification tool, protein sequence analysis using the Edman degradation is limited by the large number of subunits, more than 50% of the subunit complement, possessing modified N-terminal residues. These proteins remain undetected by this method. Although peptide mass fingerprinting using a MALDI-TOF identifies rapidly the major protein component in a sample, it is less effective at establishing the complexity of protein mixtures. PMF also has limitations in the analysis of both small proteins and hydrophobic proteins, where few tryptic peptides are generated and even fewer are recovered and appear in the mass spectrum. For these reasons, the identification of many complex I subunits with molecular masses lower than 20 kDa is reliant on MS/MS methods and the extensive characterization of single peptides. As noted earlier *(24)*, and with the characterization of modified peptides, protein identifications may require the analysis of fragments beyond those from tryptic digestion (**Table 4**; **ref. 28**).

2. The preparation of complex I samples and removal of interfering salts and detergents from samples without irreversible loss of protein is a crucial step of the analytical procedure. Selective loss of protein, particularly the more hydrophobic components, can occur during precipitation stages as a consequence of denaturation, aggregation, or dehydration, leading to irreversible insolubility. The protein content of gels, especially 2D gels, can be affected dramatically by differences in sample preparation techniques.

3. Protein samples should not be allowed to dry out during the precipitation stages of sample preparation. This will lead to difficulties in re-solubilization of the sample.

4. The presence of detergent causes problems with RP-HPLC separations unless removed by precipitation or other sample preparation techniques. Some subunits co-elute with detergent or are eluted with bound detergent. Subsequently the detergent causes difficulties with ESI-MS analyses of fractions, because the ionization of proteins is prevented and their presence is masked.

5. The SDS-polyacrylamide gel patterns of complex I and its subcomplexes are shown in **Fig. 1**. The purity of the complex I preparation can be assessed from the absence of contaminating proteins in the region of the largest subunits. The α and β subunits

of ATP synthase are indicators of contamination from other respiratory complexes, as are the largest subunits of cytochrome oxidase, although these latter examples stain poorly with Coomassie dye. Below this region, the large number of complex I components make evaluations of purity impossible. No single gel resolves all complex I components adequately and, despite the use of acrylamide gradient gels, many bands contain more than one subunit, especially in the region below 20 kDa. The Tris/Tricine gel systems of Schägger and von Jagow *(36)* offer superior resolution of smaller components, but larger proteins are less well-resolved and proteins with similar molecular masses such as the 51 and 49 kDa subunits and the 42 and 39 kDa subunits co-migrate. With care, the gel patterns shown are reproducible, but they are influenced by minor alterations in the composition of the gel. Therefore they cannot be used as a reliable basis for the precise interpretation of patterns of subunits in other gel systems. Subunits ND1–ND6 and ND4L are the most problematic. They are all highly hydrophobic proteins and most stain poorly with Coomassie dye and tend to form indistinct, diffuse bands.

6. Do not heat or boil the protein samples in loading buffers before electrophoresis. Sample heating causes aggregation of hydrophobic proteins and their loss from gels.

7. Gels containing acrylamide gradients are essential in order to resolve smaller subunits. We use 12–22% acrylamide gradients generally in a mini-gel (10 × 10 cm) format with 0.6-mm gel spacers. These gels are made "in house" and we have not found any commercially made gradient gels superior to these.

8. Adequate separations of subunits can be obtained by SDS-PAGE using either mini-gels (10 × 10 cm) or larger-format gels. A cooling fan is incorporated into the apparatus for running mini-gels (EF100, Cambridge Electrophoresis Ltd., Cambridge, UK) and the current applied to the gel is limited to 25 mA. The gels are run more slowly and at lower temperatures than is possible with this apparatus.

9. The 2D separations shown in **Fig. 2** are the outcome of a number of improvements to the 2D gel protocol. These improvements gradually increased the protein content such that 35 subunits are detected on these gels. In addition, some hydrophobic proteins (e.g., subunits B14.7 and B16.6) are resolved as spots rather than as streaks and smears seen in earlier gels. However, none of the 2D gels contain any of the seven ND subunits, the most hydrophobic constituents of the complex. Their absence, together with subunits AGGG, ESSS, and SDAP, illustrates the well-known unsuitability of 2D gels for the analysis of membrane proteins *(41–43)*. These proteins are insoluble in the solutions used for rehydration of the IPG strips and fail to enter the IEF gel. Alternatively they are readily lost during sample preparation. The multitude of small subunits, less than 25 kDa, are resolved adequately by 2DE. However the resolution of all these subunits requires the use of IPG strips with two different pH gradients, 3.0–10.0 and 6.0–11.0, as shown in **Fig. 2**. Many complex I components have isoelectric point values close to or greater than 10 and are not resolved adequately in separations within the 3.0–10.0 gradient. The 2D gel patterns were influenced to a minor extent by pre-treatment of samples with chloroform/methanol *(24)*. More major effects were noted from differences in

sample solubilization. The IEF solution used to resolve the majority of complex I subunits contained 7 M urea, 2 M thiourea, 0.15% ASB-14, 18 mM DTT, 0.5% IPG buffer, and a trace of bromophenol blue. Early experiments showed that the inclusion of thiourea allowed the 49 kDa subunit to be detected clearly. Additionally a change in detergents from CHAPS, at concentrations of up to 4% (w/v) or combinations of 2% (w/v) CHAPS and caprylyl sulfobetaine 3–10, to the detergent ASB-14 allowed more of the hydrophobic subunits to be resolved.

10. The use of high concentrations of the detergent ASB-14 (≥1%) in the IEF step resulted in distortions observed in the second dimension gel. Therefore although higher concentrations of ASB-14 are used for the protein solubilization step, the final concentration used for IEF was reduced to 0.15% (w/v) by dilution.

11. Based on the profiles of protein spots, initial experiments suggest that the detergent DDM can replace ASB-14. As in previous instances, high detergent concentrations resulted in distorted second-dimension gel profiles. The use of DDM at 2% (w/v) for solubilization followed by dilution to 0.2% for electrophoresis overcame this problem.

12. Separations using 1-mm microbore columns allow proteins to be eluted in high concentrations, most suitable for direct analysis by ESI-MS. This is beneficial for the analysis of larger subunits of complex I when higher protein concentrations are required for good-quality ESI spectra. These and hydrophobic proteins are also prone to aggregation and irreversible precipitation when HPLC eluates are dried or concentrated by vacuum centrifugation.

13. The highest recoveries of the larger subunits of complex I are achieved with new RP-HPLC columns. The recoveries of larger subunits are rapidly compromised with column usage. The most hydrophobic subunits including the seven ND subunits are not recovered at all by these methods. The recovery of these proteins and the larger subunits are not improved by elution with less polar solvents such as propan-1-ol or propan-2-ol or mixtures of propanol with ACN. We have not noted any advantage in either chromatographic resolution or subunit yields using polymeric reverse-phase supports such as (poly) styrene/divinylbenzene as alternatives to silica. Generally Aquapore RP-300 columns are used for these separations.

14. Portions of fractions from HPLC separations are dried before analysis by 1D SDS-PAGE. To avoid difficulties with re-solubilization, add SDS (e.g., 1 µL of a 1% SDS solution) to each portion before drying and dissolution in gel-loading buffer.

15. The matrix compound α-cyano 4-hydroxycinnamic acid (Sigma, Cat. no. C2020) is recrystallized from ethanol before use. The recrystallized compound is stored in the dark at −20°C. The solution (10 mg/mL in 0.1% TFA in 60% aqueous ACN) is made fresh each day. Sonication of the matrix solution at room temperature in a bath sonicator helps to dissolve the matrix compound completely.

16. Reliable protein identifications require at least five peptide matches with a mass tolerance of 50 ppm, except in the cases of smaller proteins, less than 15 kDa, where fewer tryptic peptides are observed. In these cases, information from tandem MS analyses are required to confirm the identifications (*see* supplementary information associated with **ref. 24**).

17. It is particularly important that the analytical column in the CLC system and the trap cartridge, used for desalting, contain the same reverse-phase support.
18. A desalting step is necessary to remove excess detergents, salts, and unspecified contaminants before peptide samples recovered from polyacrylamide gels can be analyzed by ESI-MS. These experiments consume a greater proportion of the protein digest than MALDI experiments. The necessary desalting step is achieved by reverse-phase capillary column chromatography on-line to the mass spectrometer using a trap cartridge as described in **Subheading 3.6.3.** Alternatively, in a technique requiring skilled manipulation, the samples are desalted manually by chromatography on a microtip containing 50–100 nL of reverse-phase support *(44)*. The desalted sample is eluted directly to a microinjection needle and the unfractionated mixture is analyzed by nano-electrospray ESI *(44)*.
19. Protein identifications made from ESI-MS molecular mass data must account for the potential for artifactual protein modifications. These modifications arise frequently from oxidation, unwanted proteolysis, and exposure to reagents commonly used during protein purification. Cysteines are modified by reaction with mercaptoethanol, 4-(2-Aminoethyl)benzene sulphonylfluoride, or phenylmethyl-sulphonyl fluoride, and lysine residues by carbamylation with cyanate from decomposition of urea.
20. Protein molecular mass data can be obtained readily by LC-ESI-MS measurements with direct transfer of LC eluent (50 µL/min from 1-mm id reverse-phase columns) to the mass spectrometer. Then proteins are analyzed immediately, and this is beneficial for some proteins (e.g., 49- and 51-kDa subunits; *see* **Note 12**). However off-line analysis of chromatography fractions, either by flow-injection analysis (as described in **Subheading 3.6.4.**) or by nano-ESI are valuable alternatives. Then MS data can be accumulated for longer periods and the spectrum quality improved, particularly when samples are introduced at nanoliter per minute flow rates from microinjection needles as with the nano-ESI interface *(44)*.
21. RP chromatography of proteins is performed in the presence of TFA in solution (0.1%). Despite the subsequent problems with suppression of ESI caused by the TFA solution, it is still used in preference to formic acid for protein chromatography. The elution of proteins in sharp peaks, smaller volumes, and therefore high concentrations, compensates for the ion-suppression properties of this acid.
22. When HPLC fractions containing TFA are examined off-line, the ESI spectra can often be improved by acidification of the analyzed sample with formic acid to a concentration of 0.2–1%.
23. The detection of sample by the mass spectrometer is monitored by plotting the total ion current vs time, a common instrument feature. Separation of components can occur during delivery through capillary tubing and so all spectra are recorded while a signal from the sample remains.

Acknowledgments

We thank Drs. J. Hirst and Richard Shannon for providing samples of bovine complex I and its subcomplexes.

References

1. Schultz, B. E. and Chan, S. I. (2001) Structures and proton-pumping strategies of mitochondrial respiratory enzymes. *Annu. Rev. Biophys. Biomol. Struct.* **30**, 23–65.
2. Nicholls, D. G. and Ferguson, S. J. (2002) *Bioenergetics 3* Academic Press, London.
3. Walker, J. E. (1992) The NADH-ubiquinone oxidoreductase (complex I) of respiratory chains. *Q. Rev. Biophys.* **25**, 253–324.
4. Friedrich, T. (2001) Complex I: a chimera of a redox and conformation driven proton pump? *J. Bioenerg. Biomembr.* **33**, 169–177.
5. Hirst, J., Carroll, J., Fearnley, I. M., Shannon, R. J., and Walker, J. E. (2003) The nuclear encoded subunits of complex I from bovine heart mitochondria. *Biochim. Biophys. Acta* **1604**, 135–150.
6. Grigorieff, N. (1999) Structure of the respiratory NADH:ubiquinone oxidoreductase (complex I). *Curr. Opin. Struct. Biol.* **9**, 476–483.
7. Grigorieff, N. (1998) Three-dimensional structure of bovine NADH:ubiquinone oxidoreductase (complex I) at 22 Å in ice. *J. Mol. Biol.* **277**, 1033–1046.
8. Guénebaut, V., Schlitt, A., Weiss, H., Leonard, K., and Friedrich, T. (1998) Consistent structure between bacterial and mitochondrial NADH:ubiquinone oxidoreductase (complex I). *J. Mol. Biol.* **276**, 105–112.
9. Chomyn, A., Mariottini, P., Cleeter, M. W. J., et al. (1985) Six unidentified reading frames of human mitochondrial DNA encode components of the respiratory chain NADH dehydrogenase. *Nature* **314**, 592–597.
10. Pilkington, S. J. and Walker, J. E. (1988) Mitochondrial NADH-ubiquinone reductase: complementary DNA sequences of import precursors of the bovine and human 24 kDa subunit. *Biochemistry* **28**, 3257–3264.
11. Runswick, M. J., Gennis, R. B., Fearnley, I. M., and Walker, J. E. (1989) Mitochondrial NADH:ubiquinone reductase: complementary DNA sequence of the import precursor of the bovine 75-kDa subunit. *Biochemistry* **28**, 9452–9459.
12. Fearnley, I. M., Runswick, M. J., and Walker, J. E. (1989) A homologue of the nuclear encoded 49 kD subunit of bovine mitochondrial NADH-ubiquinone reductase is coded in chloroplast DNA. *EMBO J.* **8**, 665–672.
13. Pilkington, S. J., Skehel, J. M., Gennis, R. B., and Walker, J. E. (1991) Relationship between mitochondrial NADH-ubiquinone reductase and a bacterial NAD-reducing hydrogenase. *Biochemistry* **30**, 2166–2175.
14. Fearnley, I. M., Finel, M., Skehel, J. M., and Walker, J. E. (1991) NADH-ubiquinone reductase from bovine heart mitochondria: cDNA sequences of the import precursors of the 39 kDa and 42 kDa subunits. *Biochem. J.* **278**, 821–829.
15. Dupuis, A., Skehel, J. M., and Walker, J. E. (1991) A homologue of nuclear-coded iron-sulfur protein subunit of bovine mitochondrial complex I is encoded in chloroplast genomes. *Biochemistry* **30**, 2954–2960.
16. Arizmendi, J. M., Runswick, M. J., Skehel, J. M., and Walker, J. E. (1992) NADH:ubiquinone oxidoreductase from bovine heart mitochondria. A fourth nuclear encoded subunit with a homologue encoded in chloroplast genomes. *FEBS Lett.* **301**, 237–242.

17. Walker, J. E., Arizmendi, J. M., Dupuis, A., et al. (1992) Sequences of twenty subunits of NADH: ubiquinone oxidoreductase from bovine heart mitochondria: application of a novel strategy for sequencing proteins using the polymerase chain reaction. *J. Mol. Biol.* **226,** 1051–1072.

18. Arizmendi, J. M., Skehel, J. M., Runswick, M. J., Fearnley, I. M., and Walker, J. E. (1992) Complementary DNA sequences of two 14.5 kDa subunits of NADH: ubiquinone oxidoreductase from bovine heart mitochondria. Completion of the primary structure of the complex? *FEBS Lett.* **313,** 80–84.

19. Skehel, J. M., Fearnley, I. M., and Walker, J. E. (1998) NADH:ubiquinone oxidoreductase from bovine heart mitochondria: sequence of a novel 17.2 kDa subunit. *FEBS Lett.* **438,** 301–305.

20. Murray, J., Zhang, B., Taylor, S. W., et al. (2003) The subunit composition of the human NADH dehydrogenase obtained by a rapid one step immunopurification. *J. Biol. Chem.* **278,** 13,619–13,622.

21. Schilling, B., Bharath, M. M. S., Row, R. H., et al. (2005) Rapid purification and mass spectrometric characterization of mitochondrial NADH dehydrogenase (complex I) from rodent brain and a dopaminergic neuronal cell line. *Mol. Cell. Proteomics* **4,** 84–96.

22. Finel, M., Skehel, J. M., Albracht, S. P. J., Fearnley, I. M., and Walker, J. E. (1992) Resolution of NADH-ubiquinone oxidoreductase from bovine heart mitochondria into two subcomplexes, one of which contains the redox centres of the enzyme. *Biochemistry* **31,** 11,425–11,434.

23. Finel, M., Majander, A. S., Tyynelä, J., De Jong, A. M. P., Albracht, S. P. J., and Wikström, M. (1994) Isolation and characterisation of subcomplexes of the mitochondrial NADH:ubiquinone oxidoreductase (complex I). *Eur. J. Biochem.* **226,** 237–242.

24. Carroll, J., Fearnley, I. M., Shannon, R. J., Hirst, J., and Walker, J. E. (2003) Analysis of the subunit composition of complex I from bovine heart mitochondria. *Mol. Cell. Proteomics* **2,** 117–126.

25. Fearnley, I. M., Carroll, J., Shannon, R. J., Runswick, M. J., Walker, J. E., and Hirst, J. (2001) GRIM-19, a cell death regulatory gene product, is a subunit of bovine mitochondrial NADH:ubiquinone oxidoreductase (complex I). *J. Biol. Chem.* **276,** 38,345–38,348.

26. Carroll, J., Shannon, R. J., Fearnley, I. M., Walker, J. E., and Hirst, J. (2002) Definition of the nuclear encoded protein composition of bovine heart mitochondrial complex I: identification of two new subunits. *J. Biol. Chem.* **277,** 50,311–50,317.

27. Chen, R., Fearnley, I. M., Peak-Chew, S. Y., and Walker, J. E. (2004) The phosphorylation of subunits of complex I in bovine mitochondria. *J. Biol. Chem.* **279,** 26,036–26,045.

28. Carroll, J., Fearnley, I. M., Skehel, J. M., et al. (2005) The post-translational modifications of the nuclear encoded subunits of complex I from bovine heart mitochondria. *Mol. Cell. Proteomics* **4,** 693–699.

29. Sazanov, L. A., Peak-Chew, S. Y., Fearnley, I. M., and Walker, J. E. (2000) Resolution of the membrane domain of bovine complex I into subcomplexes:

implications for the structural organization of the enzyme. *Biochemistry* **39**, 7229–7235.

30. Neuhoff, V., Stamm, R., and Eibl, H. (1985) Clear background and highly sensitive protein staining with Coomassie Blue dyes in polyacrylamide gels: a systematic analysis. *Electrophoresis* **6**, 427–448.

31. Wessel, D. and Flugge, U. I. (1984) A method for the quantitative recovery of protein in dilute solution in the presence of detergents and lipids. *Anal. Biochem.* **138**, 141–143.

32. Simpson, R. J. (2003) *Proteins and Proteomics.* Cold Spring Harbor Laboratory Press, Cold Spring Harbor, New York, pp. 206–207.

33. Laemmli, U. K. (1970) Cleavage of structural proteins during the assembly of the head of bacteriophage T4. *Nature* **227**, 680–685.

34. Görg, A., Postel, W., and Günther, S. (1988) The current state of two-dimensional electrophoresis with immobilised pH gradients. *Electrophoresis* **9**, 531–546.

35. Chevallet, M., Santoni, V., Poinas, A., et al. (1998) New zwitterionic detergents improve the analysis of membrane proteins by two-dimensional electrophoresis. *Electrophoresis* **19**, 1901–1909.

36. Schägger, H. and von Jagow, G. (1987) Tricine-sodium dodecyl sulfate-polyacrylamide gel electrophoresis for the separation of proteins in the range from 1 to 100 kDa. *Anal. Biochem.* **166**, 368–379.

37. Fearnley, I. M., Skehel, J. M., and Walker, J. E. (1994) Electrospray mass spectrometric analysis of subunits of NADH:ubiquinone oxidoreductase (complex I) from bovine heart mitochondria. *Biochem. Soc. Trans.* **22**, 551–555.

38. Shevchenko, A., Wilm, M., Vorm, O., and Mann, M. (1996) Mass spectrometric sequencing of proteins from silver-stained polyacrylamide gels. *Anal. Chem.* **68**, 850–858.

39. Perkins, D. N., Pappin, D. J., Creasy, D. M., and Cottrell, J. S. (1999) Probability-based protein identification by searching sequence databases using mass spectrometry data. *Electrophoresis* **20**, 3551–3567.

40. Mann, M. and Wilm, M. (1994) Error tolerant identification of peptides in sequence databases by peptide sequence tags. *Anal. Chem.* **66**, 4390–4399.

41. Santoni, V., Kieffer, S., Desclaux, D., Masson, F., and Rabilloud, T. (2000) Membrane proteomics: use of additive main effects with multiplicative interaction model to classify plasma membrane proteins according to their solubility and electrophoretic properties. *Electrophoresis* **21**, 3329–3344.

42. Santoni, V., Molloy, M., and Rabilloud, T. (2000) Membrane proteins and proteomics: un amour impossible? *Electrophoresis* **21**, 105–107.

43. Gygi, S. P., Corthals, G. L., Zhang, Y., Rochon, Y., and Aebersold, R. (2000) Evaluation of two-dimensional gel electrophoresis-based proteome analysis technology. *Proc. Natl. Acad. Sci. USA* **97**, 9390–9395.

44. Wilm, M., Shevchenko, A., Houthaeve, T., et al. (1996) Femtomole sequencing of proteins from polyacrylamide gels by nano-electrospray mass spectrometry. *Nature* **379**, 466–469.

11

Identification of Targets
of Phosphorylation in Heart Mitochondria

Roberta A. Gottlieb

Summary

Whereas most strategies in proteomics deal with changes in protein levels, posttranslational modifications often represent pathological consequences at the molecular level without changes in abundance and, therefore represent a critical phenomenon to study. The present report elucidates an approach to studying one such posttranslational modification, phosphorylation, which is an important event in a variety of signaling cascades initiated in the heart in response to physiological and pathological stimuli. Comparison of phosphorylation profiles in different conditions, including drug treatments, receptor stimulation or blockade, and so forth, may lead to identification of relevant targets of signal transduction cascades. Furthermore, it may be possible to devise similar schemes to detect the subset of proteins modified by ubiquitinylation, lipid acylation, and so forth, and to compare differential patterns. Phosphorylation of mitochondrial targets is increasingly being recognized. Mitochondria are critical targets of a number of signaling pathways and therefore it is important to develop suitable methods to detect phosphorylation of mitochondrial targets. One approach is to perform in vitro phosphorylation in a reconstituted system of cytosol and mitochondria.

Key Words: Mitochondria; phosphoproteins; ischemia; myocardium; 2D PAGE; mass spectrometry; proteomics.

1. Introduction

We used a reconstituted system to identify mitochondrial elongation factor Tu (EF-Tu) as a target of phosphorylation in the myocardium *(1)*. Abundant evidence has shown immediate activation of cytosolic mitogen-activated protein kinases, protein kinase C isoforms, and tyrosine kinases during ischemia or ischemic preconditioning *(2)*. For instance, PKC isoforms, protein kinase A, and JNK have all been reported translocate to mitochondria, and in some cases,

From: *Methods in Molecular Biology, vol. 357: Cardiovascular Proteomics: Methods and Protocols*
Edited by: F. Vivanco © Humana Press Inc., Totowa, NJ

to phosphorylate specific mitochondrial targets *(3,4)*. Comparison of phosphorylation profiles in genetically modified mice may aid in identifying relevant mitochondrial targets of protein kinases, particularly those already suspected to translocate to the mitochondria, such as PKCε.

Detection of phosphorylation can be approached through several means. The classic approach has been to use metabolic labeling, in which cells are incubated with inorganic phosphate containing ^{32}P. The cells incorporate the radioisotope into ATP, and protein kinases then use the radiolabeled ATP to phosphorylate proteins, which can be recovered from cell lysates, resolved by sodium dodecyl sulfate polyacrylamide gel electrophoresis (SDS-PAGE), and detected by autoradiography. This approach is not suitable for in vivo studies, and is technically challenging in the ex vivo isolated perfused heart model because of the large amount of ^{32}P required and the risk of radioactive exposure and contamination of the area. A second approach has utilized antibodies against phosphotyrosine, phosphoserine, and phosphothreonine. Some of these antibodies have been quite useful, particularly those directed against a specific epitope such as Ser473 on Akt *(5)*. However, these antibodies have limitations related to their epitope specificity.

The third approach (broken cell method) utilizes isolated mitochondria incubated with γ-^{32}P-ATP, and will be described in detail in this chapter. Mitochondria from stimulated hearts would be predicted to have active protein kinases associated with them, compared to mitochondria isolated from unstimulated hearts. In addition, mitochondria from unstimulated hearts can be incubated with cytosol from stimulated hearts (plus γ-^{32}P-ATP) with the expectation that cytosolic kinases may phosphorylate mitochondrial targets.

1.1. Detection of Phosphoproteins by Comparison of Two Conditions Can Be a Tool to Select Proteins of Potential Interest for Proteomic Analysis

By comparing mitochondria phosphorylated by cytosol from control and ischemic hearts, we were able to select only those proteins whose phosphorylation state changed in response to ischemia. These phosphoproteins can then be identified by two-dimensional gel electrophoresis (2DE) and mass spectrometry (MS). We assessed differential phosphorylation of mitochondria by comparing continuously perfused control hearts with hearts that had been subjected to ischemia and reperfusion.

1.2. Selection of an Organelle Simplifies the Analysis

We would not have been able to distinguish changes in the whole cell lysate, and inspection of cytosol after in vitro phosphorylation revealed too many bands to detect changes. However, relatively few proteins in the mitochondria underwent phosphorylation. Therefore the approach of identifying mitochondrial phosphoproteins that were targets of cytosolic signaling greatly simplified the analysis. In addition to organelles and cytoskeleton, it is possible to recover signaling scaffolds by immunoprecipitating one component under conditions that preserve the integrity of the entire complex.

1.3. Further Simplification of the Mixture of Proteins to be Analyzed is Desirable

The tools of cell biology (e.g., suborganellar fractionation) and biochemistry (chromatography) should be employed whenever possible. Mitochondria contain more than 1000 different polypeptides of widely varying abundance. Although detection of the most abundant proteins on 2D gels is straightforward, one is likely to miss many of the less abundant proteins. In addition, membrane proteins are problematic owing to the limiting factor of solubilization. Therefore it is useful to enrich for the subset of proteins in which one is most interested.

1.4. Selection of a Protein Separation Strategy is Essential

Although 2D gels are one widely used strategy to separate complex mixtures of proteins for MS, other approaches exist and are discussed elsewhere in this collection. One approach that shows great promise is that developed by Yates and colleagues, utilizing multidimensional liquid chromatography (LC) coupled to tandem mass spectrometry (MS/MS) *(6)*. In this approach, a complex mixture of proteins is subjected to tryptic digestion or cyanogen bromide hydrolysis, then loaded onto a microcapillary column packed with a C18 reverse-phase matrix and with a cation exchange material. Peptides are eluted through the column with an automated program using an acetonitrile (ACN) gradient and a step gradient of ammonium acetate. The eluted peptides are injected directly into the mass spectrometer an analyzed by the SEQUEST algorithm *(7)*.

1.5. Validation of the Findings in a Native System is Essential

Our studies identifying $EF-Tu_{mt}$ as a protein that was differentially phosphorylated after ischemia required careful validation after MS identification. The 46-kDa protein that we had demonstrated to be phosphorylated in the reconstituted system met the following criteria:

1. The phosphoprotein and the immunoreactive $EF-Tu_{mt}$ co-localized on 2D SDS-PAGE, in mitochondria and in submitochondrial fractions, and had identical chromatographic properties. These findings supported the conclusion that the 46-kDa phosphoprotein was indeed $EF-Tu_{mt}$ as identified by MS.
2. The protein was differentially phosphorylated (greater in ischemia than control).

3. The protein was resident in mitochondria and its phosphorylation could be demonstrated using mitochondria with intact outer membranes.
4. The protein was present in isolated cardiomyocytes.
5. Phosphorylation could be shown to occur in vivo as well as in vitro.

2. Materials

2.1. Langendorff Perfusion, Tissue Homogenization, and Mitochondrial Isolation

1. Krebs-Ringer buffer (KRB): 118.5 mM NaCl, 4.7 mM KCl, 1.2 mM MgSO$_4$, 1.2 mM KH$_2$PO$_4$, 24.8 mM NaHCO$_3$, 2.5 mM CaCl$_2$, and 10 mM glucose. KRB was gassed with 95% O$_2$, 5% CO$_2$, resulting in a pH of 7.4–7.5.
2. Homogenization buffer (MSE): 225 mM mannitol, 75 mM sucrose, 1 mM EGTA, 1 mM Na$_3$VO$_4$, 20 mM HEPES-KOH, pH 7.4.
3. Hybrid Percoll/metrizamide discontinuous gradient: 5 mL of 6% Percoll, 2 mL of 17% metrizamide, and 2 mL of 35% metrizamide, all prepared in 0.25 M sucrose and set up in 13-mL tubes.
4. Polytron homogenizer: PowerGen 125 (Fisher Scientific) with a 10-mm diameter rotor knife.

2.2. In Vitro Phosphorylation of Mitochondria and Submitochondrial Fractionation

1. 100-µg Mitochondria for pilot studies; for MS, use 7 mg mitochondria.
2. Phosphorylation buffer: MSE supplemented with 25 mM HEPES-KOH, pH 7.5, 10 mM Mg acetate, 10 µM ATP (unlabeled) and 10 µCi [γ-^{32}P]ATP.
3. Cytosol (250 µg protein) is included for reconstitution experiments; this amount can be adjusted.
4. Mitochondrial swelling buffer: 10 mM KH$_2$PO$_4$, pH 7.4.
5. Mitochondrial stabilization buffer (MC buffer): 300 mM sucrose, 1 mM EGTA, 1 mM Na$_3$VO$_4$, 20 mM MOPS, pH 7.4.

2.3. 2DE, Autoradiography, and In-Gel Digestion of Phosphoproteins

1. Isoelectric focusing (IEF) sample buffer contained 8 M urea and 20 mM Tris-HCl, pH 7.4, and appropriate ampholytes.
2. Pharmacia immobilized pH gradient (IPG) phor IEF system.
3. Gel destaining and wash buffer: 25 mM NH$_4$HCO$_3$ in 50% ACN.
4. Trypsin 0.1 µg/µL in 25 mM NH$_4$HCO$_3$.
5. Peptide extraction buffer: 50% ACN with 5% trifluoroacetic acid (TFA).

2.4. MS Analysis

1. MALDI analysis: Voyager DE-STR MALDI-TOF instrument (Applied Biosystems, Framingham, MA).
2. MS/MS analysis: Nanospray needle (Protana, Odense, Denmark); Q-Star quadrupole instrument (Sciex, Toronto, Canada).

3. Methods

3.1. Isolated Heart Perfusions and Preparation of Subcellular Fractions

1. Rats or rabbits are anesthetized with pentobarbital and the heart is rapidly excised and quickly cannulated onto the Langendorff perfusion apparatus *(8)*. The heart is perfused with oxygenated KRB for 15 min, then subjected to global no-flow ischemia for 30 min and reperfusion for 15 min. Controls are perfused continuously for the same amount of time.

2. After Langendorff perfusion, hearts are removed from the Langendorff apparatus, rapidly minced in ice-cold MSE, and disrupted by Polytron homogenization (5 s at maximal power output). The homogenate is centrifuged for 10 min at 600g, 4°C to remove nuclei and unbroken cells. The supernatant is centrifuged for 10 min at 10,000g to pellet mitochondria and lysosomes (crude mito pellet). The supernatant (crude cytosol) is centrifuged for 30 min at 100,000g to obtain particulate-free cytosol (S100).

3. The 10,000g crude mito pellet from the previous centrifugation is resuspended in 10 mL of MSE buffer and centrifuged for 10 min at 8000g *(9)*. This wash step is repeated once. The final pellet is resuspended in 3 mL of MSE buffer and layered onto the hybrid Percoll/metrizamide gradient and centrifuged for 20 min at 50,000g, at 4°C using a Beckman SW41 rotor. The mitochondrial fraction at the interface between 17 and 35% metrizamide is collected and diluted 10-fold with MSE buffer, followed by centrifugation for 10 min at 10,000g to remove metrizamide. The pellet is resuspended in 20 mL of MSE and centrifuged again. The final pellet is resuspended in 3 mL MSE.

3.2. In Vitro Phosphorylation of Mitochondria and Submitochondrial Fractionation

1. Mitochondria are incubated in phosphorylation buffer for 30 min at 30°C with or without added S100 cytosol. The reaction is terminated by centrifugation at 10,000g for 5 min followed by washing twice in 0.5 mL MC buffer.

2. We used a modification of the method of Comte and Gautheron to fractionate mitochondria *(10)*. To enrich for proteins in a particular submitochondrial compartment, mitochondria are incubated in swelling buffer on ice for 20 min. This causes swelling and rupture of the mitochondrial outer membrane. This is followed by centrifugation for 15 min at 10,000g, at 4°C to pellet mitoplasts (inner membrane and matrix). The supernatant is centrifuged 30 min at 100,000g to obtain outer membrane (pellet) and intermembrane space constituents (supernatant). The mitoplast pellet is resuspended in 0.5 mL MC buffer and sonicated on wet ice in 5 cycles of 20 s bursts and 40 s rest intervals with output settings at 8–10 watts. The sonicated mitoplast preparation is centrifuged for 10 min at 10,000g to remove any remaining intact mitoplasts or mitochondria, followed by centrifugation at 100,000g for 30 min to separate inner membrane (pellet) and matrix components (supernatant).

3. Differences in phosphorylation between samples can be verified by resolving samples by SDS-PAGE followed by autoradiography. An example result is shown in **Fig. 1**.

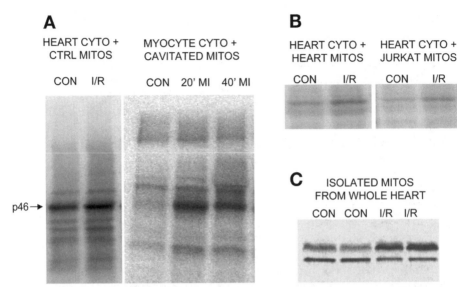

Fig. 1. Comparison of phosphoproteins in mitochondria subjected to various incubations. **(A)** Mitochondria from a normal heart (CTRL MITOS) or from cardiomyocytes disrupted by nitrogen cavitation (CAVITATED MITOS) were incubated with γ^{32}P-ATP and cytosol from a control heart (CON) or from a heart subjected to ischemia/reperfusion (I/R). **(B)** A similar reconstitution using heart mitochondria or mitochondria prepared from Jurkat cells disrupted by nitrogen cavitation. **(C)** Mitochondria were isolated from CON or from hearts subjected to I/R, then incubated with γ^{32}P-ATP without added cytosol. MI, myocardial ischemia. From **ref. _1_**, with permission.

3.3. 2DE, Autoradiography, and In-Gel Digestion of Phosphoproteins

1. Mitochondrial fractions from parallel radiolabeled and unlabeled reactions are resuspended in a minimal volume of IEF sample buffer and resolved by 2DE under identical conditions using the Pharmacia IPGphor IEF system, then stained with Coomassie blue, destained, and dried. The gel containing the radiolabeled sample is exposed to X-ray film to detect phosphoproteins. The autoradiogram is placed over the unlabeled dried gel and the spots of interest are excised for MS analysis. An example is shown in **Fig. 2**.

2. It is important to remember throughout all of these steps that great care must be taken to avoid keratin protein contamination. All materials (e.g., glass plates) should be acid-washed and gloves should be worn at all times. Avoid leaning over samples in such a way that hair or skin flakes might be introduced.

3. The gel is destained and dehydrated by washing three times (~10 min) with 25 mM NH$_4$HCO$_3$–50% ACN (or until the Coomassie stain is no longer visually detectable). The destained gel particles are then dried under vacuum for 30 min. After rehydration of the particles with a minimal amount of 25 mM NH$_4$HCO$_3$ with 0.1 µg of trypsin per µL, the protein is digested overnight at 37°C *(11)*. Recovery

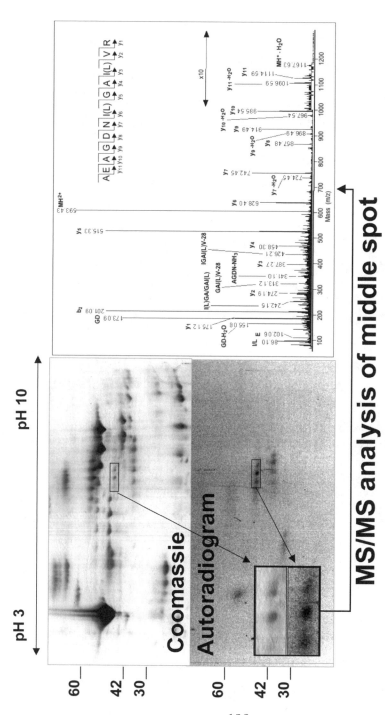

Fig. 2. Preparation of sample for identification by mass spectrometry (MS). Inner mitochondrial membrane from 7 mg of mitochondria were used for phosphorylation with "cold" or "hot" ATP and resolved by two-dimensional polyacrylamide gel electrophoresis for Coomassie staining and autoradiography, respectively. Inset shows the three spots of interest, which were removed and separately subjected to trypsin digestion and MALDI-MS. The middle spot was subsequently used for MS/MS analysis, shown on the right. From **ref. 1**, with permission.

of the peptides is accomplished by extracting the digestion mixture three times with 50% ACN–5% TFA. In an effort to reduce the amount of volatile salts (e.g., TFA and NH_4HCO_3), the recovered peptides are concentrated in a Speed-Vac vacuum centrifuge (to a final volume of ~5 µL) and rehydrated at least three times. Control digestions are performed on gel slices that did not contain any protein and are expected to yield trypsin autoproteolysis products and keratin contaminants that can be readily identified and excluded in the subsequent mass spectrometric analyses.

3.4. MS Analysis

1. MALDI analysis (11) utilizes a Voyager DE-STR MALDI-TOF instrument (Applied Biosystems) equipped with a nitrogen laser (337 nm), operated in delayed-extraction (12) and reflectron mode (13). Mass spectra are calibrated internally on the trypsin autolysis peptides.
2. For peptide sequencing MS/MS analysis, the crude peptide mixture obtained after in-gel digest is purified over a C_{18} reverse-phase Zip-Tip® (Millipore, Bedford, MA). The purified sample is supplied into a nanospray needle (Protana) and analyzed on a Q-Star quadrupole TOF instrument (Sciex) in nanospray mode. The ion spray voltage is set to 1100 V. For MS/MS experiments, the collision energy Q_0 is set to 50.

4. Additional Considerations

The following are additional considerations that can be applied to the procedures outlined in this chapter.

1. A drawback of using a reconstituted system is that because the lysates are prepared from whole heart tissue comprising cardiomyocytes, endothelial cells, and fibroblasts, it is possible that the kinase(s) could be derived from endothelial cells or fibroblasts rather than cardiomyocytes. However, the mitochondria would be predominantly derived from cardiomyocytes. One should use isolated cardiomyocytes to validate the findings.
2. A second concern is related to the fact that the mitochondria are prepared by Polytron homogenization resulting in damage to their outer membranes, thus permitting artifactual phosphorylation of target proteins in a compartment that would not ordinarily be accessible to a kinase. To address this concern, one can prepare cytosol and mitochondria from isolated cardiomyocytes that are subjected to simulated ischemia by metabolic inhibition with cyanide and 2-deoxyglucose, followed by recovery. Cell disruption by nitrogen cavitation preserves outer mitochondrial membrane integrity (14). If the same pattern of phosphorylation is observed, it argues against artifactual phosphorylation in the in vitro system.
3. Controls with no added cytosol and with varying amounts of cytosol are important to ascertain whether the kinase is derived from cytosol or mitochondria. In our experience, the phosphorylation of p46 was not directly due to a cytosolic protein kinase, because even isolated mitochondria from a control heart exhibited a cer-

tain degree of phosphorylation of p46, which was suppressed by the addition of cytosol from control hearts and enhanced by cytosol from ischemic hearts. The addition of increasing amounts of cytosol tended to suppress the phosphorylation of p46, implying the presence of an inhibitory cytosolic factor that was diminished after ischemia.

4. 2D spot mapping can be quite difficult, depending on the complexity of the sample under analysis. Although our autoradiogram of mitochondria after in vitro labeling revealed only about 20 spots, a Coomassie-stained gel of the same material was extremely complex and the region corresponding to the 46-kDa phosphoproteins had no detectable Coomassie-stained material when the gel contained as much as 2 mg of mitochondrial protein. This made it essential to further enrich the phosphoprotein.

5. Fractionation of mitochondria can further reduce the complexity and enrich for phosphoproteins of interest. In our case, submitochondrial fractionation revealed that the 46-kDa protein was present in the matrix and in the inner membrane. Because it was easier to prepare and concentrate the inner membrane, this material was used for subsequent analysis. This subfractionation was sufficient to enrich the 46-kDa protein so that it could be detected by Coomassie staining after loading inner membrane derived from 7 mg of mitochondria on a large (18 cm) preparative gel. Although having enough protein to detect by Coomassie staining is not essential, it is a reassuring indication that one will have enough material for successful identification by mass spectrometry. Moreover, it will reveal the presence of additional contaminating proteins in the neighborhood that may complicate the analysis. Verification of the identity of the phosphoprotein can be achieved by demonstrating that the immunoreactive protein co-localizes in the same subcellular and submitochondrial compartment and has the same elution profile in one or more chromatographic systems (*see* example in **Fig. 3**).

6. Analysis of the tryptic peptides by MALDI-MS revealed that the three spots had identical mass signatures, confirming that they represented multiple phosphorylated forms of the same protein. This observation is a valuable confirmation of the protein identity as a phosphoprotein, and can subsequently be used to evaluate in vivo phosphorylation events. Evidence that phosphorylation occurs in vivo is obtained by immunoblot analysis of 2D gels. In our study, EF-Tu immunoreactivity of three closely spaced spots could be superimposed on the autoradiogram of the phosphoprotein, thus confirming that EF-Tu$_{mt}$ exists as multiple isoforms that differ by isoelectric point, presumably representing multiple phosphorylated species (*see* example in **Fig. 3**). Comparison of the isoforms of EF-Tu$_{mt}$ observed in the reconstituted preparations of cytosol and mitochondria with those observed in a freshly isolated heart is shown in the inset (**Fig. 3C**, inset). The presence of multiply phosphorylated EF-Tu$_{mt}$ in vivo confirms that this is not merely an artifact of in vitro labeling.

7. A total of 12 mass fingerprints from derived rabbit heart mitochondrial protein were used to query the database, but failed to match anything in the database, thus necessitating peptide sequencing. The overall sequence of mitochondrial EF-Tu is highly conserved, demonstrating 92% homology with human and 95% homol-

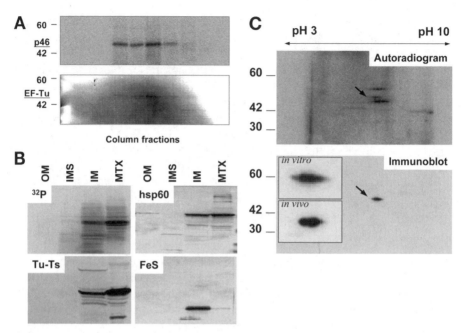

Fig. 3. Verification that p46 is elongation factor Tu (EF-Tu)$_{mt}$. (**A**) Column fractions from anion exchange were concentrated, resolved by sodium dodecyl sulfate-polyacrylamide gel electrophoresis, and subjected to autoradiography (upper panel) and immunoblotting for EF-Tu (lower panel). (**B**) Submitochondrial fractionation was performed to obtain outer membrane (OM), intermembrane space (IMS), inner membrane (IM), and matrix (MTX). Shown are the autoradiogram of p46 (^{32}P), immunoblot of mitochondrial EF-Tu/Ts (Tu-Ts), and markers for IM (Rieske iron-sulfur protein, FeS) and matrix (hsp60). (**C**) Two-dimensional gel showing autoradiogram of p46 and immunoblot for EF-Tu. Inset shows an enlargement of the immunoblot, showing two to four isoelectric variants of EF-Tu from mitochondria incubated in vitro with cytosol (in vitro) and from mitochondria rapidly isolated from a fresh heart (in vivo). From **ref.** (*1*) with permission.

ogy with bovine amino acid sequence. Despite this high degree of homology, detection by the mass fingerprint was unsuccessful, because of limitations in the current analytical software. The algorithm can tolerate one amino acid substitution in a fragment, but fails if there are two or more amino acid substitutions in a single tryptic fragment. It is further confounded if a tryptic cleavage site is missing. Because very little of the rabbit genome is in the database, one must rely on comparisons with other species such as mouse and human. Unfortunately, the rabbit genome diverges just enough to confound such comparisons. This illustrates the need to select starting material from a species that is well-represented in the database, and also emphasizes the value of sequencing the genomes of additional species that serve as important animal models, including rabbit, dog, and pig.

References

1. He, H., Chen, M., Scheffler, N. K., Gibson, B. W., Spremulli, L. L., and Gottlieb, R. A. (2001) Phosphorylation of mitochondrial elongation factor Tu in ischemic myocardium: basis for chloramphenicol-mediated cardioprotection. *Circ. Res.* **89,** 461–467.
2. Cohen, M. V., Baines, C. P., and Downey, J. M. (2000) Ischemic preconditioning: from adenosine receptor of K_{ATP} channel. *Annu. Rev. Physiol.* **62,** 79–109.
3. Vondriska, T. M., Zhang, J., Song, C., et al. (2001) Protein kinase C epsilon-Src modules direct signal transduction in nitric oxide-induced cardioprotection: complex formation as a means for cardioprotective signaling. *Circ. Res.* **88,** 1306–1313.
4. He, L. and Lemasters, J. J. (2005) Dephosphorylation of the Rieske iron-sulfur protein after induction of the mitochondrial permeability transition. *Biochem. Biophys. Res. Commun.* **334,** 829–837.
5. Alessi, D. R., Andjelkovic, M., Caudwell, B., et al. (1996) Mechanism of activation of protein kinase B by insulin and IGF-1. *EMBO J.* **15,** 6541–6551.
6. Washburn, M. P., Wolters, D., and. Yates, J. R., 3rd (2001) Large-scale analysis of the yeast proteome by multidimensional protein identification technology. *Nat. Biotechnol.* **19,** 42–247.
7. Eng, J. K., McCormack, A. L., and Yates, J. R., 3rd (1994) An approach to correlate tandem mass spectral data of peptides with amino acid sequences in a protein database. *J. Am. Soc. Mass Spectrom.* **5,** 976–989.
8. Tsuchida, A., Liu, Y., Liu, G. S., Cohen, M. V., and Downey, J. M. (1994) Alpha 1-adrenergic agonists precondition rabbit ischemic myocardium independent of adenosine by direct activation of protein kinase C. *Circ. Res.* **75,** 576–585.
9. Storrie, B. and Madden, E. A. (1990) Isolation of subcellular organelles. *Methods Enzymol.* **182,** 203–225.
10. Comte, J. and Gautheron, D. C. (1979) Preparation of outer membrane from pig heart mitochondria. *Methods Enzymol.* **LV,** 98–104.
11. Wong, D. K., Lee, B. Y., Horwitz, M. A., and Gibson, B. W. (1999) Identification of fur, aconitase, and other proteins expressed by Mycobacterium tuberculosis under conditions of low and high concentrations of iron by combined two-dimensional gel electrophoresis and mass spectrometry. *Infect. Immun.* **67,** 327–336.
12. Vestal, M. L., Juhasz, P., and Martin, S. A. (2001) Delayed extraction matrix-assisted laser desorption time-of-flight mass spectrometry. *Rapid Commun. Mass Spec.* **9,** 1044–1050.
13. Karas, M., Bahr, U., and Giessmann, U. (1991) Matrix-assisted laser desorption ionization mass spectrometry. *Mass Spectrom. Rev.* **10,** 335–357.
14. Gottlieb, R. A. and Adachi, S. (2000) Nitrogen cavitation for cell disruption to obtain mitochondria from cultured cells. *Methods Enzymol.* **322,** 213–221.

III

PROTEOMICS OF HUMAN ATHEROMA PLAQUES

12

Characterization of the Human Atheroma Plaque Secretome by Proteomic Analysis

Mari-Carmen Durán, Jose L. Martín-Ventura, Sebastián Mas,
Maria G. Barderas, Verónica M. Dardé, Ole N. Jensen,
Jesús Egido, and Fernando Vivanco

Summary

Atherosclerosis is one of the most common causes of death in developed countries. Atherosclerosis is an inflammatory process that results in the development of complex lesions or plaques that protrude into the arterial lumen. Plaque rupture and thrombosis result in the acute clinical complications of myocardial infarction and stroke. Although certain risk factors (dyslipidemias, diabetes, hypertension) and humoral markers of plaque vulnerability (C-reactive protein, interleukin-6, -10 and -18, CD-40L) have been identified, a highly sensitive and specific biomarker or protein profile, which could provide information on the stability/vulnerability of atherosclerotic lesions, remains to be identified. Recently, we have described a novel strategy consisting in the proteomic analysis of proteins released by normal and atherosclerotic arterial walls in culture. This method enables harvesting of proteins that are only secreted by pathological or normal arterial walls. By focusing only on the secreted proteins found in the tissue culture media, there is an intended bias toward those molecules that would have a higher probability of later being found in plasma. Using this approach, we have shown that carotid atherosclerotic plaques cultured in vitro are able to secrete proteins, and also that a differential pattern of protein secretion of normal arteries vs pathological ones has been observed. In this chapter, the proteomic analysis of the human atheroma plaque secretome is described.

Key Words: Secretome; atheroma plaque; atherosclerosis; two-dimensional electrophoresis (2DE), mass spectrometry (MS); biomarker.

1. Introduction

Atherosclerosis is a chronic disease that constitutes one of the main causes of death in developed countries (1). Atheroma plaque formation is the result of the interaction between several cellular and molecular species, including connective

From: *Methods in Molecular Biology, vol. 357: Cardiovascular Proteomics: Methods and Protocols*
Edited by: F. Vivanco © Humana Press Inc., Totowa, NJ

tissue matrix (collagen) produced by smooth muscle cells (SMC), lipids (into the core of macrophage foam cells or free in the tissues), and T-lymphocytes *(2)*. Both SMC and inflammatory cells (macrophages and T-lymphocytes) produce a wide range of proteins, including inflammatory cytokines, such as tumor necrosis factor-α, metalloproteinases; and cysteine proteases (Cathepsins), the last being responsible for the degradation of the collagenous matrix components or procoagulant proteins such as tissue factor *(3)*. Some of these proteins are secreted during the development and proliferation of the plaque, playing an important role in the interaction between the cellular components and allowing the connection endothelium–blood cell. This in turn promotes the proliferation and disruption of the matrix in the most complicated states, which normally results in subsequent plaque rupture and thrombosis, precipitating acute coronary events *(4,5)*. A major obstacle for applying proteomic analysis to vascular pathology is the heterogeneous cellular composition of atherosclerotic plaques (endothelial cells, SMC, leukocytes, and erythrocytes) *(6)*. In this sense, differential display experiments might reflect the heterogeneous cell composition of advanced lesions compared to normal vessels without necessarily contributing to a better understanding of vessel pathology *(7)*.

To date several attempts have been made to study the atherosclerotic phenomena by proteomic approaches *(8–10)*. Recently, we described a novel strategy consisting of the proteomic analysis of proteins released by normal and pathological arterial walls from patients affected by atherosclerosis *(11)*. We hypothesized that the patterns of protein secretion could be different between atherosclerotic plaques and normal endarteries. Thus, by culturing healthy arteries and carotid endarterectomy samples in a medium free of proteins, and later analyzing the proteins released into the supernatants by two-dimensional gel electrophoresis (2DE), we have demonstrated that carotid atherosclerotic plaques cultured in vitro are able to secrete proteins, and also that a differential pattern of protein secretion has been reported. The goal is to define new specific biomarkers released by the complicated and uncomplicated plaques in the plasma. This ex vivo system avoids technical obstacles associated with tissue extraction or with the analysis of the whole plasma, which contains large amount of nonvascular proteins that could interfere with protein identification. Progressive improvements in the methodology have allowed the characterization of an important number of proteins differentially secreted by carotid atherosclerotic plaques *(12)*. The majority of the analyzed proteins showed increased secretion levels in atheroma plaque compared with controls. In addition, this increase was parallel to the intensity of the lesion: the more complicated the lesion, the higher the secretion levels. The different secretion profile observed, with proteins promoting or preventing the proliferation of the lesion, could suggest that the balance between them finally decides the progression or not of the atherothrombotic process.

In this work, we have shown that this proteomic approach is indeed a powerful strategy for the detection of possible biomarkers in atherosclerotic disease, although more extensive analyses are necessary to determine the true role(s) that the described proteins are playing in this pathology.

2. Materials

2.1. Tissue Culture

1. Serum-free RPMI medium: RPMI, 1% HEPES, 1% penicillin, streptomycin, amphotericin.

2.2. 2D Electrophoresis

1. Precipitation of proteins: prepare a 100% trichloroacetic acid (TCA) stock solution by adding 454 mL H_2O/kg solid TCA. From this, prepare a 10% (v/v) TCA in acetone containing 0.07% β-mercaptoethanol (*see* **Note 1**).
2. Rehydration buffer: 8 *M* urea, 0.5% CHAPS, 2 m*M* TCEP, and 0.2% Pharmalyte pH 3.0–10.0 and 4.0–7.0, (broad range; Bio-Rad) (*see* **Notes 2** and **3**).
3. Ready-strip immobilized pH gradient (IPG) strips (Bio-Rad): 3.0–10.0 and 4.0–7.0 pH range strips.
4. Isoelectric focusing (IEF) System chamber (Bio-Rad).
5. Equilibration buffer: 1.5 *M* Tris-HCl pH 8.8 buffer, containing 6 *M* urea, 20% glycerol, 2% sodium dodecyl sulfate (SDS; traces of bromophenol blue can be added to track the SDS-PAGE front).
6. Dithiothreitol (DTT) equilibration buffer: add 2% (w/v) DTT (Bio-Rad; 100 mg/5 mL) in equilibration base buffer.
7. Iodoacetamide (IAA) equilibration buffer: 2.5% IAA (Bio-Rad; 125 mg/5 mL) in equilibrating base buffer.
8. Protean II system (Bio-Rad).
9. Resolving gel preparation: prepare a stock solution of acrylamide, 30% (w/v), bis-acrylamide, 0.8% (w/v), 1.5 *M* Tris base, pH 8.8, and 10% SDS. Use 12.5% acrylamide gels in 0.375 *M* Tris-HCl, pH 8.8 (final concentration).

2.3. Silver Staining

Any commercial kit (i.e., PlusOne Silver Staining Kit, Amersham Biosciences) or home-made protocol can be used (*see* **Note 4**).

3. Methods

3.1. Tissue Sampling

In this study, atherosclerotic plaques come from patients undergoing carotid endarterectomy (carotid stenosis >70%) at our institution (*n* = 28), and informed consent was obtained before enrollment. As healthy controls, we used artery fragments (mammary or radial endarteries) from cardiac surgical procedures. The study followed our institutional guidelines.

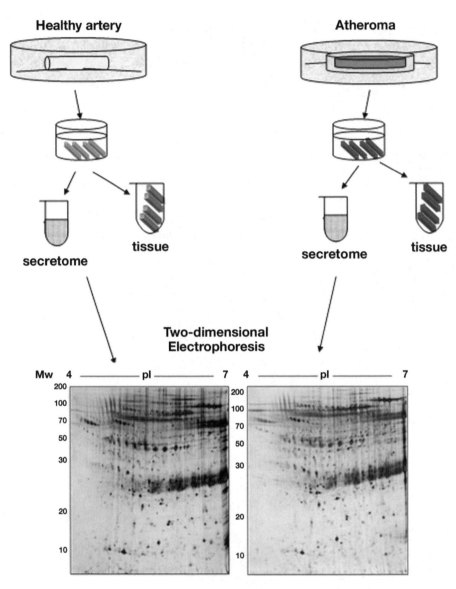

Fig. 1. Comparison of the radial and mammary artery vs the atherosclerotic plaque secretome. Human radial and mammary arteries, considered as control, as well as carotid arteries were incubated 24 h in serum-free RPMI medium at 37°C. Supernatants were collected and 600 µg of protein were precipitated with 10% trichloroacetic acid/acetone, and resuspended in 300 µL rehydration buffer. First dimension was performed using 4.0–7.0 pH range immobilized pH gradient strips. The second dimension was done with a 12.5% acrylamide gels. Gels were silver-stained.

3.2. Tissue Culture

1. Samples are collected in saline buffer and processed in a safety culture cabinet. In a sterile Petri dish, separate the stenosing complicated zone (origin of the internal carotid artery) from the adjacent plaque (common and external carotid endartery) using a surgical scalpel. Remove the regions adjacent to the noncomplicated areas. Histological analysis showed that the complicated plaques used for the study had ruptured and contained an intra-plaque hemorrhage with a variable, but important, proportion of inflammatory cells (*see* **ref. 11**). The adjacent area considered as uncomplicated plaque was composed of fibrous thickening with a variable content of vascular SMC.
2. For mammary arteries, remove the adventitia before incubation of the intima-media.
3. Cut the tissue (approx 1 cm long) into pieces of about 1 mm^3 and then incubate separately for 24 h in serum-free RPMI medium (2 mL) at 37°C (cell culture 24-well plates).
4. After 24 h, collect the supernatant and determine the protein concentration by Bradford *(13)* or another analytical method employed with that purpose.

3.3. 2D Electrophoresis

3.3.1. Precipitation with TCA/Acetone (see **Note 1**)

1. From each culture sample (2 mL supernatant), precipitate the proteins (~400–600 μg) with 5 vol of a solution containing 10% TCA in acetone and 0.07% β-mercaptoethanol, incubating 1 h at −20°C.
2. Centrifuge (2000*g*, 30 min, 4°C). Remove the supernatant and wash the pellet with 1 mL of cold acetone containing 0.07% β-mercaptoethanol. Incubate approx 20 min at −20°C.
3. Remove the residual acetone by air-drying and resuspend the pellet in 300 μL rehydration buffer to proceed with the 2DE.

3.3.2. IEF or First-Dimension Step

1. Resuspend the samples in 300 μL rehydration buffer and load them into the IEF System chamber (Multiprotean IEF cell System, Bio-Rad), employing 4.0–7.0 pH range strips.
2. The program chosen to separate the proteins in the first dimension includes the following steps: active rehydration (50 V) during 14 h, followed by a consecutive step in which voltage is applied in an step-and-hold mode: 200 V for 30 min, 500 V for 30 min, 1000 V for 60 min, 2000 V for 60 min, 5000 V for 120 min, 8000 V for 1 h, and a final gradient step of 8000 V h for 8 h. The temperature applied in all these steps is 20°C.

3.3.3. Equilibration

After IEF, strips have to be equilibrated, to reduce and alkylate the sample before separation by SDS-PAGE.

1. Incubate the IPG strips with 1% (w/v) DTT diluted in 5 mL equilibration buffer per gel, for 20 min, with agitation.
2. Remove the equilibration buffer and wash the strips for a few seconds with deionized water.
3. Finally, incubate again the strips in the same buffer but in this case with 4.8% (w/v) IAA for 20 min, also in agitation. Remove the equilibration buffer. Strips are ready to use in the second dimension. In case the strips are not going to be immediately used, we recommend storing them at −20°C (*see* **Note 5**).

3.3.4. Second Dimension: SDS-PAGE Gels

1. SDS-PAGE is performed according to Laemmli *(14)*, using a Protean II system (Bio-Rad) at 25 mA/gel at 4°C.
2. Polymerize the gels in a 12.5% acrylamide.
3. Place the strips on the surface of the polyacrylamide gel. Ensure that no air bubbles are trapped between the IPG strip and the gel surface between the gel backing and the glass plate. The addition of agarose to fix both the IPG strip and the SDS-polyacrylamide gel can be helpful. This is important to avoid proteins being lost when they migrate from one gel to another.

3.3.5. Silver Staining

For silver staining, we employed silver-staining kits (PlusOne Silver-Staining Kit, Amersham Biosciences) to increase the reproducibility of the resultant images. This kit is compatible with matrix-assisted laser desorption/ionization (MALDI) and electroscopy ionization mass spectrometry (ESI-MS) (*see* **Note 4**) and works with semipreparative loads without resulting in negative staining (*see* manufacter's protocol).

3.3.6. Image Analysis

Evaluation and processing of the 2D gels was performed by PD-Quest (Bio-Rad) gel analysis software. The protein patterns in the gels were recorded as digitalized images using a desktop scanner (Duo-Scan HiD, Agfa-Gevaert) and imported to PD-Quest.

3.4. Sample Preparation Previous to MS Analysis

3.4.1. In-Gel Digestion

1. After PD-Quest analysis, select the spots to be cut from the 2D gels, using a sterile scalpel.
2. For the in-gel trypsin digestion, we follow the same protocol as Schevchenko et al. *(15)* but with some modifications: wash the gel pieces two times with 50 μL of 50 and 100% acetonitrile (ACN); 15 min for each wash.
3. Remove the supernatants and incubate the samples with 50 μL of 10 mM DTT in 100 mM NH$_4$HCO$_3$ for 45 min at 56°C.

4. Alkylate with another 50 μL of 55 mM IAA in 100 mM NH$_4$HCO$_3$ for 30 min at room temperature in the dark.
5. Remove the supernatants and add 50 μL 100% ACN to wash the gel pieces for 15 min.
6. Remove again the supernatants and vacuum-dry the samples in a Speed-Vac for 10–15 min.
7. Gel pieces are then swollen in a digestion buffer containing 50 mM NH$_4$HCO$_3$ and 12.5 ng/μL of sequencing-grade modified trypsin (Promega) in an ice bath for 45 min.
8. Discard remaining buffer, adding and 20 μL of 50 mM NH$_4$HCO$_3$ to the gel piece. The digestion will proceed at 37°C overnight.
9. Collect the supernatants containing the tryptic peptides. Unless you are going to analyze the samples briefly, store them at −20°C. These supernatants will be employed in further MS analysis.

3.4.2. Microcolumn Purification of Digested Samples

1. Prepare a column by packing 100–300 nL of POROS R2 material (PerSeptive) in a constricted GELoader tip (Eppendorf) (*see* **Note 6**), A 1.25-mL syringe can be used to force liquid through the column by gentle applying air pressure (*see* Chapter 13).
2. Equilibrate the column with 20 μL of 5% formic acid (FA) and add the analyte solution. The volume of analyte employed will depend on the peptide concentration available. Normally, with proteins extracted from saturated spots, 5 μL or even less is enough for a good identification with MALDI-MS analysis. Wash the column with 20 μL of 5% FA.
3. Elute the bound peptides directly onto the MALDI target with 0.5 μL 2,5-dihydroxybenzoic acid (DHB) solution (20 μg/μL in ACN, 0.1% TFA 70:30 [v/v]). For external calibration, it is possible to use tryptic Lactoglobulin (5 pmol/μL), loading 1 μL of this standard onto MALDI target, together with another microliter of DHB solution.

3.5. Protein Identification and Characterization by MS

Peptide mixtures were analyzed by MALDI-MS, using a Voyager-DE STR BioSpectometry Workstation instrument (Applied Biosystems, Foster City, CA) in positive reflector mode. Typical instrument settings were: acceleration voltage 20 kV, grid 72%, delay time 300 ns, 100 shots/spectrum, and mass range 800–3500 Da. Spectra were calibrated using peaks corresponding to trypsin autolysis products, resulting in a mass accuracy of less than 50 ppm. The spectra were analyzed using Data Explorer software (Applied Biosystems).

Automated nanoflow liquid chromatography tandem mass spectrometric analysis (LC-MS/MS) was performed using a QTOF Micro mass spectrometer (Waters, Manchester, UK) employing automated DDA. A nanoflow-high-performance liquid chromatography (HPLC) system (Ultimate; Switchos2; Famos; LC Packings, Amsterdam) was used to deliver a flow rate of 200 nL/min. Chromatographic separation was accomplished by loading peptide samples onto a homemade 2-cm fused silica precolumn (75 µm internal diameter [id]; 375 µm outer diameter [od]; Zorbax® SB-C18 3 µm; Agilent, Wilmington, DE) using autosampler essentially as described by Meiring et al. *(16)*. Sequential elution of peptides was accomplished using a linear gradient from Solvent A (0% ACN in 1% FA/ 0.6% acetic acid/0.005% heptafluorobutyric acid) to 40% of solvent B (90% ACN in 1% FA/0.6% acetic acid/0.005% heptafluorobutyric acid) in 30 min over the precolumn in-line with a homemade 8-cm resolving column (75 µm id; 375 µm od; Agilent Zorbax SB-C18 3.5 µm). The resolving column was connected using a fused silica transfer line (20 µm id) to a distally coated fused silica emitter (New Objective, Cambridge, MA) (360 µm od; 20 µm id; 10 µm tip id) biased to approx 2.6 kV. The mass spectrometer was operated in the positive ion mode with a resolution of 4000–6000 full-width half-maximum using a source temperature of 80°C and a counter current nitrogen flow rate of 60 L/h. Data-dependent analysis was employed (three most abundant ions in each cycle): 1 s MS (*m/z* 350–1500) and max 3 s MS/MS (*m/z* 50–2000, continuum mode), 45 s dynamic exclusion. A charge-state recognition algorithm was employed to determine optimal collision energy for low-energy collision-induced dissociation MS/MS of peptide ions. External mass calibration using NaI resulted in mass errors of less than 50 ppm. Raw data was processed using MassLynx 3.5 Protein Lynx (smooth 3/2 Savitzky Golay and center 4 channels/80% centroid) and the resulting MS/MS data set exported in the Micromass pkl format.

Protein identification was performed by searching a nonredundant protein sequence database National Center for Biotechnology Information using the Mascot program (http://www.matrixscience.com). The following parameters were used: monoisotopic mass accuracy less than 50 ppm, 1 missed cleavage, allowed modifications carbamidomethylation of cysteine (complete), and methionine and pyroglutamic acid (partial).

4. Notes

1. Precipitation with TCA/acetone allows to remove contaminating substances from the sample, mainly lipids and salts, that would otherwise interfere with the 2DE result, and to obtain a more concentrated protein sample.
2. Caotropic agents as urea help to solubilize and denature the protein mixture. Traditional protocols use urea at 8 *M*, but the concentration can be increased to 9 or 9.8 *M* if necessary for complete sample solubilization. Thiourea, in addition to urea, can be used to further improve protein solubilization, normally as 7 *M* urea and 2 *M* thiourea.
3. Detergent election is also crucial in a good sample preparation; they solubilize hydrophobic proteins and minimize protein aggregation. The only requirement in IEF is a neutral charge: CHAPS, Triton X-100, or NP-40 in the range of 0.5–4%

are most commonly used, but novel detergents as SB 3-10 and ASB-14 can be very useful for membrane extraction.

4. Any glutaraldehyde-free silver-staining protocol can be used in detection. Cross-linking during sensitization and development (with glutaradehyde or formaldehyde) enhances detection sensibility, but interferes with the subsequent enzymatic digestion and extraction of peptides for spot identification using MS.
5. Strips can be frozen before or after equilibration. Equilibration is always performed immediately prior to the second-dimension run, never prior to storage of the IPG strips at −40°C or lower.
6. In order to use the PorosR2, this material has to be dissolved in 70% ACN, leaving it in a saturated dissolution. Store it at room temperature.

Acknowledgments

This work has been partially supported by FIS (PI02/1047) (PI 02/3093), Spanish Cardiovascular Network RECAVA (03/01), SAF-2004-06109, European Network (QLG1-CT-2003-01215), and the International Cardura Award (Pfizer, USA).

References

1. Murray, C. J. and Lopez, A. D. (1997) Global mortality, disability, and the contribution of risk factors: global burden of disease study. *Lancet* **349**, 1436–1442.
2. Hansson, G. K. (2005) Inflammation, atherosclerosis and coronary artery disease. *N. Engl. J. Med.* **352**, 1685–1695.
3. Kuang-Yuh Chyu, P. S. (2001) The role of inflammation in plaque disruption and thrombosis. *Rev. Cardiovasc. Med.* **2**, 82–91.
4. Davies, M. J. (2001) Going from immutable to mutable atherosclerotic plaques. *Am. J. Cardiol.* **88**, 2F–9F.
5. Kolodgie, F. D., Virmani, R., Burke, A. P., et al. (2004) Pathologic assessment of the vulnerable human coronary plaque. *Heart* **90**, 1385–1391.
6. Stary, H. C., Chandler, A. B., Glagov, S., et al. (1994) A definition of initial, fatty streak, and intermediate lesions of atherosclerosis. A report from the Committee on Vascular Lesions of the Council on Arteriosclerosis, American Heart Association. *Circulation* **89**, 2462–2478.
7. Mayr, M., Mayr, U., Chung, Y. L., Yin, X., Griffiths, J. R., and Xu, Q. (2004) Vascular proteomics: linking proteomic and metabolomic changes. *Proteomics* **4**, 3751–3761.
8. Stastny, J., Fosslien, E., and Robertson, A. L., Jr. (1986) Human aortic intima protein composition during initial stages of atherogenesis. *Atherosclerosis* **60**, 131–139.
9. You, S. A., Archacki, S. R., Angheloiu, G., et al. (2003) Proteomic approach to coronary atherosclerosis shows ferritin light chain as a significant marker: evidence consistent with iron hypothesis in atherosclerosis. *Physiol. Genomics* **13**, 25–30.

10. Fach, E., Garulacan, L., Gao, J., et al. (2004) In vitro biomarker discovery for atherosclerosis by proteomics. *Mol. Cell. Proteomics* **3,** 1200–1210.
11. Duran, M. C., Mas, S., Martín-Ventura, J. L., et al. (2003) Proteomic analysis of human vessels: application to atherosclerotic plaques. *Proteomics* **3,** 973–978.
12. Duran, M. C., Martin-Ventura, J. L., Mohammed, S., et al. (2005) The elucidation of atherosclerotic plaque secretome by proteomic analysis: a potential strategy for detection and characterization of proteins involved in plaque formation and rupture (in revision).
13. Bradford, M. M. (1976) A rapid and sensitive method for the quantitation of microgram quantities of protein utilizing the principle of protein-dye binding *Anal. Biochem.* **72,** 248–254.
14. Laemmli, U. K. (1970) Cleavage of structural proteins during the assembly of the head of bacteriophage T4, novel approach to testing for induced point mutations in mammals. *Nature* **227,** 680–685.
15. Shevchenko, A., Wilm, M., Vorm, O., and Mann, M. (1996) Mass spectrometric sequencing of proteins silver-stained polyacrylamide gels. *Anal Chem.* **68,** 850–858.
16. Meiring, H. D., Van der Helft, E., TenHove, G. J., and DeJong, A. P. (2002) Nanoscale LC-MS(n): technical design and application to peptide and protein analysis. *J. Sep. Sci.* **25,** 557–568.

13

Characterization of HSP27 Phosphorylation Sites in Human Atherosclerotic Plaque Secretome

Mari-Carmen Durán, Elisabetta Boeri-Erba, Shabaz Mohammed, Jose L. Martín-Ventura, Jesús Egido, Fernando Vivanco, and Ole N. Jensen

Summary

Atherosclerosis is one of the main causes of death in developed countries. Atheroma plaque formation is promoted by the interaction between the cells conforming the arterial wall, smooth muscle cells, and endothelial cells, together with lipoproteins and inflammatory cells (mainly macrophages and T-lymphocytes). These interactions can be mediated by proteins secreted from these cells, which therefore exert an important role in the atherosclerotic process. We recently described a novel strategy for the characterization of the human atherosclerotic plaque secretome, combining two-dimensional gel electrophoresis and mass spectrometry (MS). Among the identified proteins, two isoforms of heat shock protein 27 (HSP27), a protein recently described as a potential biomarker of atherosclerosis, were detected. However, the putative mechanisms in which HSP27 isoforms could be involved in the atherosclerotic process are unknown. Thus, the role that phosphorylated HSP27 could play in the atherosclerotic process is actually under study. The present work shows the strategies employed to characterize the phosphorylation in the HSP27 secreted by atheroma plaque samples. The application of liquid chromatography tandem mass spectrometry (MS/MS), as well as the combination of immobilized metal affinity chromatography methodology with matrix-assisted laser desorption/ionization MS/MS are described.

Key Words: Atherosclerosis; secretome; 2DE; mass spectrometry; MS; HSP27; stress oxidative, phosphorylation, posttranslational modification; PTM; LC-MS/MS; IMAC.

1. Introduction

Atheroma plaque formation is normally promoted by the interaction between the cells conforming the arterial wall, smooth muscle cells (SMC), and endothelial cells, together with lipoproteins and inflammatory cells (mainly macrophages

From: *Methods in Molecular Biology, vol. 357: Cardiovascular Proteomics: Methods and Protocols*
Edited by: F. Vivanco © Humana Press Inc., Totowa, NJ

and T-lymphocytes) *(1)*. Such interactions can be mediated by proteins secreted from these cells, exerting an important role in the atherosclerotic process. Until now, several attempts have been made to study the atherosclerotic phenomena by proteomic approaches *(2–4)*. Recently, we described a novel strategy consisting of the proteomic analysis of proteins released by normal and pathological arterial walls in patients affected by atherosclerosis *(5)*. We hypothesized that the patterns of protein secretion could be different between atherosclerotic plaques and normal endarteries. Thus, by culturing carotid normal arteries and carotid endarterectomy samples in a medium free of proteins, and later analysis of the proteins included in the supernatants by two-dimensional gel electrophoresis (2DE), we have shown that carotid atherosclerotic plaques cultured in vitro are able to secrete proteins, and, also that a differential pattern of protein secretion has been reported *(5)*. Among the proteins involved in the atherosclerotic process, we reported that heat shock protein (HSP)27 can be considered as a potential biomarker in atheroma plaque formation and rupture *(6,7)*. Surprisingly, HSP27 release was drastically decreased in atherosclerotic plaques and barely detectable in complicated plaque supernatants *(6)*. Although the exact participation of HSP27 in atheroma plaque formation is still unknown, we hypothesized that HSP27 could play a protective role in this pathology because it has been demonstrated that HSP27 participate in the regulation of the mobility of SMC and coordination of actin dynamics into the cells. In addition, HSP27 is able to interact with the IKK protein and inhibit the activation of nuclear factor κB, widely involved in plaque instability and rupture *(7,8)*. A similar role for HSP27 has been described recently in human heart transplantation *(9)*. Interestingly, all these phenomena are mediated by HSP27 phosphorylation/dephosphorylation. Therefore, the presence of phosphorylation in the secreted HSP27 could help to better illuminate the mechanisms of HSP27's involvement in the atherosclerotic plaque. HSP27 phosphorylation has been widely studied *(10, 11)*, presenting at least three known phosphorylation sites (Ser82, Ser15, and Ser78). In addition, the presence of phosphorylation has been related to several roles for this protein *(12,13)*. Thus, we wanted to determine if differences between both HSP27 isoforms identified in 2D gels derived from human atherosclerotic plaque samples were in fact owing to a differential phosphorylation state.

With that purpose in mind, an analysis by liquid chromatography tandem mass spectrometry (LC-MS/MS) was done to characterize the phosphorylated residues of these two HSP27 isoforms. Initially, an automated nanoflow LC-MS/MS analysis was performed using a QTOF Micro mass spectrometer (Waters, Manchester, UK) employing automated data-dependent acquisition. As result, a phosphate group was detected in the Ser82 of the most acidic isoform but not in the most basic isoform, which is not phosphorylated. The characterization of phosphoproteins is actually one of the main subjects in proteomic research, most of them supported by the application of mass spectrometric analysis *(14, 15)*. Several methods for phosphoproteins purification and enrichment have been described, including the application of phospho-serine or phospho-tyrosine antibodies *(16)*. Immobilized metal ions, such as Fe(III), bind phosphorylated peptides and some phosphoproteins with high selectivity *(17)*. Fe(III)-immobilized metal affinity chromatography (IMAC) columns have been used in combination with reverse-phase high-performance liquid chromatography (RP-HPLC) and matrix-assisted laser desorption and ionization mass spectrometry (MALDI-MS) *(17)* and with LC-MS/MS *(18,19)* for phosphopeptide purification and characterization at picomole and subpicomole levels. Fe(III)-IMAC was also combined with nanoelectrospray MS/MS for mass determination and amino acid sequencing of phosphopeptides *(20)*.

The present work includes the strategies used to characterize the differential phosphorylation states detected in the two HSP27 isoforms identified in the atherosclerotic plaque supernatants, by application of direct LC-MS/MS and further combination of IMAC columns with LC-MS/MS, is described.

2. Materials

2.1. In-Gel Digestion

1. 50 and 100 mM ammonium hydrogenocarbonate (NH_4HCO_3).
2. 10 mM dithiotreitol (DTT).
3. 55 mM iodoacetamide (Sigma, St. Louis, MO).
4. 50% (v/v) and 100% acetonitrile (ACN) (HPLC-grade; Fisher Scientific, UK).
5. 5% Formic acid (FA) (Merck KGaA, Darmstadt, Germany).
6. Digestion buffer: 12.5 ng/µL sequencing-grade modified trypsin (Promega) in 50 mM NH_4HCO_3.

2.2. Identification of HSP27 Isoforms by MALDI-MS Analysis

2.2.1. Microcolumn Purification of Digested Samples

1. GELoader tips from Eppendorf (Hamburg, Germany).
2. 1-mL Syringe, to force liquid through the column.
3. ACN (HPLC-grade; Fisher Scientific).
4. 5% FA (Merck, KGaA, Darmstadt, Germany).
5. Ortho-phosphoric acid (85%) (J. T. Baker Deventer, Holland).
   All chemicals used were ACS- or HPLC-grade.
6. Chromatographic resins: Poros 10R2 (PerSeptive Biosystems); Graphite powder (Sigma).
7. MALDI matrix: 20 mg/mL 2,5-dihydroxybenzoic acid (DHB; Aldrich Chemicals Milwaukee, WI) in 50% (v/v) ACN 1% (v/v) ortho-phosphoric acid.

2.2.2. Protein Identification and Characterization by MS

1. For MALDI-MS analysis, a Voyager-DE STR BioSpectometry Workstation instrument (Applied Biosystems, Foster City, CA) was employed, in positive ion reflector mode.
2. For external calibration: 5 pmol/µL lactoglobulin tryptic peptide mixture.
3. Data Explorer software (Applied Biosystems) was used to perform spectra analyses data, and Mascot program (http://www.matrixscience.com) for protein identification.

2.3. Phosphopeptide Characterization by LC-MS/MS SYSTEM (Q-TOF Micro)

1. Nanoflow-HPLC system (Ultimate; Switchos2; Famos; LC Packings, Amsterdam) employing the following columns:
   a. 2-cm Fused silica precolumn: 75 µm inner diameter (id); 375 µm outer diameter (od); Zorbax® SB-C18, 3 µm (Agilent, Wilmington, DE).
   b. 8-cm Resolving column (75 µm id; 375 µm od; Zorbax SB-C18, 3.5 µm; Agilent).
2. Solvents: 1% (v/v) FA (Merck) and 0.005% (v/v) heptafluorobutyric acid (HFBA; ICN Biomedicals, Aurora, OH).
3. Solvent A: 0% ACN in 1% FA, 0.6% acetic acid (AA), 0.005% HFBA.
4. Solvent B. 90% ACN in 1% FA, 0.6% AA, 0.005% HFBA.
5. Silica emitter (New Objective, Cambridge, MA) (360 µm od; 20 µm id; 10 µm tip id).

2.4. Phosphopeptide Purification by Nanoscale Fe(III)-IMAC Column

2.4.1. Preparation of Fe(III)-IMAC Resin

1. 0.6% (0.1 M) and 1.2% (0.2 M) AA (Fluka, Buchs, Switzerland).
2. 2. 0.1 M EDTA.
3. 0.1 M [Iron (III) chloride] $FeCl_3$ (Aldrich Chemicals, Milwaukee, WI). All chemicals used were ACS- or HPLC-grade.
4. Chromatographic resin: nitrilotriacetic acid (NTA)-silica (16–24 µm particle size, Qiagen, Valencia, Spain).

2.4.2. Preparation of Nanoscale Fe (III)-IMAC Columns

1. GELoader tips from Eppendorf.
2. 0.6% AA (Fluka).
3. 0.1 M Ortho-phosphoric acid (85%) (J. T. Baker Deventer).
4. Chromatographic resins: Poros 10R2 (PerSeptive Biosystems); Graphite powder (Sigma).

2.4.3. Phosphopeptide Purification

1. Nanoscale Fe(III)-IMAC, R2, and graphite columns.
2. 0.1 M AA.
3. 1-mL Syringe to force liquid through the column.

4. 5% FA.
5. 0.6% AA/30% ACN.
6. DHB matrix solution: 20 mg/mL DHB in 50% (v/v) ACN, 1% (v/v) phosphoric acid.
7. MALDI Q-TOF Ultima HT instrument (Waters/Micromass, Manchester, UK). Data acquisition and manipulation: MassLynx 3.5 program (Waters).

3. Methods

Two of the proteins detected in the 2D gels of human atherosclerotic supernatants, were identified by peptide mass fingerprinting (MALDI-MS analysis) as two distinct HSP27 isoforms *(6)*. They were subjected to further analysis in order to determine the differential phosphorylation between them (*see* **Note 1**). With that purpose in mind, two strategies were employed. Initially, the tryptic peptide mixture derived from the digested HSP27 isoforms was directly loaded on the Q-TOF Micro (LC-MS/MS) system. In the second approach, Fe(III)-IMAC columns were employed for enrichment of phosphorylated peptides from crude peptide mixtures and MALDI-MS/MS was used for phosphopeptide sequencing to determine the exact phosphorylation sites.

3.1. Sample Extraction from 2D Gels: In-Gel Digestion of the Two HSP27 Isoforms

1. Select the spots to be cut from the 2-D gels, using a sterile scalpel.
2. The protein spots are in-gel digested with trypsin, according to Schevchenko et al. *(21)* with some modifications:
   a. Wash the gel pieces twice with 30 µL of 50% and 100% ACN for 15 min in agitation. Remove the supernatant.
   b. Incubate the samples with30 µL of 10 mM DTT in 100 mM NH$_4$HCO$_3$ for 45 min at 56°C in agitation. After cooling to room temperature, remove the supernatant.
   c. Proteins are alkylated by adding 30 µL of 55 mM iodoacetamide in 100 mM NH$_4$HCO$_3$ for 30 min at room temperature in the dark. Remove the supernatants.
   d. Wash the gel pieces twice with 30 µL 50% ACN and incubate for 15 min each in agitation. Remove the supernatant.
   e. Dehydrate the gel pieces twice with 30 µL 100% ACN and incubate for 15 min each in agitation. Remove the supernatants each time.
   f. Vacuum-dry the samples in a Speed-Vac for 15 min.
   g. Gel pieces are then swollen in a freshly prepared digestion buffer, in an ice bath, for 45 min.
   h. Remove the supernatant, if any, and discard it. Add 20 µL of 50 mM NH$_4$HCO$_3$ to the gel piece. Incubate at 37°C overnight.
   i. Collect the supernatants containing the tryptic peptides. These supernatants will be employed in further MS analysis.
   j. Store the samples at −20°C if they are not going to be analyzed immediately.

3.2. Identification of HSP27 Isoforms by MALDI-MS Analysis

3.2.1. Micro-Column Purification of Digested Samples

1. The columns are prepared as described by Gobom et al. *(22)*.
2. The column body, consisting of a long and narrow pipet tip (GELoader) is carefully flattened near the end of the outlet using tweezers. Two microliters of suspension of R2 Poros chromatography medium in ACN is deposited in the loaded solution and gently pressed through. A 1.25-mL syringe can be used to force liquid through the column by gently applying air pressure. This allows the chromatography medium to form a column of 1.5 mm in length near the outlet.
3. Equilibrate the column with 20 μL of 5% FA.

3.2.1.1. Use of Nano-Reversed-Phase Columns

1. Load the sample: load 20 μL of 5% FA on the top of the pipet tip (R2 column) and add on it the tryptic peptides mixture (*see* **Note 2**). Using a 1.25-mL syringe, gently apply air pressure to force the liquid through the column.
2. Wash the column with 20 μL of 5% FA, employing the 1.25-mL syringe.
3. Elute the bound peptides directly onto the MALDI target with 1 μL DHB matrix solution.

3.2.2. Protein Identification and Characterization By MS

MALDI-TOF MS peptide mass mapping and computational analysis was used to determine the identity of the two proteins detected in the 2D gels of human atheroma plaque samples.

1. To determine the identity of the sample, peptide mixtures were first analyzed by peptide mass finger printing by MALDI-MS. We used a Voyager-DE STR BioSpectometry Workstation (Applied Biosystems).
2. Typical instrument settings were: acceleration voltage 20 kV, grid 72%, delay time 300 ns, 100 shots/spectrum, and mass range 800–3500 Da. Spectra were calibrated using peaks corresponding to trypsin autolysis products (*m/z* 842.5 and *m/z* 2211.1), resulting in a mass accuracy better than 50 ppm. The spectra were analyzed using Data Explorer software (Applied Biosystems).
3. For external calibration, it is possible to use tryptic lactoglobulin (5 pmol/μL), loading 1 μL of this standard onto the MALDI target, together with 1 μL of DHB solution.
4. The spectra were analyzed using Data Explorer software (Applied Biosystems).
5. Protein identification was performed by searching a nonredundant protein sequence database (National Center for Biotechnology information) using the Mascot program (http:// www.matrixscience.com). The following parameters were used: monoisotopic mass accuracy 50 ppm, 1 missed cleavage, as a fixed modification carbamidomethylation of cysteine; as variable modifications: phosphorylation in Ser/Thr, methionine oxidation, and pyroglutamic acid.

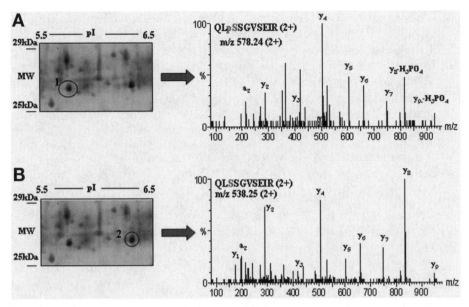

Fig. 1. Characterization of a differential phosphorylation site in two heat shock protein (HSP)27 isoforms by liquid chromatography-tandem mass spectrometry (LC-MS/MS) analysis. Two different isoforms of HSP27 were detected in the atheroma plaque supernatants analyzed by two-dimensional gel electrophoresis, presenting the same molecular weight but a difference in their isoelectric point values of 0.5 U. By LC-MS/MS analysis, a phosphate group was detected in the Ser82 of the sequence of the most acidic HSP27 isoform (**A**), but not in the basic one, considered nonphosphorylated (**B**). This figure includes a fraction of the gel were both isoforms were located, and the correspondent sequence obtained after LC-MS/MS analysis.

3.3. Characterization of Different Phosphorylated States in the HSP27 Isoforms by LC-MS/MS (see Fig. 1)

1. A nanoflow-HPLC system was used to deliver a flow rate of 200 nL/min. Chromatographic separation is accomplished by loading peptide samples onto a homemade 2-cm fused silica precolumn using autosampler according to Meiring et al. *(23)*.
2. For sequential elution of peptides, use a linear gradient from Solvent A to 40% of solvent B in 30 min over the precolumn in-line with a homemade 8-cm resolving column.
3. Connect the resolving column, using a fused silica transfer line (20 µm id), to a distally coated fused silica emitter biased to approx 2.6 kV.
4. The mass spectrometer is operated in the positive ion mode with a resolution of 4000–6000 full-width half-maximum, using a source temperature of 80°C and a counter-current nitrogen flow rate of 60 L/h.

5. Data-dependent analysis was employed (three most abundant ions in each cycle): 1 s MS (*m/z* 350–1500) and max 3 s MS/MS (*m/z* 50–2000, continuum mode), 45 s dynamic exclusion. A charge-state recognition algorithm was employed to determine optimal collision energy for low-energy collision-induced dissociation MS/MS of peptide ions. External mass calibration using NaI resulted in mass errors of less than 50 ppm. Raw data was processed using MassLynx 3.5 ProteinLynx (smooth 3/2 Savitzky Golay and center 4 channels/80% centroid) and the resulting MS/MS data set exported in the Micromass pkl format.

3.4. Characterization of Different Phosphorylated States in the HSP27 Isoforms by Nanoscale Fe(III)-IMAC Column and MALDI-MS/MS

To confirm results obtained from LC-MS/MS analysis, the presence of phosphopeptides in the HSP27 tryptic digestion mixtures was also determined using custom-made miniaturized Fe(III)-IMAC columns, according to Stensballe et al. *(17)*.

For each isoform, the peptide mixture was loaded sequentially on IMAC, R2, and graphite columns and the MALDI-MS spectra of the peptides, retained by the three distinct columns, respectively, were acquired.

The peptides present in IMAC fraction, but not the ones derived from R2 and graphite columns, were considered phosphorylated peptide candidates. MALDI-MS/MS was used to confirm the presence of a phosphoryl group and to determine the exact sites of phosphorylation.

3.4.1. Preparation of Nanoscale Fe(III)-IMAC Resin and Nanoscale Fe(III)-IMAC, R2 Columns

1. Load a volume correspondent to 40 μL of Ni-NTA material in a empty 1.5-mL Eppendorf tube.
2. Wash it with pure water (approx 1 mg/mL; *see* **Note 3**). Centrifuge and remove the water.
3. Wash with 1 mL of 50 m*M* EDTA (*see* **Note 3**; aspirate at least 5 times). Centrifuge and remove EDTA. Repeat EDTA step 3 times (*see* **Note 4**).
4. Add 600 μL 1.2% AA (aspirate at least 5 times). Centrifuge and remove AA. Repeat acid step 5 times (*see* **Note 5**).
5. Add 500 μL 1.2% AA and 500 μL FeCl$_3$ (aspirate at least 5 times). Centrifuge and remove. Repeat addition of AA and FeCl$_3$ 3 times.
6. Add 750 μL 1.2% AA. Centrifuge and remove. Repeat twice.
7. Store the resin in 1 mL of 1.2% AA.
   This ready-for-use slurry of Fe(III)-IMAC resin in can be stored at 4°C for at least 40 d without reducing the performance of the resin. Before storing Fe(III)-IMAC resin, blow argon in the tube and seal it with parafilm.
8. Nanoscale Fe(III)-IMAC columns: for each sample prepare a nanoliter bed volume (nanoscale) IMAC column by loading IMAC resin into a GELoader to pack a 15–

20 mm column (*see* **Note 6**). Use a 1-mL plastic syringe adapted to fit the GELoader tip to gently apply air pressure to pack the IMAC resin in the narrow part of the GELoader tip. It is necessary to partially constrict the narrow part, near the tip, to prevent leakage of resin.

9. Prepare an R2 column as described in **Subheading 3.2.1., step 1**.

3.4.2. Phosphopeptide Purification (see Fig. 2)

1. Load 30 μL of 0.1 *M* AA in the top of the tip containg the IMAC colum and add in them the peptidic sample (~10 μL). This volume is loaded slowly onto the IMAC column (loading time 30–60 min), using the 1-mL syringe to force liquid through the column.

2. The Fe(III)-IMAC column flow-through, containing the unretained peptides, is collected, acidified with 20 μL of 5% FA, and subjected to reverse-phase chromatography on R2 Poros resin and graphite powder column.

3. Wash the IMAC column with 3 μL of 0.6% AA, 30% ACN.

4. Wash the IMAC column with 3 μL of 0.6% AA.

5. Elute the bound peptides from the nanoscale IMAC columns with 1 μL of DHB matrix solution (*see* **Note 7**). The matrix/analyte eluate is spotted as a series of nanoliter volume droplets onto the MALDI probe.

6. Elute the bound peptides from the R2 and graphite columns with 1 μL of DHB matrix solution.

7. The matrix/analyte eluate is spotted as a series of nanoliter volume droplets onto the MALDI probe.

3.4.3. Determination of Phosphorylation Sites in the HSP27 Isoforms by MALDI-MS/MS

1. The MALDI-MS analysis is performed for each isoforms of the protein HSP27. The spectra correspondent to the peptides retained in IMAC, R2 and graphite columns, respectively, are acquired. MALDI-MS data analysis is done using the same parameters described in **Subheading 3.2.2.**

2. The spectra obtained from peptides retained in IMAC (phosphopeptides) and the spectra for the peptides retained in R2 column (nonphospholylated peptides) are compared. The peptides present in IMAC, but not in the spectra derived from R2 column, are phosphopeptide candidates. Phosphopeptide candidates are also assigned by their 79.96 amu (atomic mass unit) mass increments per phosphate moiety relative to the unmodified peptides.

3. Further MALDI-MS/MS is used to confirm the presence of a phosphoryl group and to determine the exact sites of phosphorylation. In this case, a MALDI Q-ToF Ultima HT instrument (Waters/Micromass) was employed. For MS experiments, spectra are recorded in continuum mode using the time of flight (TOF) mass analyzer; the resolving quadrupole is operated in radio frequency-only mode (mass range 800–2500 Da). For product ion analyses, the quadrupole is set to transmit an approx 4 Th window into the hexapole collision cell. The collision gas is argon and the

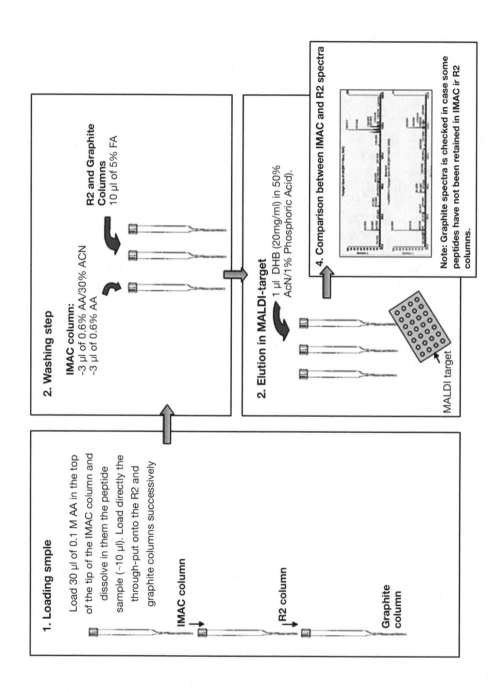

1. Loading smple

Load 30 µl of 0.1 M AA in the top of the tip of the IMAC column and dissolve in them the peptide sample (~10 µl). Load directly the through-put onto the R2 and graphite columns successively

IMAC column →

R2 column →

Graphite column

2. Washing step

IMAC column:
~3 µl of 0.6% AA/30% ACN
~3 µl of 0.6% AA

R2 and Graphite Columns
10 µl of 5% FA

2. Elution in MALDI-target

1 µl DHB (20mg/ml) in 50% AcN/1% Phosphoric Acid).

MALDI target

4. Comparison between IMAC and R2 spectra

Note: Graphite spectra is checked in case some peptides have not been retained in IMAC ir R2 columns.

collision energy is determined by the level of fragmentation observed and is typically between 60–150 eV. Data acquisition and manipulation is performed using MassLynx 3.5 (Waters).

4. Notes

1. One of the advantages of using 2DE is that it allows detection of the presence of posttranslational modifications (PTMs) because normally PTMs are associated with isoforms that present the same molecular weight and different isoelectric points or different molecular weights and same isoelectric point. In the case of proteins with different phosphorylation states, it is quite characteristic that diverse isoforms are present that have quite similar molecular weight values but different isoelectric points owing to the presence of the phosphate group (negative charge). In this case, these isoforms will appear separate in a row of spots, often called "trains" of proteins, where the most phosphorylated isoform will be located in the left (acidic part), whereas the nonphosphorylated isoform will appear in the right side (basic part; *see* **Fig. 1**).
2. The volume of analyte employed will depend on the peptide concentration available. Normally, with proteins extracted from saturated spots, 5 μL or less is enough for a good identification with MALDI-MS analysis.
3. Do not vortex; this could damage the resin.
4. During the activation of the IMAC resin, EDTA is employed to capture all possible cations presents in the medium that later could interfere with the interaction between NTA silica resin and the cation Fe(III).
5. This step is important to remove the EDTA, which could interfere with the binding of the Fe^{3+} to the resin. The treatment with $FeCl_3$ is to bind the Fe^{3+} to the resin.
6. This volume is necessary to ensure that most part of the phosphopeptides will be in contact with the resine and, therefore, will be retained in the IMAC column, whereas the nonphosphorylated peptides will be washed away.
7. It has been demonstrated recently that phosphoric acid in combination with DHB matrix significantly enhanced phosphopeptide ion signals in MALDI mass spectra of crude peptide mixtures derived from the phosphorylated proteins *(24)*.

Fig. 2. *(Opposite page)* Characterization of different phosphorylated states in the heat shock protein (HS)P27 isoforms by nanoscale Fe(III)-immobilized metal affinity chromatography (IMAC) column and matrix-assisted laser desorption/ionization tandem mass spectrometry (MALDI-MS/MS). An schematic protocol for characterization of phosphopeptides in the HSP27 isoforms by using Nanoscale Fe(III)-IMAC columns, is presented. The peptide mixture extracted from the gel is loaded directly onto the IMAC column, collecting the flow-through directly on a R2 column and graphite powder column, respectively, in order to capture those peptides that are not retained on the IMAC column. Columns are washed and finally the retained peptides from each column are loaded directly onto a MALDI target for further analysis. By comparison of the IMAC and the R2 columns, it is possible to detect the presence of phosphorylated peptides.

Acknowledgments

This work has been partially supported by FIS (PI02/1047) (PI 02/3093), Spanish Cardiovascular Network RECAVA (03/01), SAF-2004-06109, European Network (QLG1-CT-2003-01215), and the International Cardura Award (Pfizer, USA).

References

1. Hansson, G. K. (2005) Inflammation, atherosclerosis and coronary artery disease. *N. Engl. J. Med.* **352,** 1685–1695.
2. Stastny, J., Fosslien, E., and Robertson, A. L., Jr. (1986) Human aortic intima protein composition during initial stages of atherogenesis. *Atherosclerosis* **60,** 131–139.
3. You, S. A., Archacki, S. R., Angheloiu, G., et al. (2003) Proteomic approach to coronary atherosclerosis shows ferritin light chain as a significant marker: evidence consistent with iron hypothesis in atherosclerosis. *Physiol. Genomics* **13,** 25–30.
4. Fach, E., Garulacan, L., Gao, J., et al. (2004) In vitro biomarker discovery for atherosclerosis by proteomics. *Mol. Cell. Proteomics* **3,** 1200–1210.
5. Duran, M. C., Mas, S., Martín-Ventura, J. L., et al. (2003) Proteomic analysis of human vessels: application to atherosclerotic plaques. *Proteomics* **3,** 973–978.
6. Martin-Ventura, J. L., Duran, M. C., Blanco-Colio, L. M., et al. (2004) Identification by a differential proteomic approach of Heat Shock Protein 27 as a potential marker of atherosclerosis. *Circulation* **110,** 2216–2219.
7. Vivanco, F., Martin-Ventura, J. L., Duran, M. C., et al. (2005) Quest for novel cardiovascular biomarkers by proteomic analysis. *J. Proteom. Res.* **4,** 1181–1191.
8. Martin-Ventura, J. L., Blanco-Colio, L. M., Muñoz-García, B., et al. (2004) NF-kappaB activation and Fas ligand overexpression in blood and plaques of patients with carotid atherosclerosis: potential implication in plaque instability. *Stroke* **35,** 458–463.
9. De Souza, A. I., Wait, R., Mitchell, A. G., et al. (2005) Heat shock protein 27 is associated with freedom from graft vasculopathy after human cardiac transplantation. *Cir. Res.* **97,** 192–198.
10. Larsen, J. K., Yamboliev, I. A., Weber, L. A., and Gerthoffer, W. T. (1997) Phosphorylation of the 27-kDa heat shock protein via p38 MAP kinase and MAPKAP kinase in smooth muscle. *Am. J. Physiol.* **273,** L930–940.
11. Zhou, M., Lambert, H., and Landry, J. (1993) Transient activation of a distinct serine protein kinase is responsible for 27-kDa heat shock protein phosphorylation in mitogen-stimulated and heat-shocked cells *J. Biol. Chem.* **268,** 35–43.
12. Hirano, S., Rees, R. S., and Gilmont, R. R. (2002) MAP kinase pathways involving hsp27 regulate fibroblast-mediated wound contraction. *J. Surg. Res.* **102,** 77–84.
13. McGregor, E., Kempster, L., Wait, R., Gosling, M., Dunn, M. J., and Powell, J. T. (2004) F-actin capping (CapZ) and other contractile saphenous vein smooth muscle proteins are altered by hemodynamic stress: a proteonomic approach. *Mol. Cell. Proteomics* **3,** 115–124.

14. Steen, H. and Mann, M. (2002) A new derivatization strategy for the analysis of phosphopeptides by precursor ion scanning in positive ion mode. *J. Am. Soc. Mass Spectrom.* **13**, 996–1003.
15. Bennett, K. L., Stensballe, A., Podtelejnikov, A. V., Moniatte, M., and Jensen, O. N. (2002) Phosphopeptide detection and sequencing by matrix-assisted laser desorption/ionization quadrupole time-of-flight tandem mass spectrometry. *J. Mass Spectrom.* **37**, 179–190.
16. Gronborg, M., Kristiansen, T. Z., Stensballe, A., et al. (2002) A mass spectrometry-based proteomic approach for identification of serine/threonine-phosphorylated proteins by enrichment with phospho-specific antibodies: identification of a novel protein, Frigg, as a protein kinase A substrate. *Mol. Cell Proteomics* **1**, 517–527.
17. Stensballe, A., Andersen, S., and Jensen, O. N. (2001) Characterization of phosphoproteins from electrophoretic gels by nanoscale Fe(III) affinity chromatography with off-line mass spectrometry analysis. *Proteomics* **1**, 207–222.
18. Watts, J. D., Affolter, M., Krebs, D. L., Wange, R. L., Samelson, L. E., and Aebersold, R. (1994) Identification by electrospray ionization mass spectrometry of the sites of tyrosine phosphorylation induced in activated Jurkat T cells on the protein tyrosine kinase ZAP-70. *J. Biol. Chem.* **269**, 29,520–29,529.
19. Figeys, D., Gygi, S. P., Zhang, Y., Watts, J., Gu, M., and Aebersold, R. (1998) Electrophoresis combined with novel mass spectrometry techniques: powerful tools for the analysis of proteins and proteomes. *Electrophoresis* **19**, 1811–1818.
20. Cleverley, K. E., Betts, J. C., Blackstock, W. P., Gallo, J. M., and Anderton, B. H. (1998) Identification of novel in vitro PKA phosphorylation sites on the low and middle molecular mass neurofilament subunits by mass spectrometry. *Biochemistry* **37**, 3917–3930.
21. Shevchenko, A., Wilm, M., Vorm, O., and Mann, M. (1996) Mass spectrometric sequencing of proteins silver-stained polyacrylamide gels. *Anal. Chem.* **68**, 850–858.
22. Gobom, J., Nordhoff, E., Mirgorodskaya, E., Ekman, R., and Roepstorff, P. (1999) Sample purification and preparation technique based on nano-scale reversed-phase columns for the sensitive analysis of complex peptide mixtures by matrix-assisted laser desorption/ionization mass spectrometry. *J. Mass Spectrom.* **34**, 105–116.
23. Meiring, H. D., Van der Helft, E., TenHove, G. J., and DeJong, A. P. (2002) Nanoscale LC-MS(n): technical design and application to peptide and protein analysis. *J. Sep. Sci.* **25**, 557–568.
24. Kjellstrom, S. and Jensen, O. N. (2004) Phosphoric acid as a matrix additive for MALDI-MS analysis of phosphopeptides and phosphoproteins. *Anal. Chem.* **76**, 5109–5117.

14

Western Array Analysis of Human Atherosclerotic Plaques

Wim Martinet

Summary

High-throughput immunoblotting or Western array screening of tissue is a unique advancement in proteomics that may help researchers in their quest to elucidate the proteins and signaling pathways involved in complex human pathologies such as atherosclerosis. The technique entails polyacrylamide gel electrophoresis and Western blotting followed by screening of the blots with hundreds of high-quality antibodies that are combined into unique cocktails. Because each monoclonal antibody identifies a unique target among the array of thousands of proteins displayed on the Western blot, subnanogram quantities of proteins with altered expression can be readily detected. This approach of proteome analysis avoids the limitations of traditional high-throughput protein screening techniques such as two-dimensional gel electrophoresis and mass spectrometry.

Key Words: Polyacrylamide gel electrophoresis; Western blotting; high-throughput proteome analysis; antibody cocktails; atherosclerosis.

1. Introduction

Atherosclerosis is a complex, chronic inflammatory disease of the arterial vessel wall that is characterized by smooth muscle cell migration and proliferation, accumulation of cholesterol, calcification, infiltration of inflammatory cells, and induction of cell death (*1,2*). According to recent evidence, each of the distinct steps in atherogenesis is driven, or at least influenced, by a set of differentially expressed gene products (*3,4*). Hence, we are faced with the challenge of identifying these genes and elucidating their function, in order to obtain a firm understanding of the molecular basis of atherosclerosis. During the last decade, much progress in our search for pivotal genetic factors has been made predominantly at the mRNA level using DNA chip array technology or other high-throughput genomics methods (e.g., representational difference analysis,

From: *Methods in Molecular Biology, vol. 357: Cardiovascular Proteomics: Methods and Protocols*
Edited by: F. Vivanco © Humana Press Inc., Totowa, NJ

serial analysis of gene expression) *(5,6)*. As marvelous as these assays may be, many researchers are not aware of their potential pitfalls, including chip errors *(7)* and difficulties with interpretation of transcript profiling data *(8)*. Moreover, the correlation between mRNA levels and protein concentration is often poor. The use of proteomics-based strategies—examining the protein content of cells or tissues in a high-throughput fashion—is expanding rapidly and offers a powerful complementary approach, overcoming some of the limitations of RNA analysis.

Two-dimensional gel electrophoresis (2DE) is currently the most popular technique for high-throughput proteome analyses and has already offered much insight into the molecular mechanisms underlying cardiovascular disease *(9)*. Despite standardization of this technique, several technical issues remain unsolved. These issues include difficulties with isoelectric focusing and the inability to resolve all the proteins on gel (especially lower-abundance, basic, and membrane proteins). Other problems are inherent to the technique itself. Indeed, 2D gel electrophoresis is very difficult to automate and requires considerable hands-on time. Furthermore, although it is the highest-resolving technique for very complex protein mixtures, multiple proteins are frequently found in one 2D gel spot so that quantitative and comparative analysis of 2D gels is prone to mistakes. One technique that may compete with 2DE is high-throughput Western blotting, also called Western array screening of total cell or tissue lysate. This type of technology includes polyacrylamide gel electrophoresis (PAGE) and Western blotting followed by screening of the blots with hundreds of high-quality antibodies that are combined into unique cocktails. Antibodies detect sub-nanogram quantities of protein and can distinguish closely related members of many protein families. Because each monoclonal antibody (MAb) identifies a unique target among the array of thousands of proteins displayed on the Western blot, this approach of proteome analysis eliminates the need for further protein identification such as is necessary with 2DE. Interestingly, a Western array screening service, called PowerBlot, is offered by BD Biosciences so that the technique is accessible to anyone, thus also to researchers who are not familiar with protein analyses. Although high-throughput Western blotting is relatively new in proteomics, the number of reports from different research disciplines using this type of technology is increasing rapidly *(10–15)*.

2. Materials

2.1. Preparation of Samples

1. Container with liquid nitrogen.
2. Lysis solution: 10 mM Tris-HCl pH 7.4, 1 mM sodium orthovanadate, 1% sodium dodecyl sulfate (SDS; *see* **Note 1**). Prepare this solution fresh.

3. Frozen human blood vessels (*see* **Note 2**). Store at –80°C.
4. NuPAGE LDS sample buffer (4X) and NuPAGE reducing agent (both from Invitrogen, Carlsbad, CA). Store at room temperature and 4°C, respectively.
5. (Optional) BCA Protein Assay kit (Pierce, Rockford, IL).

2.2. PAGE and Western Blotting

1. NuPAGE Novex 4–12% *bis-tris* gels (1 well per gel, 1 mm thick), NuPAGE MOPS (or MES) SDS running buffer (20X), NuPAGE antioxidant, and NuPAGE transfer buffer (20X). All products are purchased from Invitrogen. Gels and buffers are stored at room temperature, with the exception of NuPAGE antioxidant, which should be stored at 4°C.
2. Immobilon-P Transfer membranes, pore size 0.45 μm (Millipore, Bedford, MA) and 3 MM chromatography paper (Whatman, Maidstone, UK).
3. *Tris*-buffered saline: 20 mM Tris-HCl, pH 7.6, 137 mM NaCl, supplemented with 0.05% Tween-20 (TBS-T).
4. Blocking buffer: 5% blotting grade nonfat dry milk (Bio-Rad, Richmond, CA) in TBS-T.

2.3. Western Array Analysis

1. Mini-PROTEAN II multiscreen apparatus (Bio-Rad).
2. Antibody dilution buffer: 1% blotting grade nonfat dry milk in TBS-T.
3. Cocktails of high-quality monoclonal and/or polyclonal primary antibodies.
4. Horseradish peroxidase (HRP)-conjugated secondary antibodies (DAKO, Glostrup, Denmark).
5. SuperSignal West Pico Chemiluminescent substrate (Pierce).

3. Methods

3.1. Preparation of Samples

These instructions assume the use of a SPEX CertiPrep 6750 Freezer/Mill and grinding vials (Metuchen, NJ) for homogenization of tissue samples (*see* **Note 3**).

1. Heat lysis solution to boiling on a hot plate. Meanwhile, fill the SPEX CertiPrep Freezer/Mill with liquid nitrogen.
2. Place each sample in a separate grinding vial along with a steel impactor. Then insert the vial in the coil assembly of the mill and lower it into the liquid nitrogen.
3. When the vial is thoroughly chilled (usually within 30 s), activate the grinding cycle for 1 min. During this step, the tissue is pulverized into a fine powder. If small pieces remain, repeat the grinding cycle for another minute.
4. Transfer the sample to a polypropylene tube and immediately add 3 mL of boiling lysis buffer for 0.2 g tissue (*see* **Note 4**). After homogenization, you may take a 50-μL aliquot to determine protein concentration using the BCA Protein Assay kit (*see* **Note 5**). To the remaining lysate, add NuPAGE LDS sample buffer and NuPAGE reducing agent to 1X final concentration (*see* **Note 6**) and mix well. Heat again at 70°C for 10 min.

3.2. PAGE and Western Blotting

Instructions are provided below for electrophoresis of one NuPAGE Novex *bis-tris* gel (*see* **Note 7**) using the Xcell SureLock Mini-Cell (Invitrogen). Blotting is performed with the Mini Trans-Blot Electrophoretic Transfer Cell (Bio-Rad).

1. Remove the NuPAGE gel from the pouch and rinse with deionized water. Peel off the tape from the bottom of the gel cassette.
2. Prepare 1000 mL of 1X NuPAGE SDS running buffer by diluting 50 mL NuPAGE SDS running buffer (20X) in 950 mL deionized water.
3. In one smooth motion, gently pull the comb out of the cassette. Rinse the sample wells with 1X NuPAGE SDS running buffer. Invert the gel and shake to remove the buffer. Repeat two more times.
4. Orient the gel in the Mini-Cell such that the notched "well" side of the cassette faces inwards toward the buffer core. Seat the gels on the bottom of the Mini-Cell and lock into place with the Gel Tension Wedge.
5. Immediately prior to electrophoresis, add 500 µL of NuPAGE antioxidant to 200 mL of 1X NuPAGE SDS running buffer and fill the upper (inner) buffer chamber with a small amount of the running buffer to check for tightness of seal. If you detect a leak from upper to the lower buffer chamber, discard the buffer, reseal the chamber, and refill.
6. Once the seal is tight, fill the upper buffer chamber with 200 mL of 1X NuPAGE SDS running buffer supplemented with NuPAGE antioxidant. The lower buffer chamber is filled with 600 mL 1X NuPAGE SDS running buffer.
7. Load approx 200 µg of protein in one big well across the entire width of the gel and run the gel for 35 min (MES buffer) or 50 min (MOPS buffer) at 200 V.
8. During electrophoresis, prepare 1000 mL of 1X NuPAGE transfer buffer by diluting 50 mL NuPAGE transfer buffer (20X) in 849 mL deionized water. Supplement solution with 1 mL NuPAGE antioxidant and 100 mL methanol (*see* **Note 8**).
9. Cut an Immobilon-P transfer membrane to the dimension of the gel (7 × 8 cm). Pre-wet the membrane for 30 s in methanol. Briefly rinse in deionized water, then place in a shallow dish with 50 mL of 1X NuPAGE transfer buffer for at least 5 min (*see* **Note 9**). To avoid membrane contamination, always use forceps or wear gloves when handling membranes.
10. After electrophoresis, remove the gel from the Mini-Cell and open the gel cassette. Rinse the gels in 1X NuPAGE transfer buffer prior to blotting to facilitate the removal of electrophoresis buffer salts and detergents (*see* **Note 10**).
11. Soak two 3MM filter papers briefly in 1X NuPAGE transfer buffer and place on top of the gel. Turn the plate over gloved hand or clean flat surface so that the gel and filter papers are facing downwards. Place the pre-soaked transfer membrane on the gel followed by another pair of wetted filter papers.
12. Remove all trapped air bubbles by gently rolling over the surface using a glass pipet as a roller.
13. Carefully pick up the gel/membrane assembly, place on blotting pads (pre-soaked in transfer buffer), and slide the assembly into the guide rails of the blotting tank

so that the transfer membrane is between the gel and the anode (+) core of the blot module. It is vitally important to ensure this orientation or the proteins will be lost from the gel into the transfer buffer rather than transferred to the membrane.

14. Place the lid on the unit and connect the electrical leads to the power supply. Perform transfer using 100 V constant for 45–60 min (*see* **Note 11**). To avoid buffer breakdown, make sure that you have adequate cooling (*see* **Note 12**). Therefore, place a frozen cooling unit in the transfer tank and circulate the buffer by placing the transfer tank on a magnetic stirrer.

15. Once the transfer is complete, the cassette is taken out of the tank and carefully disassembled.

3.3. Western Array Analysis

The protocol here describes immunodetection of blotted proteins using HRP-conjugated secondary antibodies and a Lumi-Imager from Roche Diagnostics (Mannheim, Germany) for detection of chemiluminescence.

1. Block nonspecific binding sites by immersing the membrane in blocking buffer for 1 h at room temperature on a rotary platform shaker (*see* **Note 13**).

2. Prepare antibody cocktails by diluting two to six antibodies per cocktail in 600 μL antibody dilution buffer (*see* **Note 14**).

3. Clamp the membrane with the Mini-PROTEAN II multiscreen apparatus that isolates 20 leak-proof channels across the membrane. In each channel, add 600 μL antibody cocktail and incubate for at least 1 h at room temperature (*see* **Note 15**).

4. Remove the blot from the Mini-PROTEAN II multiscreen apparatus and place it in an appropriate dish. Wash the membrane 3×5 min with antibody dilution buffer (*see* **Note 16**). During the washing step, dilute peroxidase-conjugated secondary antibody in 20 mL antibody dilution buffer (*see* **Note 17**).

5. Incubate the membrane for 1 h with diluted HRP-conjugated secondary antibody on a rotary platform shaker.

6. Wash the membrane using three changes of antibody dilution buffer (5 min each), then wash twice for 5 min with fresh changes of TBS-T. During the final wash step, mix the two substrates from the Supersignal West Pico Chemiluminescent Substrate kit (Pierce) at a 1:1 ratio to prepare a substrate working solution (approx 0.125 mL per cm^2 of membrane) (*see* **Note 18**).

7. Incubate blot 1–5 min in substrate working solution. Then drain excess reagent and cover the blot with clear plastic wrap.

8. Expose the blot to a Lumi-Imager for 5–30 min (*see* **Note 19**). Analyze data using LumiAnalyst software. To define differential protein expression, we generally use a fivefold threshold value (*see* **Note 20**). An example of a Western array comparison between pooled carotid endarterectomy specimens and mammary arteries is shown in **Fig. 1**. The proteins that were screened in this experiment are listed in **Table 1**.

9. After identification of the differentially expressed proteins, run additional blots with multiple samples using one antibody per blot (instead of an antibody cocktail) to confirm array results (**Fig. 2**; *see* **Note 21**).

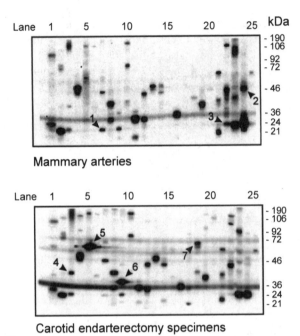

Fig. 1. Comparative analysis of protein expression in pooled nonatherosclerotic mammary arteries (*n* = 7) and pooled carotid endarterectomy specimens (*n* = 12) by Western array technology. The entire screening included 141 protein targets (*see* **Table 1**) spread over 25 lanes. Proteins were detected by carefully formulated monoclonal antibody combinations (all purchased from BD Biosciences Transduction Laboratories, Lexington, KY). Upregulation (fivefold) of seven protein spots as shown in three different Western array runs is indicated: p23 (spot 1), GSK-3β (spot 2), caveolin 3 (spot 3), apoE (spot 4), cyclin B (spot 5), cdk2 (spot 6), and PTP1C (spot 7). Differential expression of caveolin 3, cyclin B, and cdk2 was not confirmed by conventional immunoblot assays and could therefore represent false-positive signals.

4. Notes

1. Addition of protease inhibitors to the lysis solution is not required because proteins (including proteases) are denatured immediately in hot SDS.
2. Blood vessels should be frozen in liquid nitrogen immediately after surgery to prevent protein degradation and/or changes in posttranslational protein modifications. Analysis of postmortem tissue material is not recommendable.
3. SPEX CertiPrep Freezer/Mills are cryogenic laboratory mills that chill samples in liquid nitrogen and pulverize them with a magnetically driven impactor. Each sample is placed in a separate grinding vial that is immersed in a liquid nitrogen bath inside the mill. Chilling tissue in liquid nitrogen (at temperatures approaching −200°C) has two important advantages for sample preparation: it allows thorough homogenization of the tissue and it prevents proteolytical degradation of proteins.

Table 1

Overview of the Proteins With Indication of Their Expected Molecular Weight That Were Screened in the Western Array Experiment Shown in Fig. 1

Lane	MW	Protein ID	Lane	MW	Protein ID	Lane	MW	Protein ID
1	180	ROK α	8	220	EIF-4 γ	16	66	TRF2
1	120	PRK1	8	150	Symplekin	16	55	SKAP 55
1	92	Stat3	8	113	PARP	16	27	eIF6
1	54	SRP54	8	89	Stat4 cl. 8	17	240	CRIK
1	36	RACK1	8	76	SLP-76	17	110	Exportin-t
1	24	Rab8	8	55	p55Cdc	17	66	SNX 1
2	180–220	CD45	8	34	TRADD	17	38	Mona
2	110–120	FYB	8	24	GRB2	17	11	Annexin II LC
2	97	Karyopherin B	9	357	Mitosin	18	450	AKAP 450
2	72	PTP1D/SHP2	9	180	Integrin α L/LFA-1	18	150	Ataxin 2
2	56	lck	9	120	c-Cbl	18	68	PTP1C/SHP1
2	36	PCNA	9	76	Cul-2	18	38	Jab 1
2	20	VHR	9	46	IAK1	18	15	cytochrome c
3	250	RPTP β	9	33	CDK2	19	340	Smrt
3	200	L1	9	22	CDC42	19	110	PI3-Kinase p110 α
3	110	LAMP-1	10	160	DSIF	19	65	TLS
3	80	BMX	10	110	rSec8	19	33	TRAX
3	48	RBBP	10	70	Annexin VI	20	150	Btf
3	36	apoE	10	55	PDI	20	110	PI4-Kinase β
3	21	Rho	10	35	HAX-1	20	60	BAF60A
4	125	DNA pol δ	11	150	CA150	20	40	Hsp40 cl.5
4	100	Cas	11	87	Dystrobrevin	21	160	TopBP1

(continued)

Table 1 (Continued)

Lane	MW	Protein ID	Lane	MW	Protein ID	Lane	MW	Protein ID
4	78	Moesin	11	70/75	ZAP70 Kinase	21	103	DNA ligase III
4	59	fyn	11	45	Fas/CD95/APO-1	21	56	Annexin XI
4	42	Flotillin-2/ESA	11	27/30	ERAB	21	35/36	α/β-SNAP
4	21	DHFR	12	160	AIβ-1	21	18	eIF-5a
5	200–220	MAP4	12	95	ARNT-1	22	170	Sos-1
5	130	Rb2	12	60	Cam kinase IV	22	110/95	Striatin
5	106	Adaptin β	12	42	p38 Map kinase	22	85/56/44/42	panERK
5	79	PKC t	12	24	Ral A	22	33/300	Cdk4
5	62	Cyclin B	13	130	RPTP α	22	18	Caveolin 3
5	46	Nek2	13	104	Adaptin γ	23	400	Utrophin
5	24	hsMAD2	13	68	G3BP	23	160	MSH6
6	320	c-NAP1	13	48	Flotillin-1	23	80	L-Caldesmon
6	160	TIF2	13	23	PMF-1	23	56	B56a
6	110	Rb	14	154	Rad50	23	21	RBP
6	70	TAP	14	120	NFAT-1	24	140	p140 mDia
6	53	CART1	14	76	HEC	24	102	MSH2
6	32	Caspase-3	14	60	p54nrb	24	75	FIN13
6	15	GS15	14	45	ERP	24	46	GSK-3β
7	350	DNA-PKcs/p350	14	32	LAIR-1	24	34	Cdk1/Cdc2
7	220	ZO-1	15	200	GMAP-210	24	22	Caveolin 1
7	105	KRIP-1	15	66	hPrp17	25	130	mSin3A
7	69	ERp72	15	36	Rnase HI	25	93	Gephyrin
7	48	Caspase-2	15	19	Stathmin/Metablastin	25	70	Cox-2
7	31	Syntaxin 6	16	280	NPAT	25	51	ETS-1
7	23	p23	16	140	TTF-I	16	66	TRF2

MW, molecular weight.

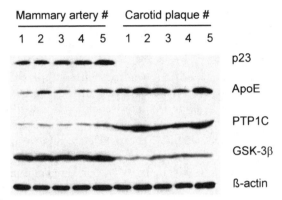

Fig. 2. Confirmation of Western array results shown in **Fig. 1** by conventional immunoblotting (one antibody per Western blot). The figure shows differential expression of p23, apolipoprotein E (apoE) protein-tyrosine phosphatase 1C (PTP1C) and glycogen synthase kinase-3β (GSK-3β) in 5 different mammary arteries and 5 different carotid endarterectomy specimens. Expression of the housekeeping gene β-actin is shown in the bottom panel and served as a loading control.

In case a Freezer/Mill is not available, tissue may be ground with mortar and pestle in liquid nitrogen. Do not allow the frozen tissue samples to thaw at any time because this may lead to protein degradation.

4. Most advanced human atherosclerotic plaques are extremely acellular and contain a large necrotic core that is composed of cell debris and oxidized lipids. When fibrous plaques are examined, use 5–10 times more tissue (1–2 g tissue for 3 mL lysis buffer) to compensate for reduced protein content. Although large amounts of lipid are present in advanced human plaques, we do not observe significant lane artefacts after electrophoresis. However, if desired, protein samples can be purified from contaminants using the PAGEprep Protein Clean-Up and Enrichment Kit (Pierce).

5. There is no absolute need to measure protein concentration from homogenized plaques, because this will not give a fair representation of cellular proteins.

6. For best results, reducing agent should be added immediately prior to electrophoresis.

7. In theory, any type of polyacrylamide gel can be used. For many years, the Laemmli system has been the standard method used to perform SDS-PAGE and subsequent Western blotting. Although Laemmli-type gels are useful for a broad range of protein separations, they are subject to several potential problems. First, the gel is cast at pH 8.7. At this pH, polyacrylamide undergoes gradual hydrolysis, which eventually causes band distortion and/or loss of resolution. Second, the pH of the separating region of the gel is about 9.5 during electrophoresis through which proteins are potentially subjected to chemical modifications such as deamination and alkylation. Third, the redox state of the gel is not well controlled. This means that reduced disulfides are more prone to reoxidation, giving rise to diminished band sharpness

and reduced transfer efficiency, particularly for cysteine-containing proteins. Fourth, the traditional Laemmli-style sample buffer (pH 6.8) changes to a pH of 5.2 when heated. This lower pH environment is known to induce Asp-Pro peptide bond cleavage, putting the protein at risk of degradation. NuPAGE gels, which are used in this protocol, eliminate the problems associated with the traditional Laemmli system. Both NuPAGE *bis-tris* (for small and mid-sized proteins) and NuPAGE *tris*-acetate gels (for large proteins) are discontinuous SDS-PAGE systems that operate in the same way as the traditional *tris*-glycine system, but are cast at a lower pH (pH 6.4 in *bis-tris* gels and pH 7.0 in *tris*-acetate gels), resulting in a dramatically improved gel stability, better protein stability during the run, and a higher protein capacity. In combination with optimized sample, running, and transfer buffers, both the *bis-tris* and *tris*-acetate gels greatly minimize protein oxidation during electrophoresis and contributes to a much better transfer efficiency by eliminating inter-protein disulfide formation. Because of the optimal band resolution and higher transfer efficiencies, NuPAGE gels also offer a much higher sensitivity (5- to 10-fold difference) than traditional gels *(16)*. For Western arrays, we always prefer gradient gels so that a wide range of proteins can be detected on one gel. For example, proteins ranging from 15–250 kDa can be separated on 4–12% Bis-Tris gels with MOPS buffer. NuPAGE *tris*-acetate gels are recommended for analysis of large proteins (up to 400 kDa).

8. Methanol is usually added to prevent or reduce swelling of the gel matrix during transfer. Methanol, however, can remove SDS from the proteins. This may lead to their precipitation, which impairs their electrotransfer. This effect is more pronounced with larger proteins and thus, if working with proteins greater than 100 kDa, methanol should be omitted. If in such cases methanol is still included, it might be necessary to supplement the buffer with SDS. On the other hand, methanol facilitates binding of proteins to the membrane. Small proteins, which are prone to blowthrough, are transferred efficiently even at elevated methanol concentrations (up to 40%). Blotting of small proteins, therefore, is best done in the presence of methanol. In general, NuPAGE transfer buffer with 10–20% methanol provides optimal transfer of a wide range of proteins.

9. Membranes used for protein blotting obviously must have a high protein binding and retention capacity, should be reasonably mechanically resistant, and should not interfere with the probing process (e.g., binding of the antibodies to their epitope). Nitrocellulose (NC) and polyvinylidene difluoride (PVDF) membranes are the most widely used today. NC is a long-established membrane for protein blotting, because it has a high protein binding capacity and good protein retention. The main drawback of NC is its comparatively low mechanical strength. We prefer to use PVDF membranes (such as Immobilon-P membrane, used in this protocol) because it combines high protein-binding capacity (170–200 µg/cm^2 vs 80–100 µg/cm^2 for NC) with excellent mechanical resistance and good staining properties. The protein binding capacity also is higher than NC so that a much better sensitivity can be obtained. If you are using NC instead of PVDF, wet membranes in water (not in methanol) and proceed as described in **Subheading 3.**

10. If the salts are not removed, they will increase the conductivity of the transfer buffer and the amount of heat generated during the transfer. Also, low-percentage gels (<12%) will shrink in methanol buffers. Equilibration in transfer buffer allows the gel to adjust to its final size prior to electrophoretic transfer.

11. Transfer efficiency is affected by the size of the protein, the percentage of acrylamide in the gel, the strength of the electric field, the duration of transfer, and the pH of the buffer. Generally speaking, the larger the protein, the more slowly it will transfer. The best way to transfer large proteins is to blot with a high field strength. However, smaller proteins may be forced through the membrane if blotted for a long time with a high field strength. It is therefore extremely difficult for Western arrays to provide ideal transfer conditions. In most cases, 100 V constant for 45–60 min offers satisfactory results. If too much blow-through occurs (this is especially true for proteins <15 kDa), reduce the transfer time or use PVDF membranes with 0.2-µm pore size (for example, Immobilon-P^{SQ}, Millipore, Bedford, MA).

12. Transferring in a cold room, or placing the transfer chamber in an ice bath, does not give sufficient cooling. This is because most transfer cell buffer chambers are made of plastic, which does not allow efficient heat transfer.

13. There is a plethora of protocols and compounds for blocking free protein-binding sites of the transfer membrane. The blocking buffer should improve the sensitivity of the Western array by reducing background interference. In our hands, 5% nonfat dry milk gives excellent blocking with low background and without impairing immunoreactivity of the blotted proteins. It is also the most commonly used blocking agent today. However, nonfat dry milk contains endogenous biotin and should not be used with avidin–biotin systems. Alternative blocking compounds are bovine serum albumin and Tween-20, although they have a tendency to give elevated background. In addition, Tween-20 may mask or even detach immunoreactive proteins in some occasions. For true optimization of the blocking step, empirical testing is essential. Pierce offers a complete line of blocking buffers for this purpose.

14. Besides good protein samples, the quality and specificity of the antibodies is the most important factor influencing the sensitivity, reproducibility, and interpretation of the Western array pattern *(17)*. The primary antibodies should not only have a high binding affinity for the protein of interest but also display low background staining. Therefore, the use of affinity-purified antibodies is recommended. A huge number of pimary antibodies are commercially available and can be identified quickly by searching sites such as www.antibodyresource.com or www.sciquest.com on the internet. Both monoclonal and polyclonal antibodies can be used as long as they meet the aforementioned criteria. If there is some doubt about the quality of an antibody, run a Western blot with a test sample to verify good binding affinity and background staining. Also note that antibodies degrade with age and will break down very rapidly with repeated freeze–thaw cycles.

    When preparing antibody cocktails, the molecular weight of the different protein targets should differ at least 10 kDa (preferably >20 kDa) so that the individual bands on the Western blot can be easily distinguished from each other. For large proteins (>100 kDa) this difference should be at least 30 kDa (*see* **Table 1**). Also try to anticipate potential variations in molecular weight of the protein targets as a result of posttranslational modifications (e.g., glycosylation, proteolytical cleavage) or alternative splicing of mRNA. Phospho-specific antibodies can be used. However, be aware that phospho-proteins migrate with an apparent higher molecular weight because less SDS binds to the protein.

Antibodies for Western blotting are typically used as dilute solutions, ranging from 1/100–1/500,000 dilutions beginning from a 1 mg/mL stock solution. The optimal dilution of a given antibody with a particular detection system must be determined experimentally. More sensitive detection systems require that less antibody be used, and this can result in substantial savings on antibody costs. It also produces side benefit of reduced background because the limited amount of antibody shows increased specificity for the target with the highest affinity.

15. To obtain maximum sensitivity, it may be advisable to probe the membrane with antibodies overnight. In that case, cover the channels with parafilm and incubate at 4°C to prevent evaporation of the buffer solution.
16. Washing steps are necessary to remove unbound reagents and reduce background, thereby increasing the signal-to-noise ratio. As a general rule, the volume of the washing buffer that is used each time should be as large as possible.
17. Sodium azide is a powerful inhibitor of HRP. Therefore, do not use azide in any of the solutions used in peroxidase-conjugate Western blotting.
18. Exposure to the sun or any other intense light can harm the working solution. For best results, keep the working solution in an amber bottle and avoid prolonged exposure to any intense light. Typical laboratory lighting will not harm the working solution. In case you may expect low-intensity signals, use Supersignal West Femto Maximum Sensitivity Substrate from Pierce to increase sensitivity.
19. Chemiluminescent signals from Western blots can be captured using either X-ray film or cooled charge-coupled device cameras. We prefer the latter option because it offers the advantages of instant image manipulation, higher sensitivity, greater resolution, and a larger dynamic range than film. It also eliminates the need for a darkroom and film processing equipment. The Western array screening facility of BD Biosciences hybridizes blots with secondary antibodies conjugated to fluorescent dyes. Blots are then scanned with a fluorescence imaging system, which is a valuable alternative. Exposition of X-ray films and scanning densitometry of the bands can be done, providing that care is taken to ensure that the signal has not saturated on film.
20. We have realized that the signal intensity from samples run in duplicate or triplicate can vary considerably. Therefore, a differential gene expression threshold of at least fivefold is highly recommendable to avoid false-positive results.
21. Although we always apply a differential gene expression threshold of fivefold, we generally experience a high rate of false-positive results *(13)*. Presumably, this is the most important pitfall of the Western array technique. False-positive protein spots on Western arrays may result from proteolytic degradation of the protein target during sample preparation (or inherent in certain tissues) so that fragments

of a specific protein may overlap with proteins of lower molecular weight. Alternatively, antibodies may crossreact with nonspecific proteins. We therefore believe that it is essential that all Western array results are confirmed by conventional Western blotting.

Acknowledgments

W. Martinet is a postdoctoral fellow of the Fund for Scientific Research-Flanders (Belgium).

References

1. Lusis, A. J. (2000) Atherosclerosis. *Nature* **407,** 233–241.
2. Glass, C. K. and Witztum, J. L. (2001) Atherosclerosis: the road ahead. *Cell* **104,** 503–516.
3. Monajemi, H., Arkenbout, E. K., and Pannekoek, H. (2001) Gene expression in atherogenesis. *Thromb. Haemost.* **86,** 404–412.
4. Laukkanen, J. and Ylä-Herttuala, S. (2002) Genes involved in atherosclerosis. *Exp. Nephrol.* **10,** 150–163.
5. Cook, S. A. and Rosenzweig, A. (2002) DNA microarrays: implications for cardiovascular medicine. *Circ. Res.* **91,** 559–564.
6. Patino, W. D., Mian, O. Y., and Hwang, P. M. (2002) Serial analysis of gene expression: technical considerations and applications to cardiovascular biology. *Circ. Res.* **91,** 565–569.
7. Knight, J. (2001) When the chips are down. *Nature* **410,** 860–861.
8. Liu, T.-J., Lai, H.-C., Wu, W., Chinn, S., and Wang, P. H. (2001) Developing a strategy to define the effects of insulin-like growth factor-1 on gene expression profile in cardiomyocytes. *Circ. Res.* **88,** 1231–1238.
9. Arrell, D. K., Neverova, I., and Van Eyk, J. E. (2001) Cardiovascular proteomics: evolution and potential. *Circ. Res.* **88,** 763–773.
10. Yoo, G. H., Piechocki, M. P., Ensley, J. F., et al. (2002) Docetaxel induced gene expression patterns in head and neck squamous cell carcinoma using cDNA microarray and Powerblot. *Clin. Cancer Res.* **8,** 3910–3921.
11. Malakhov, M. P., Kim, K. I., Malakhova, O. A., Jacobs, B. S., Borden, E. C., and Zhang, D. E. (2003) High-throughput immunoblotting. Ubiquitin-like protein ISG15 modifies key regulators of signal transduction. *J. Biol. Chem.* **278,** 16,608–16,613.
12. Lorenz, P., Ruschpler, P., Koczan, D, Stiehl, P., and Thiesen, H. J. (2003) From transcriptome to proteome: differentially expressed proteins identified in synovial tissue of patients suffering from rheumatoid arthritis and osteoarthritis by an initial screen with a panel of 791 antibodies. *Proteomics* **3,** 991–1002.
13. Martinet, W., Schrijvers, D. M., De Meyer, G. R. Y., Herman, A. G., and Kockx, M. M. (2003) Western array analysis of human atherosclerotic plaques: downregulation of apoptosis-linked gene 2. *Cardiovasc. Res.* **60,** 259–267.
14. Gross, K. L., Cioffi, E. A., and Scammell, J. G. (2004) Increased activity of the calcineurin-nuclear factor of activated T cells pathway in squirrel monkey B-lymphoblasts identified by Powerblot. *In Vitro Cell. Dev. Biol. Anim.* **40,** 57–63.

15. Kim, H. J. and Lotan, R. (2004) Identification of retinoid-modulated proteins in squamous carcinoma cells using high-throughput immunoblotting. *Cancer Res.* **64,** 2439–2448.
16. Martinet, W., Abbeloos, V., Van Acker, N., De Meyer, G. R. Y., Herman, A. G., and Kockx, M. M. (2004) Western blot analysis of a limited number of cells: a valuable adjunct to proteome analysis of paraffin wax-embedded, alcohol-fixed tissue after laser capture microdissection. *J. Pathol.* **202,** 382–388.
17. Martinet, W., Schrijvers, D. M., De Meyer, G. R. Y., Herman, A. G., and Kockx, M. M. (2004) Cytosolic prostaglandin E2 synthase/p23 but not apoptosis-linked gene 2 is downregulated in human atherosclerotic plaques. *Cardiovasc. Res.* **61,** 360–361.

IV

THE PROTEOME OF CELLS OF THE CARDIOVASCULAR SYSTEM

15

The Proteome of Endothelial Cells

Jesús González-Cabrero, Mayte Pozo, Mari-Carmen Durán,
Rosario de Nicolás, Jesús Egido, and Fernando Vivanco

Summary

Endothelial cells form a continuous monolayer lining the inside face of all blood vessels, and present the ability to selectively control vascular permeability. The endothelium is involved in a wide variety of normal physiological and pathological processes. The endothelial dysfunction occurs under activation conditions, with the acquisition of many new functional, inflammatory, and immune properties, and as a consequence, endothelial cells display many different transcription profiles. We describe here the isolation and culture of the most useful model of human umbilical vein endothelial cells, and undertake the proteomic analysis under both basal quiescent condition and activated by stimulation with a proinflammatory cytokine. Series of two-dimensional electrophoresis have allowed us to detect a total of close to 600 polypeptide spots using 4.0–7.0 pH range in both culture conditions. We have selected 233 proteins by cross-matching the gels, and found that 70% showed an increase and 30% a decrease of expression levels in activated cells. Subsequent identification of 35 altered peptides is made by matrix-assisted laser desorption and ionization time-of-flight mass spectrometry, as well as a study of posttranslational modifications. These global findings may contribute to understand the effects of pathological stimuli and the mechanisms that regulate vascular diseases.

Key Words: Endothelial cells; proteomic analysis; vascular pathology; inflammation; cytokines; cellular quiescence; cellular activation.

1. Introduction

The endothelium is a continuous and flattened monolayer of polygonal cells lining the inner face of all vessels in the vascular and lymphatic tissues. It regulates a wide variety of biological responses and exerts a complex array of specialized physiological functions, depending on the organ in which it is located. Vascular endothelial cells are considered constituents of an extraordinarily active interface between the blood and the tissues, and a dynamic and semipermeable

From: *Methods in Molecular Biology, vol. 357: Cardiovascular Proteomics: Methods and Protocols*
Edited by: F. Vivanco © Humana Press Inc., Totowa, NJ

barrier that forms a heterogeneous organ. Moreover, the endothelium can be considered one the largest multifunctional metabolic, endocrine, and paracrine tissues of the body *(1–7)*.

1.1. The Biology of the Quiescent Vascular Endothelium

Vascular endothelial cells are in a quiescence state in normal physiological conditions. Thus, normal vascular endothelium presents a slow renewal population of cells, controls the medial smooth muscle cell growth, allows free blood circulation without activation of intrinsic defense systems, and does not constitutively express activation antigens.

The endothelium plays a critical role in the regulation of the permeability for solutes, macromolecules, and blood cells between the vessel lumen and the surrounding tissue. The formation and maintenance of this selective barrier require specific interendothelial junctional complexes of membrane proteins (cadherins, platelet/endothelial-cell adhesion molecule [PECAM]-1, CD99) *(5,8)*, cytoskeleton components (F-actin, vimentin) *(5,9)*, and associated signaling molecules (mitogen-activated protein kinases [MAPKs], and protein kinases C [PKCs]) *(10)*.

Endothelial cells also contribute to the control of vascular tone by releasing several potent vasodilators (nitric oxide, prostacyclin, bradykining) and vasoconstrictors (endothelin-1, angiotensin-II, thromboxane A2, prostaglandins). Furthermore, the vascular endothelium contributes to normal physiological homeostasis, regulates the coagulation and fibrinolysis, and maintaines the nonthrombogenic intimal surface *(2,6,11–13)*.

Vascular endothelial cells also have the ability to proliferate and form new blood vessels from existing vasculature. This fundamental process of angiogenesis occurs in normal physiological conditions in restricted places (wound healing, development, ovulatory cycle), and transiently in which transcription activation is necessary for cell structural reorganization and acquisition of an invasive phenotype *(2,7)*.

1.2. The Active Surface and the Complexity
of the Vascular Endothelium in Pathological Processes

The endothelium seems to be the main target of pathogenetic mechanisms for vascular diseases. Endothelial dysfunctions have been implicated in atherosclerosis, vascular inflammation, immune response, dyslipidemias, essential hypertension, diabetes, tumor angiogenesis, metastasis, and chronic allograft rejection *(3,11,13–17)*. Some of the endothelial markers are constitutive, but many other molecules are inducibles by inflammatory cytokines, growth factors, and hormones in those pathological processes, given the regulatory role of the endothelium on the inflammatory and immune responses of particular interest *(3,4,6,12,18)*. Thus, the ability of vascular endothelial cells to express a wide array of adhesion molecules (vascular cell adhesion molecule, intercellular adhesion molecule, selectins) and recognition structures initiates changes in membrane permeability, involves the recruitment of circulating leukocytes, and controls their transendothelial migration *(4,19–22)*.

Therefore, endothelial cells are an important source of transcriptional gene expression. This display of many different protein profiles includes, besides adhesion molecules, cytoskeleton components, extracellular matrix-modulating proteins, metalloproteinases, human leukocyte antigen (HLA), and receptors of chemokines and growth factors, among many others.

1.3. The Culture of Endothelial Cells

Isolation and culture of endothelial cells has contributed significantly to the development of vascular biology in the last two decades, and still remain a powerful in vitro model with which to study a range of important pathophysiological processes of the endothelial dysfunction *(1)*. Endothelial cells can be harvested from a variety of vessels, but most of our knowledge in vascular pathology comes from the study of human umbilical vein endothelium. This source of endothelial cells has several advantages: it is a large nonbranching vessel; the cannulation, flushing, and recovery of effluents are not very complicated; and it is available premortem.

Primary endothelial cells derived from human umbilical veins (human umbilical vein endothelial cells [HUVECs]) were first successfully cultured in vitro, and unequivocally identified by morphological, immunohistological, biochemical, and functional criteria in 1973 *(23)*. These assays were confirmed one year later, together with the first studies on cell growth behavior and DNA synthesis *(24)*.

It is imperative to have a full characterization of these cells before using them in different studies. Cultures of endothelial cells from umbilical veins grow as a homogeneous and confluent monolayers, which show a polygonal cobblestone shape (**Fig. 1**; refs. *2,3,12,18*). Disorganization of this monolayer occurs when it is maintained for long periods in the confluent state. The typical cobblestone appearance is also modified after exposure to dynamic flow or shear stress *(2)*. Moreover, endothelial cells present the capacity to organize in capillary-like microtubules when seeded on collagen gels or three-dimensional Matrigel *(12)*.

Endothelial cells express specific and constitutive markers, present in essentially all types of endothelium, which are very helpful to unequivocally identify these cells in vivo and in culture by immunohistochemical studies. The cytoplasm of these cells contains numerous secretory elongated granules called Weibel-Palade bodies, serving as a storage compartment for a variety of proteins, including von Willebrand factor (vWF) antigen or Factor VIII, serving

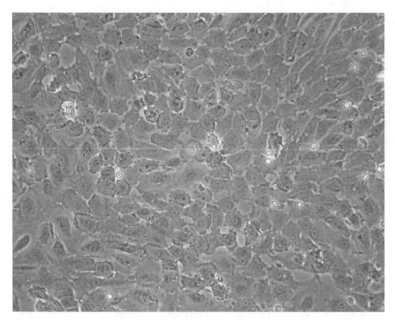

Fig. 1. Morphology of endothelial cells. Primary culture of human umbilical vein endothelial cells after the first passage, showing a confluent monolayer with polygonal cobblestone shape. Photomicrography was performed using a Nikon Eclipse E400 microscope with a Nikon Coolpix 990 digital camera. Original magnification ×200.

as a traditional and universal marker (*see* **Fig. 2**; **refs.** *12,18*). Other constitutive classical markers are also the PECAM-1 (CD31), CD34 and vascular endothelial cadherin (CD144). Many endothelial molecules are expressed only after stimulation with inflammatory cytokines or growth factors, and among these inducible markers are VCAM-1 (CD106), ICAM-1 (CD54), and E-selectin (CD62E) *(3,11,18,20,25,26)*. Lastly, the most useful biochemical and functional marker to unequivocally identify vascular endothelial cells is their ability to incorporate acetylated low-density lipoproteins (LDLs) *(12,18,27)*.

1.4. The Proteome of Endothelial Cells

The development of proteomics and its application to vascular research has been very successful *(28–34)*, and specifically the study of the molecular diversity of the endothelium has taken on increased importance in the last 2 yr *(35–39)*. The complexity of endothelial cells is a promising challenge to provide a more complete picture of the proteins that are modified during the transition from a quiescent phenotype into an activated situation induced by multiple stimuli. Moreover, activated endothelial cells can be major targets for pharmaco-delivery in various clinical states, including inflammatory diseases and cancer.

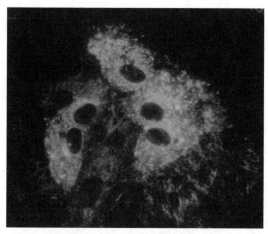

Fig. 2. Endothelial cells express the specific, constitutive, and universal marker (vWF) antigen or Factor VIII. Cells are cultured on 8-well glass slides (Lab-Tek, Chamber slide system, Nalge Nunc), and fixed with 2% paraformaldehyde (Sigma) for 20 min, and permeabilized with 0.1% Triton X-100 (Sigma) for 10 min at room temperature. Preparations are washed with PBS, and cells are specifically stained for vWF antigen or Factor VIII (Boehringer Mannheim) diluted 1:100 in PBS, for 30 min at 4°C. After washing with phosphate-buffered solution, samples are incubated with a fluorescein isothiocyanate-conjugated anti-mouse IgG (Sigma) as a secondary antibody in the dark for 30 min at 4°C. Cover slips are mounted with 90% buffered glycerin. Fluorescent micrographs are acquired as described in **Fig. 1**. Original magnification ×400.

The purpose of the present study is to develop a representative experimental model of an endothelial culture system, to characterize the whole proteome of HUVECs in a quiescent state, and to identify differentially expressed proteins under activated conditions with a well-known proinflammatory cytokine, such as interleukin (IL)-1β.

2. Materials

2.1. Cell Culture

1. Medium 199 (M199) with Hank's Balanced Salt Solution (HBSS), with L-Glutamine, and with 25 mM HEPES (BioWhittaker–Cambrex).
2. Fetal calf serum (FCS; BioWhittaker–Cambrex). It has to be heat-inactivated at 56°C for 30 min before use.
3. Antibiotics: 5000 U/mL penicillin and 5000 μg/mL streptomycin (BioWhittaker–Cambrex).
4. Fungizone (Amphotericin B; Sigma). Stock solution is prepared at 20 mg/mL in sterile saline, and should be stored at −20°C. Working solution is prepared at 2 μg/mL.
5. Trypsin-EDTA: trypsin 500 mg/mL 1:250 and Versene (EDTA 200 mg/mL) (Bio Whittaker–Cambrex).

6. Versene (EDTA, 200 mg/mL; BioWhittaker–Cambrex).
7. Endothelial cell growth supplement (ECGF), from bovine pituitary glands (Sigma). Working solution is prepared in culture medium at 50 µg/mL.
8. Heparin. Sodium salt, 10,000 U. Grade I-A, from porcine intestinal mucosa (Sigma). Working solution is prepared culture medium at 100 µg/mL.
9. Collagenase type II from *Clostridium histolyticum* (Sigma). Working solution is prepared at 0.1% (w/v) in M199 with antibiotics and filtration to sterilize.
10. Gelatin, type B, from bovine skin, approx 225 Bloom (Sigma). Coating plastic with gelatin is prepared by adding an autoclaved solution of 0.5% (w/v) in phosphate-buffered solution (PBS), for at least 2 h at 37°C, and remove it just before adding the cells.
11. IL-1β (PeproTech).
12. Sequencing-grade modified porcine trypsin (Promega).
13. Dithiothreitol (DTT), iodoacetamide, EDTA, formaldehyde, acetonitrile (ACN), formic acid, trifluoroacetic acid (TFA), and tributylphosphine (TBP) (Sigma).
14. Urea, thiourea, CHAPS, acrylamide/bisacrylamide mixture solution, sodium dodecyl sulphate (SDS) (Bio-Rad).
15. 2, 5-Dihydroxybenzoic acid (DHB).

2.2. Buffers and Solutions

1. Lysis solution for HUVECs: 7 *M* urea, 2 *M* thiourea, 4% (w/v) CHAPS, 1% (w/v) DTT.
2. Bradford Protein Assay kit (Bio-Rad).
3. Immobilized pH gradient (IPG) strip gel, 180 mm, pH 3.0–10.0 and pH 4.0–7.0 linear gradient (Immobiline DryStrip, Amersham Biosciences).
4. Rehydration buffer for isoelectric focusing (IEF): 8 *M* urea, 0.5% (w/v) CHAPS, 1% (w/v) TBP, and 0.2% (w/v) Pharmalite pH 3.0–10.0 and 4.0–7.0 (Bio-Rad).
5. Equilibration buffer: 1.5 *M* Tris, pH 8.8, containing 6 *M* urea, 30% (v/v) glycerol, 2% (w/v) SDS, and 0.003% (w/v) bromophenol blue.
6. SDS-Electrophoresis running buffer: 25 m*M* Tris, 192 m*M* glycine, 0.1% (w/v) SDS.
7. PlusOne Silver Staining Kit (Amersham Biosciences).

3. Methods

3.1. Culture of HUVECs

1. HUVECs are obtained from veins of six to eight pooled umbilical cords by collagenase enzymatic digestion (*see* **Note 1**).
2. Segments of umbilical cords, at least 20 cm long, are wiped with a clean gauze. Make sure that they are not injured, and place them inside the tissue culture hood.
3. Cut one end with a sterile surgical blade, and look for the vein (the widest vessel). Introduce a cannula or thick blunt needle (14G or 16G) into the vein, tightly fix it or clamp it with a surgical clip or hemostatic forcep, and screw it up into a stopcock three-way luer-lock.

4. Each umbilical cord is washed, applying a gentle flow through the vein, with 60 mL of saline and 20 mL of M199 with antibiotics; using syringes fit them onto the top of the three-way luer-lock.
5. The vein is filled with collagenase solution, and when the free end of the cord is leaking out, tightly clamp it with an umbilical clamp or surgical clip, inject more enzyme solution until the cord is moderately swollen, and close the three-way luer-lock.
6. Place the umbilical cords onto a tray and leave them into the incubator at 37°C for 20 min (*see* **Note 2**).
7. After this incubation time, the clamped end with the umbilical clamp is cut with a sterile scissor, the cord is massaged and squeezed gently to help cell detachment.
8. The collagenase solution containing endothelial cells is flushed from the cord by perfusion with M199 with 20% FCS, antibiotics, and fungizone (*see* **Note 3**).
9. The effluent is collected into a 50-mL tube and the vein is washed with 20 mL of M199 with 20% FCS, antibiotics, and fungizone. The cell suspension is then centrifuged at 180*g* for 8 min at room temperature. The supernatant is carefully discarded, and the cell pellet is resuspended in culture medium containing M199 with 20% FCS, antibiotics, and fungizone.
10. Cells from four to six cords are transferred into a polystyrene T25 flask, and incubated at 37°C in a 5% CO_2 95% air-humidified atmosphere overnight (*see* **Note 4**). Nonadherent cells are removed the following day by changing culture medium, with medium replacement occurring every 2 d until confluence is reached.

3.2. Propagation of HUVECs

The number of passages that it is possible to perform on a culture of endothelial cells depends of the batch; a good yield is between four and six.

1. Propagation of HUVECs in culture requires an artificial support for cell matrix deposition and attachment, usually gelatin (*see* **Note 5**).
2. Confluent primary cultures are harvested by treatment with trypsin-EDTA for 2 min at 37°C, and then centrifuged at 180*g* for 8 min at room temperature (*see* **Note 6**).
3. Cells are split at a ratio about 1:2 from each T25 flask, and seeded on gelatin-coated tissue-culture T75 flasks, plates, or Petri dishes (*see* **Note 7**). The culture medium M199 with 20% FCS and antibiotics is now supplemented with ECGF and heparin (*see* **Notes 8** and **9**). The cells are fed twice a week with complete changes of fresh culture medium throughout the entire culture period.
4. The purity of endothelial cell cultures is checked by their typical cobbelstone morphology and the expression of vWF or Factor VIII antigen, and it must be more than 95% positive (*see* **Fig. 2**).
5. Routinely, we used confluent cells between the first and fourth passage for these experiments.
6. The treatment of the endothelial cells in the assays described in this chapter is done, for example, with the proinflammatory cytokine IL-1β, at a concentration

of 50 ng/mL, for 24 h at 37°C. The different results are compared with a situation without treatment or quiescent state. For this purpose, the cells are seeded in 6-well plates, and used three wells per treatment, or in one Petri dish.

7. All the next experiments are repeated between four and six times with different batches of cultures. Some protein samples are applied to two-dimensional electrophoresis (2DE) and MS at least four times to improve the reproducibility.

3.3. Lysis of HUVECs

1. Endothelial cell monolayers at confluence are washed twice with PBS, and then released by treatment with versene (EDTA) for 30 min at 37°C.
2. Cells are collected into an Eppendorf tube and centrifuged at 200g for 10 min.
3. The pellet is suspended in lysis solution and incubated for 2 h at 4°C (*see* **Note 10**).
4. The lysate is cleared by centrifugation at 10,000g for 15 min at 4°C.
5. The pellet is discarded and the supernatant is stored frozen at −20°C until use.
6. The protein concentration is quantified by the Bradford reagent kit.

3.4. Two-Dimensional Electrophoresis

3.4.1. First Dimension: IEF

1. The cellular lysate containing 250 µg of protein is added to the rehydration buffer in a final volume of 350 µL.
2. The sample is applied near the cathode on 180-mm dry IPG strips, and proteins are separated either pH 3.0–10.0 or 4.0–7.0 linear gradients, and loaded in the IPGPhor IEF System (Amersham Biosciences).
3. Rehydration is performed at 50 V for 12 h at room temperature without pause, followed by the IEF in a stepwise fashion at 500 V for 1 h, 1000 V for 1 h, and 8000 V for 7–9 h *(30,35,39)*.
4. The focused strips are then incubated twice with the equilibration buffer for 20 min at room temperature each step, in which 1% (w/v) DTT is added in the first incubation, and replaced by 4.8% (w/v) iodoacetamide in the second incubation *(30,35)* (*see* **Note 11**).

3.4.2. Second Dimension: SDS-PAGE Gels

1. The equilibrated IPG strip is transferred for the second dimension onto the top of a 12.5% acrylamide gel (separating proteins in the range 14–200 kDa) (17 × 20 cm) from a 30% acrylamide–0.8% bisacrylamide mixture solution in 1.5 M Tris-HCl pH 8.8, and 10% SDS (*see* **Note 12**).
2. Electrophoresis is carried out at 25 mA/gel at 4°C for about 2 h, using a Protean II System (Bio-Rad) *(30,40,41)*.

3.4.3. Visualization Procedure by Silver Staining

1. Proteins are revealed by silver nitrate staining using the methodology of the PlusOne Silver Staining Kit, following the manufacturer's instruction (Amersham Biosciences). This method is compatible with MS *(35)* (*see* **Note 13**).

2. Gels are fixed overnight with the fixing solution (30% [v/v] ethanol, 5% [v/v] acetic acid). Then, gels are washed three times with double-distilled water for 5 min each time (*see* **Note 14**).
3. Gels are treated with sensitizing solution (30% [v/v] ethanol, 5% [v/v] sodium thiosulphate, and 6.8% [w/v] sodium acetate) for at least 30 min, and washed three times for 5 min each with distilled water.
4. The silver nitrate solution (2.5% [v/v]) is added to the gel and the staining proceeds for 20 min. This solution is removed and the gel is washed four times with double-distilled water for 1 min each.
5. The gel is soaked in the developing solution (2.5% [w/v] sodium carbonate, and 37% [w/v] formaldehyde added just before use) for 2–8 min, until protein spots show up and before background becomes dark.
6. Finally, the stopping solution (EDTA 15% [w/v]) is added for 10 min, and the gel is washed three times with double-distilled water for 5 min each time. The gel is sealed in a plastic bag with a little of distilled water, and stored at 4°C *(41,42)*.
7. All steps should be performed with gentle shaking in a tray with a volume of 250 mL of each solution per gel of 17 × 20 cm.

3.5. Evaluation of 2D Data by Gel Image Analysis

1. Scanning of gels and digitalization of images are accomplished with the DuoScan HiD (Agfa-Gevaert). The evaluation, normalization of the image, and compensation of the background variation are performed by the PDQuest gel analysis software version 6.2 (Bio-Rad). Differences in protein spot intensities among gels under several conditions are quantified and comparative processed. All these programs obviously allow an automated detection system, but generally the peptide detection is also manually edited, and checked to ensure proper identification.
2. Preliminary results can be obtained with broad-range IPGs, such as immobilized pH 3.0–10.0 gradient, but the superior ability of narrow-range IPGs to separate protein species and isoforms of two different states for comparative studies has been demonstrated *(43)*. Thus, the procedure of 2DE and the processing by PD-Quest resulted in a resolution of more than 500 protein spots from whole lysates of HUVECs, separated by their pI 4.0–7.0 IPG strips, in both quiescent (**Fig. 3A**) and activated conditions (**Fig. 3B**). Four to six gels of each condition are performed, revealing the presence of an average of 536 polypeptide spots in quiescent cells and 548 in stimulated with IL-1β.
3. We then create a synthetic gel representing the basal quiescent condition of most of the cultures as a standard reference. The comparison of the central areas chosen from both groups shows strong matches of 233 protein species. The analyses of these spots demonstrate that 160 (69%) present an increase of the expression levels in cells activated with IL-1β (**Fig. 4A**), and 66 spots (28%) show a decreased expression pattern with the inflammatory cytokine (**Fig. 4B**). Moreover, three proteins are exclusively expressed in stimulated cultures (**Fig. 4C**), and two are not detected in basal conditions (**Fig. 4D**).

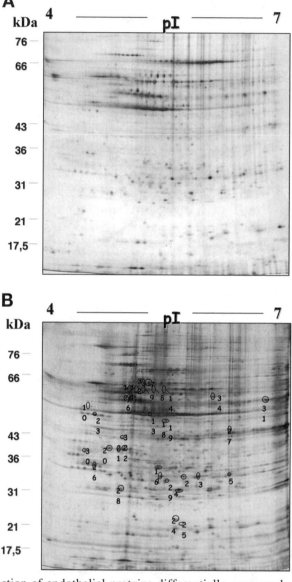

Fig. 3. Detection of endothelial proteins differentially expressed under activated conditions. Comparative two-dimensional gel electrophoresis patterns between whole cell lysate protein expression in quiescent human umbilical vein endothelial cells (**A**), and stimulated with 50 ng/mL of IL-1β for 24 h at 37°C (**B**). Endothelial proteins are separated using 4.0–7.0 immobilized pH gradient strips in the first dimension, and a 12.5% acrylamide sodium dodecyl sulfate-polyacrilamide gel electrophoresis in the second dimension. Spots are visualized with silver staining. Protein spots differentially expressed are circled.

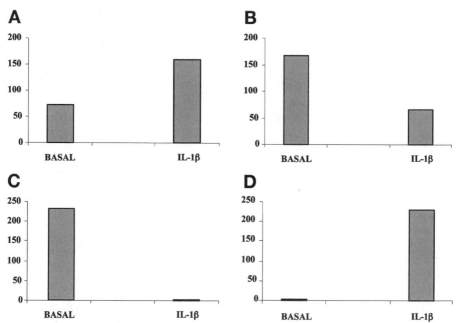

Fig. 4. PDQuest gel analysis shows that inflammatory conditions promote alterations in the endothelial protein expression levels. Differences in spot intensities of the central areas of both groups of gels, quiescent and interleukin (IL)-1β-activated conditions, are quantified and evaluated. This comparison demonstrates that **(A)** 69% (66 of 233 spots) present an increase and **(B)** 28% (66 spots) a decrease in the expression pattern of peptides in cells activated with IL-1β. **(C)** Three proteins are expressed exclusively in stimulated cultures, and **(D)** two in quiescent cells.

3.6. Preparation of Peptide Samples

3.6.1. Tryptic Digestion of Proteins

1. Protein spots of interest are excised from the gel with a sterile blade, and washed with 100 μL of double-distilled water. Gel spots are dehydrated with two cycles of 50 μL of 50% and 100% ACN for 15 min each incubation.
2. The liquid is removed and samples are reduced with 50 μL of 10 mM DTT in 100 mM of NH$_4$HCO$_3$ buffer for 45 min at 56°C.
3. Proteins are then alkylated with 50 μL of 55 mM of iodoacetamide in 100 mM of NH$_4$HCO$_3$ buffer for 30 min at room temperature in the dark.
4. Supernatants are removed and the pieces of gel are treated again with 50 μL of 100% ACN for 15 min. The samples are briefly dried under vacuum in a SpeedVac centrifuge (Savant, Fisher Scientific).
5. The dry gel spots are rehydrated with 50 mM of NH$_4$HCO$_3$ digestion buffer containing 12.5 ng/μL sequencing-grade modified porcine trypsin for 45 min at 4°C.

6. The excess of protease solution is removed, and replaced with 20 µL of NH_4HCO_3 buffer without enzyme. The digestion of the protein gel spots is continued at 37°C overnight *(35,38,39,42)*.
7. The supernatant containing the tryptic peptide mixtures is collected and purified before matrix-assisted laser desorption/ionization analysis.

3.6.2. Extraction and Purification of Digested Proteins

The concentration and desalting of peptide mixtures on reversed-phase (RP) microcolumns prior to mass spectrometric analysis have frequently resulted in an increase of signal-to-noise ratio and sensitivity *(44)*.

1. The mixture of digested peptides is extracted and purified in a custom-made chromatographic microcolumns filled with 0.5 µL of Poros R2 resin (PerSeptive Biosystem), using GELoader micropipet tips (Eppendorf; *see* **Note 15**).
2. This matrix is first equilibrated with 20 µL of 5% formic acid. Then the sample with the digested peptides is added to the resin, and washed with 20 µL more of 5% formic acid.
3. The tryptic peptides retained on the columns are eluted using 0.5 µL of DHB at a concentration of 20 µg/µL in a mixture of 70% ACN:30% of 0.1% TFA *(44,45)*.
4. The extract of purified peptides can be loaded directly onto the MALDI target.

3.6.3. Identification of Peptides by MS

The peptide solution is spotted in a Voyager DE-STR MALDI time-of-flight (TOF) mass spectrometer (Applied Biosystem).

1. Spectra are obtained in positive reflector and linear modes, with a mass range of 800–3500 Da, using an accelerating voltage of 20 kV, with an extraction delay time of 300 ns, and a grid voltage of 72%. The trypsin autoproteolysis products are used as internal calibration.
2. Proteins are identified by mass fingerprinting using the Data Explorer software (Applied Biosystem), against the NCBI and Swiss-Prot databases using the MASCOT program, with the parameters of human species, a mass tolerance setting of less than 50 ppm, and one missed cleavage site by trypsin. Partial chemical modifications (oxidation of methionine, carboxiamidomethylation of cysteine) are taken into consideration.
3. Endothelial proteins are successfully identified after MALDI peptide mass fingerprinting of a subset of 35 spots from the 233 differentially expressed in HUVECs previously analyzed by PD-Quest. Thus, we found that 21 proteins from these 35 present an upregulated expression (*see* **Table 1**), whereas 9 show a downregulation (*see* **Table 2**) under inflammatory conditions in comparison with the quiescent cells.
4. Our proteomics approach also enables the study of potential posttranslational modifications. Some peptides are identified in several spots, corresponding to different isoforms of the same protein. Among them, we have detected several isoforms of

Table 1
List of Identified Endothelial Proteins With an Increased Expression From Cultures Stimulated With IL-1β, Obtained After MALDI-TOF-MS

Identified protein	HUVEC-Basal	HUVEC + IL-1β	Function
HSP27	20292 ± 14.8	20779.4 ± 9.0	Stress defense
HSP27	28359.2 ± 14.5	35328 ± 23.9	Stress defense
Glutation-S transferase	29299.7 ± 16.7	39866.8 ± 44.1	Enzyme
ATP synthase B chain	8648.1 ± 33.1	Saturate	Enzyme
Vimentin	7054.4 ± 17.6	21697.7 ± 33.4	Structural
Unknown/Vimentin	6682.6 ± 56.6	9489.9 ± 8.7	Structural
Unknown/b Actin	41545.1 ± 07.7	Saturate	Structural
Vimentin	6296.3 ± 23.1	6878.9 ± +−8	Structural
Prohibitin	6018.6 ± 43.9	17856 ± 1.6	Signal transduction
HSPC108	3063.9 ± 56.1	3321 ± 6.1	Unknown
Stomatin-like protein 2	3063.9 ± 56.1	3321 ± 6.1	Signal transduction
Annexin V, Chain A	520.3 ± 24.4	1329.1 ± 1.2	Structural
Laminin-binding protein	2789.8 ± 3.6	3661.3 ± 24.8	Structural
Serum albumin	3542.9 ± 335.6	4113.5 ± 47.8	
Signal sequence receptor delta	660.6 ± 59.1	1894.9 ± 13.9	Signal transduction
Tyrosine 3-monooxygenase	14978 ± 70.8	2202.1 ± 2.3	Signal transduction
Fortilin, tumor protein	11392.7 ± 48.1	12157.3 ± 20.9	Signal transduction
Serum albumin	11295.7 ± 57.5	1586.6 ± 55.88	
Tropomyosin isoform	1294.5 ± 49	2346.8 ± 38.5	Structural
Dynactin complex 50 kDa	12318.9 ± 14.9	14787.4 ± 35.2	Structural
Nuclear ribonucleoprotein H1	1066.6 ± −41.1	3028.5 ± 46.6	Signal transduction

Table 2
**List of Identified Endothelial Proteins With a Decreased Expression
From Cultures Stimulated With IL-1β, Obtained After MALDI-TOF-MS Analysis**

Identified protein	HUVEC-Basal	HUVEC + IL-1β	Function
HSP27	13600 ± 14	10684.4 ± 13.2	Stress defense
HSP27	16402 ± 17.7	12816.8 ±.0.4	Stress defense
HSP60	10111.7 ± 0.1	9556.6 ± 48.7	Stress defense
Vimentin	1917.8 ± 14.1	1156.1 ± 28.2	Structural
Unknown/Vimentin	2705.2 ± 11.4	1973.8 ± 31.5	Structural
Vimentin	4255.7 ± 6.8	267.2 ± 11.9	Structural
Annexin V	1420 ± 8.3	1372.9 ± 88.6	Structural
Vimentin	2508.3 ± 20.5	2051.8 ± 77.4	Structural
Enolase 1	7234.2 ± 17.4	5308 ± 42.9	Metabolic enzyme

    vimentin, and two different phosphorylation states in two of the identified isoforms of heat shock protein (HSP)27, which is also secreted by atherosclerotic plaques *(31)*.

5. These data indicate a partial association between the differential protein expression and their function. Identification of upregulated proteins under activation condition shows that they are mainly isoforms of structural and stress defense, as well as signal transduction factors. A downregulated protein expression under inflammatory conditions is found in other isoforms of HSPs.

3.7. Conclusions

    The association of 2DE with MALDI-TOF-MS allows the identification of proteins differentially expressed during pathological or environmental stimuli exposure, in order to provide a more complete picture of activation markers of human endothelial cells. Moreover, the comparative analyses that we perform in this chapter show the usefulness of a proteomic approach in identifying quantitative and qualitative variations in protein levels associated with specific vascular dysfunctions. These findings may lead to the detection of diagnostic markers and novel drugs targets.

4. Notes

1. The cords can be placed in sterile plastic containers after delivery, without any liquid, and kept for 2–3 d at 4°C.
2. It is strongly recommended to limit the incubation time with the collagenase solution to 20 min to avoid as much as possible the contamination with other cells types (vascular smooth muscle cells, fibroblasts).
3. The addition of fungizone (amphotericin B) is recommended in primary cultures at 2 µg/mL.
4. It is important to use the best plastic for primary endothelial cell cultures, such as Corning, Falcon, or Nunc.

5. The propagation and subculturing of endothelial cells is usually performed in plastic coated with gelatin, but can also be done by coating the plastic with other extracellular matrix proteins, such as collagen, fibronectin, or vitronectin. It is important to take into account that these proteins are much more expensive and, moreover, that the activity of some endothelial integrins can be more affected than with gelatin.

6. Endothelial cells are very sensitive to trypsinization. Thus, it is important to limit the incubation with trypsin-EDTA to 2 min at 37°C, and then shake the flask several times, to keep as much as possible the adhesive properties in the next passage.

7. The best propagation of these cultures to form a monolayer is at a ratio of 1:2 or 1:3, avoiding high dilutions.

8. The concentration of 10% of FCS for subculturing and propagation purposes is usually sufficient, but some batches of cells grow faster with an amount of 20% of FCS.

9. The ECGS and the heparin dissolved in culture medium are stored at 4°C, and must be used in the next 10 days.

10. The lysis buffer solution for HUVECs must be prepared fresh every 2–3 mo.

11. IPG strips can be stored at −20°C after the equilibration buffer until further used if they are not immediately added to the second dimension.

12. IPG strips placed on top of the gel can be sealed if it is necessary with 0.5% (w/v) agarose in SDS electrophoresis buffer.

13. Visualization of protein spots can also be done with similar protocols compatible with MS and with our purposes, such as Quick CBB reagent silver-staining kit (Wako) *(36)*, and with fluorescent staining, such as SYPRO Ruby protein gel kit (Molecular Probes) *(39)*.

14. After the second dimension, gels can be fixed for 30 min in 50% ethanol, 7% acetic acid instead of 30% ethanol, 5% acetic acid overnight. Follow with the silver-staining procedure.

15. An alternative to the microcolumns filled with Poros R2 matrix to clean the digested peptides is to use ZipTip C18 pipet tip resin (Millipore) *(35,38)*.

Acknowledgments

This work was supported by grants from the Ministerio de Ciencia y Tecnología (BFI2002-03892) to Jesús González Cabrero, from the Ministerio de Sanidad (FIS, PI021047), Ministerio de Educación (SAF 2004/0619), and Comunidad de Madrid (GR/SAL/0411/2004) to Jesús Egido, and from the Ministerio de Educacion y Ciencia (BFU2005-0838/BMC) and FIS (RECAVA) to Fernando Vivanco. Mayte Pozo was a recipient of a predoctoral fellowship of Fundación Conchita Rábago de Jiménez Díaz.

References

1. Nachman, R. L. and Jaffe, E. A. (2004) Endothelial cell culture: beginnings of modern vascular biology. *J. Clin. Invest.* **114,** 1037–1040.

2. Cines, D. B., Pollak, E. S., Buck, C. A., et al. (1998) Endothelial cells in physiology and in the pathophysiology of vascular disorders. *Blood* **91,** 3527–3561.
3. Pasyk, K. A. and Jakobczak, B. A. (2004) Vascular endothelium: recent advances. *Eur. J. Dermatol.* **14,** 209–213.
4. Nash, G. B., Buckley, C. D., and Ed Rainger, G. (2004) The local physicochemical environment conditions the proinflammatory response of endothelial cells and thus modulates leukocyte recruitment. *FEBS Lett.* **569,** 13–17.
5. Stevens, T., Garcia, J. G., Shasby, D. M., Bhattacharya, J., and Malik, A. B. (2000) Mechanisms regulating endothelial cell barrier function. *Am. J. Physiol. Lung Cell. Mol. Physiol.* **279,** L419–L422.
6. Sumpio, B. E., Riley, J. T., and Dardik, A. (2002) Cells in focus: endothelial cell. *Int. J. Biochem. Cell. Biol.* **34,** 1508–1512.
7. Sato, Y. (2001) Current understanding of the biology of vascular endothelium. *Cell Struct. Funct.* **26,** 9–10.
8. Bazzoni, G. and Dejana, E. (2004) Endothelial cell-to-cell junctions: molecular organization and role in vascular homeostasis. *Physiol. Rev.* **84,** 869–901.
9. Lee, T. Y. and Gotlieb, A. I. (2003) Microfilaments and microtubules maintain endothelial integrity. *Microsc. Res. Tech.* **60,** 115–127.
10. Yuan, S. Y. (2002) Protein kinase signalling in the modulation of microvascular permeability. *Vasc. Pharmacol.* **39,** 213–223.
11. De Caterina, R. (2000) Endothelial dysfunctions: common denominators in vascular disease. *Curr. Opin. Lipidol.* **11,** 9–23.
12. Bachetti, T. and Morbidelli, L. (2000) Endothelial cells in culture: a model for studying vascular functions. *Pharmacol. Res.* **42,** 9–19.
13. Blann, A. D. (2004) Assessment of endothelial dysfunction: focus on atherothrombotic disease. *Pathophysiol. Haemost. Thromb.* **33,** 256–261.
14. Libby, P., Ridker, P. M., and Maseri, A. (2002) Inflammation and atherosclerosis. *Circulation* **105,** 1135–1143.
15. Cook-Mills, J. M. and Deem, T. L. (2005) Active participation of endothelial cells in inflammation. *J. Leukoc. Biol.* **77,** 487–495.
16. Endemann, D. H. and Schiffrin, E. L. (2004) Endothelial dysfunction. *J. Am. Soc. Nephrol.* **15,** 1983–1992.
17. Landmesser, U., Hornig, B., and Drexler, H. (2004) Endothelial function: a critical determinant in atherosclerosis? *Circulation* **109,** II27–II33.
18. Garlanda, C. and Dejana, E. (1997) Heterogeneity of endothelial cells. Specific markers. *Arterioscler. Thromb. Vasc. Biol.* **17,** 1193–1202.
19. McIntyre, T. M., Prescott, S. M., Weyrich, A. S., and Zimmerman, G. A. (2003) Cell-cell interactions: leukocyte-endothelial interactions. *Curr. Opin. Hematol.* **10,** 150–158.
20. Muller, W. A. (2003) Leukocyte-endothelial-cell interactions in leukocyte transmigration and the inflammatory response. *Trends Immunol.* **24,** 327–334.
21. Steeber, D. A., Venturi, G. M., and Tedder, T. F. (2005) A new twist to the leukocyte adhesion cascade: intimate cooperation is key. *Trends Immunol.* **26,** 9–12.

22. Blankenberg, S., Barbaux, S., and Tiret, L. (2003) Adhesion molecules and atherosclerosis. *Atherosclerosis* **170**, 191–203.
23. Jaffe, E. A., Nachman, R. L., Becker, C. G., and Minick, C. R. (1973) Culture of human endothelial cells derived from umbilical veins. Identification by morphologic and immunologic criteria. *J. Clin. Invest.* **52**, 2745–2756.
24. Gimbrone, M. A., Jr., Cotran, R.S., and Folkman, J. (1974) Human vascular endothelial cells in culture: growth and DNA synthesis. *J. Cell Biol.* **60**, 673–684.
25. Raab, M., Daxecker, H., Markovic, S., Karimi, A., Griesmacher, A., and Mueller, M. M. (2002) Variation of adhesion molecule expression on human umbilical vein endothelial cells upon multiple cytokine application. *Clin. Chim. Acta* **321**, 11–16.
26. Daxecker, H., Raab, M., Markovic, S., Karimi, A., Griesmacher, A., and Mueller, M. M. (2002) Endothelial adhesion molecule expression in an in vitro model of inflammation. *Clin. Chim. Acta* **325**, 171–175.
27. Tan, P. H., Chan, C., Xue, S. A., et al. (2004) Phenotypic and functional differences between human saphenous vein (HSVEC) and umbilical vein (HUVEC) endothelial cells. *Atherosclerosis* **173**, 171–183.
28. Jungblut, P. R., Zimny-Arndt, U., Zeindl-Eberhart, E., et al. (1999) Proteomics in human disease: cancer, heart and infectious diseases. *Electrophoresis* **20**, 2100–2110.
29. Arrell, D. K., Neverova, I., and Van Eyk, J. E. (2001) Cardiovascular proteomics: evolution and potential. *Circ. Res.* **88**, 763–773.
30. Durán, M. C., Mas, S., Martín-Ventura, J. L., et al. (2003) Proteomic analysis of human vessels: application to atherosclerotic plaques. *Proteomics* **3**, 973–978.
31. Martín-Ventura, J. L., Durán, M. C., Blanco-Colio, L. M., et al. (2004) Identification by a differential proteomic approach of heat shock protein 27 as a potential marker of atherosclerosis. *Circulation* **110**, 2216–2219.
32. Mayr, M., Mayr, U., Chung, Y. L., Yin, X., Griffiths, J. R., and Xu, Q. (2004) Vascular proteomics: linking proteomic and metabolomic changes. *Proteomics* **4**, 3751–3761.
33. Zerkowski, H. R., Grussenmeyer, T., Matt, P., Grapow, M., Engelhardt, S., and Lefkovits, I. (2004) Proteomics strategies in cardiovascular research. *J. Proteome Res.* **3**, 200–208.
34. Oh, P., Li, Y., Yu, J., et al. (2004) Subtractive proteomic mapping of the endothelial surface in lung and solid tumours for tissue-specific therapy. *Nature* **429**, 629–635.
35. Bruneel, A., Labas, V., Mailloux, A., et al. (2003) Proteomic study of human umbilical vein endothelial cells in culture. *Proteomics* **3**, 714–723.
36. Kamino, H., Hiratsuka, M., Toda, T., et al. (2003) Searching for genes involved in arteriosclerosis: proteomic analysis of cultured human umbilical vein endothelial cells undergoing replicative senescence. *Cell. Struct. Funct.* **28**, 495–503.
37. Traxler, E., Bayer, E., Stockl, J., Mohr, T., Lenz, C., and Gerner, C. (2004) Towards a standardized human proteome database: quantitative proteome profiling of living cells. *Proteomics* **4**, 1314–1323.

38. Sprenger, R. R., Speijer, D., Back, J. W., De Koster, C. G., Pannekoek, H., and Horrevoets, A. J. (2004) Comparative proteomics of human endothelial cell caveolae and rafts using two-dimensional gel electrophoresis and mass spectrometry. *Electrophoresis* **25,** 156–172.

39. Scheurer, S. B., Rybak, J. N., Rosli, C., Neri, D., and Elia, G. (2004) Modulation of gene expression by hypoxia in human umbilical cord vein endothelial cells: a transcriptomic and proteomic study. *Proteomics* **4,** 1737–1760.

40. McGregor, E., Kempster, L., Wait, R., et al. (2001) Identification and mapping of human saphenous vein medial smooth muscle proteins by two-dimensional polyacrylamide gel electrophoresis. *Proteomics* **1,** 1405–1414.

41. Yan, J. X., Wait, R., Berkelman, T., et al. (2000) A modified silver staining protocol for visualization of proteins compatible with matrix-assisted laser desorption/ ionization and electrospray ionization-mass spectrometry. *Electrophoresis* **21,** 3666–3672.

42. Shevchenko, A., Wilm, M., Vorm, O., and Mann, M. (1996) Mass spectrometric sequencing of proteins silver-stained polyacrylamide gels. *Anal. Chem.* **68,** 850–858.

43. Westbrook, J. A., Yan, J. X., Wait, R., Welson, S. Y., and Dunn, M. J. (2001) Zooming-in on the proteome: very narrow-range immobilised pH gradients reveal more protein species and isoforms. *Electrophoresis* **22,** 2865–2871.

44. Larsen, M. R., Cordwell, S. J., and Roepstorff, P. (2002) Graphite powder as an alternative or supplement to reversed-phase material for desalting and concentration of peptide mixtures prior to matrix-assisted laser desorption/ionization-mass spectrometry. *Proteomics* **2,** 1277–1287.

45. Gobom, J., Nordhoff, E., Mirgorodskaya, E., Ekman, R., and Roepstorff, P. (1999) Sample purification and preparation technique based on nano-scale reversed-phase columns for the sensitive analysis of complex peptide mixtures by matrix-assisted laser desorption/ionization mass spectrometry. *J. Mass Spectrom.* **34,** 105–116.

16

Proteomic Study of Caveolae and Rafts Isolated From Human Endothelial Cells

Richard R. Sprenger and Anton J. G. Horrevoets

Summary

Caveolae and rafts are specialized microdomains of the endothelial cell plasma membrane, which play an important role in signal transduction, transcellular transport, and cholesterol homeostasis. The dynamic protein composition of these subcellular lipid domains has been implicated in a variety of patho-physiological states of the vasculature, and is receiving increased attention. As a result of the membranous composition and abundance of insoluble intrinsic and membrane-associated proteins, determination of the raft/caveolae subproteome composition requires specially adapted methods. In this chapter, we present a straightforward protocol to obtain comprehensive and reliable peptide mixtures from this subproteome by subcellular fractionation and both one-dimensional and two-dimensional gel electrophoresis. These mixtures allow dynamic monitoring of composition and posttranslational modification of the raft/caveola subproteome using peptide mass fingerprinting and direct peptide sequencing tandem mass spectrometry.

Key Words: Endothelial cells; lipid rafts; caveolae; 2D gel electrophoresis; subproteomics; subcellular fractionation; mass spectrometry.

1. Introduction

In the post-genome era, various proteomics approaches are being developed to elucidate protein function and interaction, both in fundamental and disease-related research. The methods being developed to study the whole cellular or tissue proteome are still in their infancy with regard to sensitivity and robustness. In contrast, the study of subproteomes has already demonstrated its relevance in (disease-related) functional proteomics *(1–3)*. Among these subproteomes are rafts and caveolae, specialized lipid domains of the plasma membrane, enriched in cholesterol and glycosphingolipids. These lipid microdomains harbor a variety of, often strongly membrane-associated, proteins that are involved in many of the key functions of the vascular endothelium *(4,5)*. Furthermore,

From: *Methods in Molecular Biology, vol. 357: Cardiovascular Proteomics: Methods and Protocols*
Edited by: F. Vivanco © Humana Press Inc., Totowa, NJ

the actual protein composition of these domains with regard to signaling molecules (receptors) varies greatly during disease and pathological situations, often in the absence of changes at the mRNA transcription level *(6–9)*. This makes subcellular fractionation followed by proteomics the method of choice to study the functional implications of the dynamic rafts/caveolae composition. Because the practical implications of specifically studying membrane-associated proteins are complex, we will provide a robust protocol, yielding pure unmodified peptide mixtures for mass spectrometry (MS).

2. Materials

2.1. Cell Culture and Subcellular Fractionation

1. Medium 199 (Gibco BRL, Paisley, Scotland) supplemented with 20% (w/v) fetal bovine serum, 12.5 µg/mL endothelial cell growth supplement (Sigma, St. Louis, MO), 50 µg/mL heparin (Sigma), and 100 U/mL penicillin/streptomycin (Gibco BRL) (*see* **Note 1**).
2. MES-buffered saline (MBS): 25 m*M* MES (Sigma), pH 6.5, 150 m*M* NaCl. Filter through 0.45 µm and store at 4°C. Add 1:20 mammalian protease inhibitor cocktail (Sigma) prior to use (*see* **Note 2**).
3. Teflon cell scrapers (Costar Corning Life Sciences, Acton, MA).
4. 10% (w/v) Triton X-100 (Sigma) in MBS. Store at 4°C (*see* **Note 3**).
5. 2-mL glass Dounce homogenizers with tight fitting glass pestle type "B."
6. 80% (w/v) sucrose in MBS. Prepare at room temperature and dilute with MBS to obtain 30% and 5% sucrose in MBS. Store at 4°C.
7. Ultracentrifuge tubes with 12-mL capacity to fit in a Beckman SW41 (Beckman Instruments, Palo Alto, CA) or Kontron TST41.14 rotor (Kontron Instruments, Milan, Italy) (*see* **Note 4**).

2.2. Isoelectric Focusing

1. 1% (w/v) sodium dodecyl sulfate (SDS) in water, filtered through 0.22 µm. Store at room temperature (*see* **Note 5**).
2. Rehydration sample buffer: 7 *M* urea (Amersham Biosciences, Uppsala, Sweden), 2 *M* thiourea (Sigma), 4% (v/v) Triton X-100 (Sigma), 2% (v/v) carrier ampholytes pH 4.0–7.0 (Amersham Biosciences), 1% DeStreak reagent (Amersham Biosciences), 0.003% (w/v) bromophenol blue (*see* **Note 6**).
3. 18-cm Immobilized pH gradient (IPG) strips pH 4.0–7.0 (Amersham Biosciences).
4. Equilibration buffer: 50 m*M* Tris-HCl, pH 6.8, 6 *M* urea, 30% (v/v) glycerol, 2% SDS. Make fresh or store aliquots at −20°C in 50-mL tubes. Freshly dissolve 1% (w/v) dithiothreitol (DTT) (Amersham Biosciences) or 2.5% (w/v) iodoacetamide (Sigma) prior to use. Keep the iodoacetamide solutions in the dark.

2.3. SDS-Polyacrylamide Gel Electrophoresis

1. Standard Laemmli-compatible polyacrylamide gels of desired size and composition, ready-made or prepared according to standard protocols (*see* **Note 7**).

2. Running buffer: 25 mM Tris-HCl, 192 mM glycine, 0.1% SDS.
3. Embedding solution: 1% (w/v) low-melting point agarose (Gibco BRL) in running buffer (*see* **Note 8**).
4. Precision Plus molecular-weight standards (Bio-Rad, Hercules, CA).

2.4. Silver Staining (see Note 9)

1. Fixative: 50% (v/v) methanol, 7% (v/v) acetic acid.
2. Wash solution: 50% (v/v) ethanol.
3. Sensitizer: 0.02% (w/v) sodium thiosulfate. Prepare freshly.
4. Silver solution: 0.1% (w/v) silver nitrate. Prepare freshly and cool to 4°C.
5. Developer: 2% (w/v) sodium carbonate, 0.015% (v/v) formaldehyde. Prepare freshly and add 400 µL/L 37% formaldehyde just before use.
6. Stopper: acetic acid.
7. Storage solution: 1% acetic acid.

2.5. Destaining and In-Gel Digestion

1. A: 30 mM K$_3$Fe(CN)$_6$ and B: 100 mM Na$_2$S$_2$O$_3$. Store both solutions at room temperature in the dark.
2. Destaining solution: mix A and B in a 1:1 ratio just before use.
3. 6 M Guanidine hydrochloride. Store at room temperature. Freshly dissolve 0.2% (w/v) DTT or 1% (w/v) iodoacetamide prior to use. Keep the prepared iodoacetamide solution in the dark.
4. Ammonium bicarbonate buffer: 100 mM NH$_4$HCO$_3$. Prepare fresh with high-performance liquid chromatography (HPLC)-grade water; can be stored at 4°C for up to 24 h.
5. LiChrosolv-grade acetonitrile (ACN; Merck, Whitehouse Station, NJ).
6. 100 µg/mL sequence grade trypsin (Roche Diagnostics, Indianapolis, IN). Dissolve the contents of one vial trypsin (25 µg) in 250 µL 1 mM HCl (25% HCl is 7.7 M: dilute 7700 times; 10 µL concentrated HCl in 990 µL HPLC-grade water, then 10 µL of this solution in 760 µL water). Vortex well and keep on ice. Any trypsin solution that is not used for the digestion buffer can be stored at −20°C for several weeks.
7. Trypsin digestion buffer: combine 1 part 100 µg/mL trypsin with 4 parts ammonium bicarbonate buffer and 3 parts HPLC-grade water. Optionally add 0.05% (v/v) β-octylglucoside (Ultra-grade, Sigma).
8. Digestion buffer without trypsin: dilute ammonium bicarbonate buffer 1:1 with HPLC-grade water.
9. 50% (v/v) ACN, 5% (v/v) formic acid (Suprapure-grade, Merck). Prepare using HPLC-grade water.
10. 60% (v/v) ACN, 1% (v/v) formic acid. Prepare using HPLC-grade water.

3. Methods

Caveolae and raft fractions have a specific lipid composition and are enriched for membrane proteins and therefore require specific adaptations of existing protocols. The various, often subtle, modifications as combined in our optimized procedure will result in a successful and more complete analysis of this interesting membrane compartment. Therefore, we will describe a robust, reproducible, and reliable protocol for isolating this membrane fraction from cultured endothelial cells using extraction and density gradient centrifugation (based on **refs. 10** and **11**) In the next step, the proteins are extracted by a special protocol allowing full representation of membrane proteins that frequently remain insoluble when using standard proteomics protocols. Next, proteins will be separated by either one-dimensional (1D) or two-dimensional (2D) gel electrophoresis, preceding peptide generation for MS. Electrophoresis in two dimensions is very informative with regard to (changes in) posttranslational modifications (PTMs), such as phosphorylation and glycosylation, and inspection of relative abundance of the individual proteins. Unfortunately, the 2D approach is relatively unsuitable for very basic or acidic proteins and for large proteins (>150 kDa), which includes many receptors. These proteins are poorly represented on 2D gels, mostly because of their inability to efficiently enter the IPG strip during rehydration, or poor transfer from first to second dimension. Therefore, an alternative 1D approach is also described, used to preserve and prepare large proteins for further analysis. As a consequence, however, more proteins will be represented per gel slice/protein band, as compared to a single protein spot from a 2D gel, thereby necessitating additional separation steps. Fortunately, most proteins associated with caveolae and rafts prepared using this method have an isoelectric point between 4.5 and 6.0, allowing the use of strips with a pH range of 4.0–7.0 (instead of 3.0–10.0) yielding enhanced horizontal resolution.

The described protocols yield pure, unmodified peptides representing the full raft/caveolae proteome, allowing a comprehensive map of changing protein composition of this microdomain. Once the peptides have been obtained, standard peptide mass fingerprinting using a matrix-assisted laser desorption/ionization time-of-flight (MALDI-TOF) instrument can be used to obtain the identity of the isolated proteins spots from 2D gels. As the 1D gel segments usually contain peptides from a mixture of proteins, additional separation techniques are needed for which we use 2D liquid chromatography tandem mass spectrometry (LC-MS/MS), allowing direct sequence determination of individual peptides and/or PTMs. Because the actual MS techniques are usually performed in dedicated proteomics facilities and heavily rely on the exact instrumentation, they fall outside the scope of the present chapter (for more information on MS technology and various applications, *see* introductory reviews by, e.g., R. Aebersold, M. Mann, or P. Roepstorff).

3.1. Preparation of Endothelial Rafts and Caveolae

1. Grow the human umbilical vein endothelial cells (HUVEC) in fibronectin-coated 150-mm dishes *(12)* and refresh the culture medium upon reaching confluency.

Allow the cells to develop their characteristic cobblestone appearance for at least 24–48 h before starting the procedure (*see* **Note 10**).

2. For six 150-mm dishes (two sets of three dishes), freshly prepare 2.5 mL MBS with protease inhibitors and chill together with all other required working solutions and consumables, including the ultracentrifuge/rotor and buckets (*see* **Note 11**).

3. Starting with the first set of three dishes, pour off the culture medium and wash each dish twice with 25 mL MBS (*see* **Note 12**).

4. To collect the cells from the three 150-mm dishes in approx 1.4 mL, first pipet 400 μL of MBS with protease inhibitors at the lower edge of each slightly tilted plate and scrape the cells toward this point using a Teflon cell scraper, then pool the scraped cells from the dishes together in a standard 1.5-mL Eppendorf tube.

5. Pipet 150 μL 10% Triton X-100 in the chilled Dounce homogenizer and mix gently with 1350 μL of the cell suspension (*see* **Note 13**). Leave the mixture incubating on ice for approx 10 min.

6. In the meantime, repeat the procedure (**steps 3–5**) for the second set of dishes.

7. After incubation on ice, homogenize the cells with 15 strokes of the tight-fitting glass pestle (labeled "B") and repeat with 15 strokes after a short pause.

8. Fill an Ultraclear ultracentrifuge tube with 1.5 mL 80% sucrose in MBS and add the homogenized cells (1.5 mL). Mix gently but thoroughly by short interval vortexing, 2–5 s each time.

9. Leave the tube on ice while repeating **steps 7** and **8** for the second set of cells.

10. By using a 5-mL pipet and pipetting balloon, gently overlay the lysate in both tubes with 5 mL of 30% sucrose/MBS followed by 4 mL of 5% sucrose/MBS (*see* **Note 14**).

11. Wipe the outside of the tubes with a tissue, mark the two visible interfaces with a water-resistant marker (because the interfaces will no longer be visible after centrifugation), place the tubes in the pre-cooled ultracentrifuge buckets, and tarrate carefully (within 0.005 g) by removing liquid from the heavier tube.

12. Spin at 36,000 rpm (160,000g) for 18 h at 4°C in a pre-cooled SW41 rotor or equivalent (such as the Kontron TST 41.14). Apart from the pellet, a thin light-scattering band should be visible around the former 5–30% sucrose interface (indicated by the top marker), representing the region highly enriched for rafts and caveolae (*see* **Fig. 1**).

13. To harvest the raft/caveolae-enriched fractions, carefully remove the fluid above the membrane band without disturbing it, collect the band (~1 mL) in a 2-mL Eppendorf tube on ice, add 1 mL MBS and centrifuge at 4°C for 1 h at 20,000g. After removal of the supernatant, the pellet can be snap-frozen in liquid nitrogen and processed directly or stored at −80°C (up to several months).

14. To monitor the quality and enrichment of the procedure, routinely collect twelve 1-mL fractions from the top of each gradient (label "1-12"), resuspend the pellet in 1 mL MBS and collect (label "P") then, before proceeding, save aliquots (50–100 μL) of all samples, including "I" from **step 5**, for later analysis by 1D SDS-PAGE and caveolin-1 immunoblotting (*see* **Note 15**; **Fig. 1**).

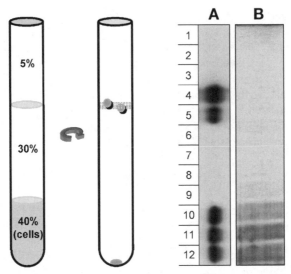

Fig. 1. Isolation of endothelial rafts and caveolae using detergent resistance at 4°C and (buoyant density) flotation ultracentrifugation. As schematically depicted, endothelial cell homogenate, prepared at 4°C in the presence of 1% Triton X-100 was adjusted to 40% sucrose and overlayed with 30% and 5% sucrose. After overnight ultracentrifugation, membranous floating material is observed around the former 5–30% sucrose interface. Equal aliquots from each 1-mL fraction, taken from the top down, were separated by sodium dodecyl sulfate polyacrylamide gel electrophoresis. Caveolin-1 and total protein levels were determined by **(A)** Immunoblotting and **(B)** Coomassie staining, respectively. Caveolin-1 is more than 2000-fold enriched in the floating fractions 4 and 5, because approx 99% of total cellular protein is retained in fractions 10–12.

15. For the 1D SDS-PAGE analysis (optional for analysis of large proteins) only, proceed to **Subheading 3.3.1.**; for extensive 2D analysis (standard), continue with **Subheading 3.2.**

3.2. First Dimension: Isoelectric Focusing

1. These instructions assume the use of a MultiPhor II and Iso-Dalt system (Amersham Biosciences), but can be adapted to other systems if specific alterations, as described in notes, are also applied (detailed description of machine operation can be found in supplier manual).
2. Presolubilize the pelleted caveolae/raft membranes by addition of 10 μL 1% SDS (*see* **Note 16**) and a 15-min incubation in a sonication bath at 25°C or in a waterbath at 25°C with occasional vortexing (*see* **Note 17**).
3. In the meantime, prepare a sufficient amount of sample rehydration buffer (*see* **Note 18**) required for the desired number of samples (for an 18-cm strip, each sample requires 340 μL sample buffer) (*see* **Note 19**).

4. Add 340 µL sample rehydration buffer to each 10 µL of presolubilized membranes and incubate while shaking for 1 h at room temperature. Afterward, pellet any insoluble material by centrifugation for 2 min at 12,000g.

5. Apply the samples to the slots of the reswelling tray and carefully apply the IPG strips, without trapping any air bubbles or pushing any fluid on top of the strip. Overlay each strip with 2 mL IPG cover fluid and allow to rehydrate overnight at room temperature (*see* **Note 20**).

6. Prepare all required isoelectric focusing (IEF) equipment according to the manufacturer's instructions and switch on the thermostatic circulator with the temperature set to 20°C. Program the power supply to start with 200 V for 30 min, followed by two prefocusing steps of 500 V for 45 min and 750 V for 1 h, ramping to 3500 V in 2.5 h and continue at 3500 V for 20 h. The complete sequence takes 25 h and corresponds to a total focusing time of 75 kVh (*see* **Note 21**).

7. Prepare two 11-cm IEF electrode strips by cutting up one large strip. Place them on a clean flat surface (a large Petri dish, for example), wet each strip with 500 µL distilled water, and then gently blot with thin filter paper to remove excess water.

8. Prepare a strip of thick filter paper (roughly the size of the reswelling tray), soak it with water, and blot with filter paper to remove excess water. Position the DryStrip aligner sheet next to the damp filter.

9. Lift the first IPG strip from the tray using a pair of thin forceps and rinse both sides briefly with water (use a squeeze bottle). First place the strip on its edge on the damp filter for a few seconds, then transfer to the aligner, gel side up, with its acidic end (marked with "+") at the top. Repeat this step for all remaining strips, making sure they are aligned.

10. Place the prepared electrode strips over the aligned acidic and basic ends of the strips, making sure to contact the gel surface. Carefully place the aligner with the strips onto the DryStrip tray with the acidic ends pointed toward the red, positive electrode.

11. Align the electrodes above the electrode strips and press them down to make contact. Submerge the strips with cover fluid, apply the lid, and start the program (*see* **Notes 22** and **23**).

12. Focused IPG strips can be processed directly or stored at −80°C for weeks without any noticeable deterioration.

3.3. Gel Electrophoresis

3.3.1. Single-Dimension SDS-PAGE

1. For 1D analysis of large proteins, first incubate the harvested membranes in 5X concentrated sample buffer for 15 min at room temperature with shaking, before diluting with the required amount of water. Heat the protein solution for 5 min at 95°C, cool on ice for 5 min, and spin for 1 min at high speed in an Eppendorf centrifuge to pellet any insoluble material. Separate the proteins on a 7.5% SDS-PAGE gel using MS-compatible materials and standard protocols.

2. Proceed to **Subheading 3.4.** for visualization of protein bands by silver staining.

3.3.2. Second-Dimension SDS-PAGE

1. Prepare running buffer and the required number of large 10% SDS-PAGE gels, using, for example, a multicaster; alternatively, pre-ordered ready-made gels can be used (*see* **Note 24**). Meanwhile, melt the embedding solution using a microwave, cool it down under a stream of cold tap water, and place the solution on a heater set at approx 45°C.
2. Put the strips in individual tubes, with the plastic backing against the tube wall, add 10 mL of equilibration buffer with 1% DTT (dissolved just prior to use) to each tube and incubate for 15 min at room temperature with gentle horizontal rocking.
3. Pour of the solution, replace with equilibration buffer containing 2.5% iodoacetamide (also dissolved just prior to use), and incubate for another 15 min at room temperature in the dark with gentle horizontal rocking.
4. Meanwhile, prepare the desired number of molecular-weight markers. Cut 5 × 5 mm pieces of thin filter paper, place them on a small glass plate, deposit 5 μL diluted marker (unstained or prestained) on each piece and add one or two drops of melted agarose using a disposable glass pipet. Allow the agarose to solidify (storing the pieces in a refrigerator for a few minutes speeds up the process), flip the pieces around, again add a few drops of agarose, allow to solidify, and trim the pieces if necessary (*see* **Note 25**).
5. Briefly wash both sides of each strip with water and place them on one edge on a moistened piece of filter paper. Dip the strip in electrophoresis buffer and position the strip on the surface of the second-dimension gel ensuring no air bubbles are trapped between the strip and the gel surface (*see* **Note 26**).
6. Apply the prepared markers to the surface of the gel, alongside the acidic end of the IPG strip (marked "+") and seal them both into place by pipetting the melted agarose until the space between the glass plates is completely filled. Allow sufficient time for the agarose to solidify.
7. Assemble the electrophoresis unit and start the run. For optimal stacking and migration, the current should be approximately half of the required normal value during the first 30–60 min (*see* **Note 27**).

3.4. Visualization

1. To adapt the following 2D gel-staining procedure to smaller 1D gels, use 50 mL of each solution and reduce the 50% ethanol washing time to 3 × 30 min.
2. Prepare the needed amount of fixative for the silver stain and stop the run when the dye-front has reached the bottom of the gel. Remove the gels from their cassettes and mark each gel identifying the acidic end before removing the sealing agarose, filter marker, and IPG strip. Rinse each gel briefly with 250 mL of water to remove any contaminants or debris.
3. Fix the gels for 30 min with 250 mL of fixative.
4. Wash three times with 250 mL 50% ethanol for 60 min each (*see* **Note 28**). Prepare the sensitizer, silver (pre-chill) and pre-develop solutions during the last wash (*see* **Note 29**).

5. Pour off the wash solution and add 200 mL sensitizer. Gently shake the gels by hand from side to side for exactly 60 s, pour off the solution, and wash three times for exactly 60 s each with 250 mL of water (*see* **Note 30**). Replace water with 200 mL of pre-chilled silver solution and incubate on an orbital shaker for 20 min at room temperature. This step can now be repeated for another set of gels.

6. Add the required amount of formaldehyde (*see* **Note 31**) to the pre-developing solution and wash the gels three times by hand for exactly 60 s. Replace with 250 mL of developer under sufficient agitation to prevent the formation of brown precipitate. Replace the solution after a few minutes when the solution has turned slightly yellow.

7. Stop the reaction by addition of 25 mL acetic acid and incubate until the formation of gas bubbles has stopped. Replace the solution with 250 mL 1% acetic acid, scan the gels, and store at 4°C until further analysis (*see* **Note 32**). In **Figs. 2A,B** examples of silver-stained 2D gels are shown with total HUVEC lysate and prepared caveolae/rafts fractions, respectively.

3.5. Generation of Peptides for MS

3.5.1. Excision and Destaining

1. Wash the gel with water and excise protein bands or spots of interest from the silver-stained gel using a scalpel and tweezers. Cut the gel slabs in small cubes of about 2×2 mm, transfer to 0.5-mL Eppendorf tubes containing 100 μL water, and vortex briefly.

2. Freshly prepare destaining solution by mixing both components in a 1:1 ratio and incubate the gel pieces with 100 μL destaining solution for 30 min at room temperature on a shaker. The gel pieces will turn bright yellow (*see* **Note 33**).

3. Wash four times for 10 min with 500 μL of water while shaking. After washing, the brown color from the silver stain will be gone and the gel pieces should look transparent.

4. For excised 1D protein bands, continue with **Subheading 3.5.2.**; for excised 2D protein spots, proceed to **Subheading 3.5.3.**

3.5.2. Modification (for 1D Protein Bands Only)

1. Incubate the gel pieces with 50 μL 6 *M* guanidine hydrochloride with freshly dissolved DTT, for 30 min at 56°C in an oven or a closed water bath (*see* **Note 34**).

2. Cool the tubes to room temperature and replace the solution by 50 μL 6 *M* guanidine hydrochloride with freshly dissolved iodoacetamide and incubate on a shaker for 30 min at room temperature in the dark and proceed to **Subheading 3.4.3.**

3.5.3. In-Gel Digestion and Extraction of Peptides

1. Wash the gel pieces for 15 min with 100 μL ammonium bicarbonate buffer with shaking, then dehydrate the pieces in 50% ACN by adding 100 μL 100% ACN and continue the incubation for another 15 min.

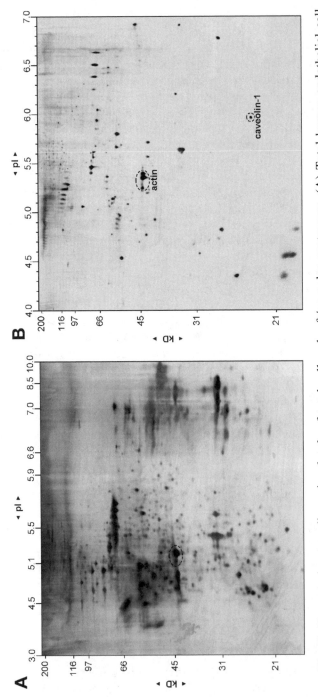

Fig. 2. Comparative two-dimensional gels of total cell and raft/caveolae proteomes. **(A)** Total human endothelial cell proteins, separated using pH 3.0–10.0 nonlinear immobilized pH gradient (IPG) strips in the first dimension and 10% sodium dodecyl sulfate polyacrylamide gel electrophoresis (SDS-PAGE) in the second dimension, and **(B)** caveolae and raft proteins, horizontally separated using pH 4.0–7.0 linear IPG strips and vertically by 10% SDS-PAGE. The encircled protein cluster identified and marked as actin is the only corresponding marker between both gels. This is in contrast to the raft/caveolae marker protein caveolin-1 (encircled in [B] but undetectable in [A]) that is only detectable after applying both subcellular fractionation and protocols specifically adapted to visualize the proteins within these specialized membrane microdomains, as described in this chapter.

208

2. Remove all liquid and incubate the gel pieces with 50 µL ammonium bicarbonate buffer. After shaking for 15 min, dehydrate the pieces in 75% ACN by adding 150 µL 100% ACN and continue shaking until the gel pieces turn white (usually 15–30 min).

3. Shortly spin the tubes, remove all traces of liquid, and dry the gel pieces for about 60 min in a vacuum centrifuge without heating. Prepare the needed amount of trypsin digestion buffer and save on ice until use (*see* **Note 35**).

4. Add 25 µL of trypsin digestion buffer to each tube and allow the dried gel pieces to swell on ice for 45 min.

5. Remove any unabsorbed liquid and add 10 µL of digestion buffer without trypsin to keep the gel pieces moist and incubate overnight at 37°C in an oven or a closed water bath (*see* **Note 36**).

6. Transfer the solution from the incubated vials to a fresh Eppendorf vial (first eluate) and incubate the gel pieces with 50 µL 5% formic acid in 50% ACN for 30 min at room temperature with shaking.

7. Combine the second eluate with the first eluate and further extract the gel pieces with 50 µL 100% ACN for 15 min. Collect the last extract and thoroughly dry the combined eluates in a vacuum centrifuge.

8. Resuspend the dried peptides in the appropriate solution for further mass spectrometric analysis. For most applications, peptides can be dissolved in 5 µL 60% acetonitril with 1% formic acid.

9. For initial standard PMF applications, MALDI analysis using 0.5 µL of the dissolved peptides is usually sufficient for unambiguous protein identification (*see* **Note 37**).

4. Notes

1. Refrain from adding antifungals like Fungizone to the culture medium because this will negatively influence the isolation yield owing to their cholesterol chelating abilities.

2. Unless stated otherwise, prepare all solutions with deionized water, such as Milli-Q (Millipore, Bedford, MA) also referred to as "water" in this text.

3. As a general method for accurately pipetting viscous solutions such as Triton X-100 and 80% sucrose, cut off the first millimeters from a pipet tip.

4. Use Ultraclear centrifuge tubes (Beckman) for better inspection of any floating material.

5. All materials used for 1D and 2D analysis should be ultrapure and/or MS-grade.

6. First make a solution of 7.7 M urea + 2.2 M thiourea. For two IPG strips: combine 629 µL of this solution with 28 µL Triton X-100, 14 µL carrier ampholytes, 7 µL DeStreak, and 2 µL 1% (w/v) 0.22 µm-filtered bromophenol blue.

7. It is preferable to use 1-mm instead of 1.5-mm thick gels.

8. Low melting point (LMP) agarose is used to minimize the chance of introducing urea and protein modifications that generally occur at higher temperatures.

9. An alternative for silver staining is, for instance, Sypro Ruby (Bio-Rad), which requires no fixation and destaining prior to mass spectrometric analysis. However, this

promotes the loss of proteins of sizes less than 15 kDa during washing procedures, and fluorescent stains require specialized scanning and visualization equipment.

10. For each isolation gradient, 3 dishes are required, yielding about 30 µg of protein.

11. All procedures should be performed on ice or in a cold room, with samples and both homogenizers kept on ice at all times.

12. Use a 50-mL tube for quick pouring and dispensing. Working quickly is more important than getting the exact volume of 25 mL. To prevent loss of cells when using an automatic pipet, position the stream along the sides of the culture plate, never directly on the cells.

13. The total volume should be 1.5 mL. If the cell suspension volume is less then 1350 µL, supplement with MBS. Any remaining cell lysate can be saved as input control, labeled "I."

14. To avoid mixing of the sucrose layers, pipet slowly with constant speed, while keeping the tip of the pipet a few millimeters above the meniscus.

15. For the detection of caveolin-1 levels, use a 1:5000 dilution of an anti-caveolin-1 polyclonal antibody (Transduction Laboratories, Lexington, KY). (*See* **Fig. 1** for a typical result.)

16. SDS is a very effective solubilizer of (membrane) proteins, but is also charged and forms complexes with proteins. To cancel out the negative effects of SDS on IEF, the final concentration of SDS in the sample buffer should be below 0.25% and the ratio of non-ionic detergent to SDS should be at least 8:1, which is the case for the used sample buffer.

17. To avoid keratin contamination, wear powder-free gloves when handling samples and IPG strips.

18. Although CHAPS is a potent detergent, Triton X-100 is much more efficient than CHAPS for solubilizing membrane proteins when used in combination with both urea and thiourea *(13,14)*.

19. The prepared buffer and samples should be kept at room temperature well below 37°C at all times to prevent modification of proteins by urea products.

20. Sufficient rehydration takes about 10 h, but overnight is usually more practical.

21. Prefocusing by slowly ramping the voltages is absolutely essential for getting rid of small charged molecules such as salt ions and SDS prior to the focusing at high voltage.

22. It is very important to ensure that the current check of the power supply is switched off before starting IEF.

23. Document the start and end current values for future reference. The starting current at 200 V is approx 0.08 mA/strip; after focusing, the final current at 3500 V should be around 15 µA/strip.

24. It is preferable to use 1-mm instead of 1.5-mm thick gels. Because older gels tend to give better identification results then fresh gels, prepare them the day before. Freshly cast gels tend to contain more free radicals and acrylamide, which can induce unwanted protein modifications.

25. Proper embedding of the filter markers greatly reduces any marker streaking visible after silver-staining.

26. For convenient mounting, stick the plastic backing of the strip against the glass plate and use gel spacers to push down and position the strip.

27. A typical run at maximum current takes about 6 h, but sufficient cooling is required. Although active temperature control generally improves gel-to-gel reproducibility, alternatively the gels can be run overnight at much lower currents, causing less heat generation.

28. Washing is sufficient when the gels start to look opaque, usually during the third wash. To save time, change the wash solution every 30 min until the gels turn opaque, usually observed after four to five solvent changes.

29. For the final steps, process two gels in parallel to reduce gel-to-gel variability. The other gels can be left in the washing solution for an extended period of time, without any problem.

30. Take extra care to keep the surface of the gels wet during the short incubations, because the gels are quite water-repellent after the extensive ethanol washing.

31. Never cool formaldehyde solutions because this will result in the formation of para-formaldehyde.

32. Although gels can be stored this way for weeks, it is recommended to proceed as soon as possible to avoid accumulation of contaminants, such as keratin.

33. Removal of silver-ions prior to in-gel digestion was shown to generally enhance protein identification *(15)*, and at the same time to be particularly useful for the recovery of hydrophobic peptides *(13)*. For an example, *see* **Fig. 3**.

34. Reduction and alkylation of iso-electric focused proteins always has to occur under denaturing conditions, whereas proteins excised from 1D gels usually are not denatured. Modification under reducing conditions is more homogenous and complete, which improves protein identification.

35. Do not dry excessively, and check every 15 min. The gel pieces are sufficiently dry when they have a nonsticky "plastic" appearance.

36. Hydrophobic peptides at low concentration (below ~10 fmol/µL), particularly those generated from membrane proteins, adhere to the vial during digestion or fail to dissolve after drying the peptide extract. To improve the recovery of such peptides, the trypsin digestion buffer can be supplemented with 0.05% β-octyl-glucoside, although this may interfere with later procedures. The detergent can be removed using SCX tips (Millipore), but to avoid the tedious work of having to clean up all samples, use this extended method only for proteins that prove difficult to identify.

37. When MS ionization is hampered owing to contamination of the peptides with salts, cleaning of the sample with a C18-ZipTip (Millipore) can be very helpful. Cleaning is usually not necessary when peptides are to be analyzed using a nano-LC system, which is equipped with an online desalting column.

Acknowledgments

This work was made possible by a Research Support Grant (902-26-201) from the Netherlands Organization for Scientific Research (MW), The Hague, Netherlands.

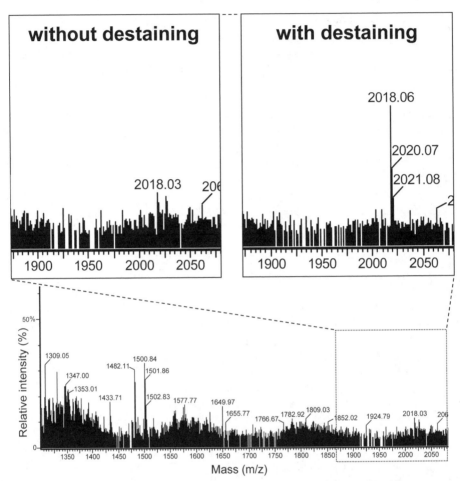

Fig. 3. Improved detection and identification of hydrophobic peptide-containing proteins by removal of silver ions prior to in-gel digestion. Two identical protein spots from two different silver-stained caveolae/raft two-dimensional gels were excised and processed with or without destaining prior to in-gel digestion, respectively. The effect of destaining on the peak intensity of a specific hydrophobic peptide is clearly demonstrated. Despite the low-quality spectrum, the appearance of at least one additional major peak allowed definite identification of the protein by peptide mass fingerprinting.

References

1. Aebersold, R. and Mann, M. (2003) Mass spectrometry-based proteomics. *Nature* **422,** 198–207.
2. Brunet, S., Thibault, P., Gagnon, E., Kearney, P., Bergeron, J. J., and Desjardins, M. (2003) Organelle proteomics: looking at less to see more. *Trends Cell Biol.* **13,** 629–638.

3. Huber, L. A., Pfaller, K., and Vietor, I. (2003) Organelle proteomics: implications for subcellular fractionation in proteomics. *Circ. Res.* **92**, 962–968.

4. Cohen, A. W., Hnasko, R., Schubert, W., and Lisanti, M. P. (2004) Role of caveolae and caveolins in health and disease. *Physiol. Rev.* **84**, 1341–1379.

5. van Deurs, B., Roepstorff, K., Hommelgaard, A. M., and Sandvig, K. (2003) Caveolae: anchored, multifunctional platforms in the lipid ocean. *Trends Cell Biol.* **13**, 92–100.

6. Razani, B., Woodman, S. E., and Lisanti, M. P. (2002) Caveolae: from cell biology to animal physiology. *Pharmacol. Rev.* **54**, 431–467.

7. Matveev, S. V. and Smart, E. J. (2002) Heterologous desensitization of EGF receptors and PDGF receptors by sequestration in caveolae. *Am. J. Physiol. Cell Physiol.* **282**, C935–946.

8. Legler, D. F., Micheau, O., Doucey, M. A., Tschopp, J., and Bron, C. (2003) Recruitment of TNF receptor 1 to lipid rafts is essential for TNFalpha-mediated NF-kappaB activation. *Immunity* **18**, 655–664.

9. Di Guglielmo, G. M., Le Roy, C., Goodfellow, A. F., and Wrana, J. L. (2003) Distinct endocytic pathways regulate TGF-beta receptor signalling and turnover. *Nat. Cell Biol.* **5**, 410–421.

10. Brown, D. A. and Rose, J. K. (1992) Sorting of GPI-anchored proteins to glyolipid-enriched membrane subdomains during transport to the apical cell surface. *Cell* **68**, 533–544.

11. Lisanti, M. P., Tang, Z., Scherer, P. E., and Sargiocomo, M. (1995) Caveolae purification and glycosylphosphatidylinositol-linked protein sorting in polarized epithelia. *Methods Enzymol.* **250**, 655–668.

12. Jaffe, E. A., Nachman, R. L., Becker, C. G., and Minick, C. R. (1973) Culture of human endothelial cells derived from umbilical veins. Identification by morphologic and immunologic criteria. *J Clin. Invest.* **52**, 2745–2756.

13. Sprenger, R. R., Speijer, D., Back, J.-W., De Koster, C. G., Pannekoek, H., and Horrevoets, A. J. (2004) Comparative proteomics of human endothelial cell caveolae and rafts using two-dimensional gel electrophoresis and mass spectrometry. *Electrophoresis* **25**, 156–172.

14. Luche, S., Santoni, V., and Rabilloud, T. (2003) Evaluation of nonionic and zwitterionic detergents as membrane protein solubilizers in two-dimensional electrophoresis. *Proteomics* **3**, 249–253.

15. Gharahdaghi, F., Weinberg, C. R., Meagher, D. A., Imai, B. S., and Mische, S. M. (1999) Mass spectrometric identification of proteins from silver-stained polyacrylamide gel: a method for the removal of silver ions to enhance sensitivity. *Electrophoresis* **20**, 601–605.

17

Proteomic Identification
of *S*-Nitrosylated Proteins in Endothelial Cells

Antonio Martínez-Ruiz and Santiago Lamas

Summary

Nitric oxide (NO) produced in endothelial cells exerts important roles in the vascular system. In recent years, posttranslational modifications induced by NO have been increasingly studied and, among them, cysteine modification by *S*-nitrosylation (also called *S*-nitrosation) has been hypothesized to represent a relevant mechanism for cell signaling. Thus, knowledge of the proteins that can be *S*-nitrosylated in endothelial cells will help to better understand the possible role of this modification. We describe a protocol to identify the *S*-nitrosylome or *S*-nitrosoproteome of endothelial cells, based on the specific derivatization of the *S*-nitrosylation, substituting it by a biotinylation, and the purification of the biotinylated proteins.

Key Words: Nitric oxide; *S*-nitrosylation; posttranslational modification; nitrosative stress.

1. Introduction

It is well-known that nitric oxide (NO) produced in endothelial cells is implied in controlling vascular tone through paracrine signaling involving cGMP synthesis. In the last years, posttranslational modifications induced by NO have also been postulated to have a role in cell signaling and, among them, *S*-nitrosylation has been thoroughly studied *(1,2)*. In endothelial cells, basal submicromolar quantities of NO are produced in very tightly controlled ways by endothelial nitric oxide synthase and are sufficient to exert the tone control function. However, endothelial cells can be exposed to higher quantities of NO and related reactive nitrogen species produced by surrounding cells, especially in pathophysiological situations such as inflammation or atherosclerosis. To better understand the role of this modification, it would be important to identify the proteins that can be *S*-nitrosylated in endothelial cells, i.e., the so called "*S*-nitrosylome" or "*S*-nitrosoproteome" of these cells *(3,4)*.

From: *Methods in Molecular Biology, vol. 357: Cardiovascular Proteomics: Methods and Protocols*
Edited by: F. Vivanco © Humana Press Inc., Totowa, NJ

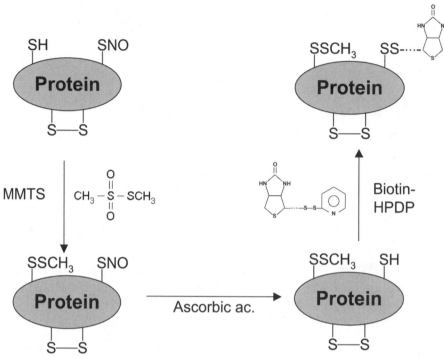

Fig. 1. Schematic representation of the three chemical steps involved in the biotin-switch technique, which derivatizes *S*-nitrosylated proteins to biotinylated proteins. The spacer arm included in the biotinylating reagent, biotin-HPDP, is depicted with a dotted line.

There are some specific features inherent to this modification that have hampered the application of proteomic techniques to the study of the modification. On the one hand, the lability of the nitrosothiol *S-N* bond does not allow some of the common manipulations that are performed to study the proteomes. Among these, light sensitivity and lack of stability under laser ionization used in the matrix-assisted laser desorption/ionization (MALDI) technique *(5)* are significant hindrances. On the other hand, there are no methods that can be reliably used to directly detect the modification (although there are commercially available antibodies, their specificity has been debated), or to enrich modified proteins or peptides by affinity purification (as is done with phosphopeptides).

However, an indirect method paved the way to study this modification in the proteomes. It was named the "biotin-switch" method by their authors *(6)*, and it uses three chemical steps to specifically derivatize the nitrosylated thiols to biotinylated residues (**Fig. 1**):

1. Blocking of free thiols, using methylmethanethiosulfonate (MMTS).
2. Specific reduction of nitrosothiols, using ascorbate.
3. Labeling of newly formed thiols with a biotinylating agent, using N-[6-(biotin-amido)hexyl]-3'-(2'-pyridyldithio)propionamide (biotin-HPDP), which incorporates to the thiol by forming a disulfide bridge.

In the original report, the authors checked that these reactions did not derivatize cysteines modified in other ways and corroborated that they were detecting NO-related protein modification by using neuronal nitric oxide synthase-deficient mice *(6)*. In our opinion, specificity of the technique is acceptable, provided that all the adequate controls are performed. The proteins identified so far in the published reports of S-nitrosylomes in different systems *(3,4,6–11)* correlate well with the proteins that have been studied individually or with proteins that potentially become S-nitrosylated.

However, sensitivity of the technique in its present stage of development is very low. Some reports treated cell extracts with S-nitrosothiols or NO donors to obtain the S-nitrosylome *(6–8)*. Others have used the same reagents, which promote S-nitrosylation, but in intact endothelial cells *(3,4)* or in macrophages *(10,11)*. In these conditions, the amount of S-nitrosylated proteins is very high, but seems to be far from physiological conditions, according to quantitative assessments *(11,12)*. However, in some cases the concentrations of NO used may fall within those assumed to be present in some pathophysiological conditions *(9)*.

We provide the protocol we have used to study the S-nitrosylome of human endothelial proteins treated with S-nitroso-L-cysteine *(3)*, a nitrosothiol that is transported into the cell cytoplasm, producing a higher amount of S-nitrosylated proteins compared with other nitrosothiols *(13)*. Because this nitrosothiol is not commercial and is very unstable, a succinct description of its synthesis is provided. The second part of the protocol describes the treatment of endothelial cells with S-nitroso-L-cysteine. The third part of the protocol describes the biotin-switch treatment itself, which is the crux of the whole protocol: substitution of the nitrosylated cysteine thiols by biotinylated residues. After this step, the degree of biotinylation of the protein extracts should be ready for its detection by Western blot of the biotinylated proteins. For the proteomic identification, the next step involves the separation of biotinylated proteins. This is performed by capture with immobilized avidin resins in stringent conditions. Elution is performed by reducing the disulfide bridges that link the biotin moieties to the cysteine residues. After this step, the proteins obtained should be those which had been previously S-nitrosylated, and common proteomic techniques can be applied for protein identification: gel or liquid separation, digestion before or after separation, MALDI or electrospray ionization, peptide mass fingerprinting or peptide fragmentation. Also, the presence of specific proteins can be checked in this proteome by using Western blot with antibodies against the protein of interest.

2. Materials

2.1. Synthesis of S-Nitroso-ʟ-Cysteine

1. 200 mM ʟ-cysteine in 1 M HCl (prepared fresh). When storing ʟ-cysteine, we treat the bottle with N_2 to prevent oxidation.
2. 200 mM $NaNO_2$ in water (prepared fresh).
3. 1 M Potassium phosphate, pH 7.4.

2.2. Treatment of Cells With S-Nitroso-ʟ-Cysteine and Extract Preparation

1. Dulbecco's modified Eagle's medium (DMEM) with HAT supplement, 20% fetal bovine serum, 100 U/mL penicillin, 100 µg/mL streptomycin, and 5 µg/mL gentamicin.
2. Phosphate-buffered saline (PBS) or other iso-osmotic solution to wash the cells.
3. Nondenaturing lysis solution: 50 mM Tris-HCl, pH 7.4, 300 mM NaCl, 5 mM EDTA, 0.1 mM neocuproine, 1% Triton X-100. Add protease inhibitors.

2.3. Biotin-Switch Treatment

1. HEN buffer: 250 mM HEPES, pH 7.7, 1 mM EDTA, 0.1 mM neocuproine.
2. Blocking buffer: 225 mM HEPES, pH 7.7, 0.9 mM EDTA, 90 µM neocuproine, 2.5% sodium dodecyl sulfate (SDS), 20 mM MMTS (*see* **Note 1**).
3. HENS buffer: HEN buffer with 1% SDS.
4. Biotin-HPDP solution: 4 mM in N,N-dimethylformamide (DMF).
5. ʟ-ascorbic acid: 100 mM.

2.4. Avidin Capture

1. Neutralization buffer: 20 mM HEPES, pH 7.7, 100 mM NaCl, 1 mM EDTA, 0.5% Triton X-100.
2. Equilibration buffer: 20 mM HEPES, pH 7.7, 100 mM NaCl, 1 mM EDTA.
3. Immobilized avidin resin. We use UltraLink® Immobilized NeutrAvidin™ Plus Gel, from Pierce.
4. Washing buffer: 20 mM HEPES, pH 7.7, 600 mM NaCl, 1 mM EDTA, 0.5% Triton X-100.
5. Elution buffer: 20 mM HEPES, pH 7.7, 100 mM NaCl, 1 mM EDTA, 100 mM 2-mercaptoethanol (freshly added).

3. Methods

3.1. Synthesis of S-Nitroso-ʟ-Cysteine

S-nitroso-ʟ-cysteine is an unstable nitrosothiol, and is not available commercially. We use it as a nitrosylating agent because it is a naturally occurring nitrosothiol and we and others have observed that it is able to raise the nitrosothiol content in the cell *(3,13)*, because it is transported across the plasma membrane and it transnitrosates other low-molecular-weight and protein thiols. We follow a protocol previously described *(14)*.

1. Mix 1 vol of 200 mM L-cysteine with 1 vol of 200 mM NaNO$_2$. The solution becomes red almost immediately (*see* **Note 2**).
2. Incubate 30 min at room temperature.
3. Add 2 vols (respect to L-cysteine) of 1 M potassium phosphate, pH 7.4. Immediately place on ice.
4. Aliquot conveniently and store at −80°C (*see* **Note 3**).
5. With the remaining of the solution, make a 1/100 dilution and perform a wave scan. Determine S-nitroso-L-cysteine concentration from absorbance at 338 nm, using the molar absorption coefficient $\varepsilon_{338} = 900\ M^{-1}\ cm^{-1}$ (*15*; *see* **Note 4**).

3.2. Treatment of Cells With S-Nitroso-L-Cysteine and Extract Preparation

We have used the EA.hy926 cell line (kindly provided by Dr. Cora-Jean S. Edgell, University of North Carolina, Chapel Hill, NC) derived from human endothelial cells *(16)* for the proteomic studies, because high quantities of starting material are needed and the human origin helps in the proteomic identification. It is maintained in DMEM medium with HAT supplement, 20% fetal bovine serum, 100 U/mL penicillin, 100 µg/mL streptomycin, and 5 µg/mL gentamicin.

1. Culture cells in Petri dishes until they are just confluent.
2. Wash cells with PBS and add DMEM medium, without serum, or PBS.
3. Add S-nitroso-L-cysteine to 1 mM concentration. As a control, incubate cells with just 1 mM L-cysteine, apart from basal control of untreated cells. Incubate at 37°C in the cell incubator for 15 min (*see* **Note 5**).
4. After treatment, wash cells with PBS and place on ice. Add nondenaturing lysis solution. We add 1 mL to a 100-mm-diameter dish.
5. Scrape cells, harvest them, and incubate for 15 min in ice.
6. Centrifuge at 10,000g, 4°C for 15 min and collect supernatant.
7. Quantify protein concentration. Adjust to 0.5 mg/mL with lysis solution, if necessary (*see* **Note 6**).
8. Protein extracts can be processed immediately or stored at −80°C, although we try not to keep them frozen for many days.

3.3. Biotin-Switch Treatment

The starting material is the protein extract obtained as explained, at a concentration of 0.5 mg/mL or less. For the proteomic identification, several mg of protein extracts will be needed (*see* **Note 7**). If you just want to perform the biotin detection after the biotin switch, the amount of protein can be significantly reduced. In the case of specific protein detection after avidin capture, the amount can also be reduced, depending on the sensitivity of the specific antibody.

1. Add 4 vol of blocking buffer, and incubate at 50°C for 20 min, with frequent or constant agitation.
2. Precipitate with acetone. Add 2 vol of acetone stored at −20°C. Leave at −20°C for more than 10 min. Centrifuge at 2000g, 4°C, for 5 min. Discard the supernatant. Wash with cold acetone and centrifuge again. Discard all the supernatant without affecting the pellet, and let it dry.
3. Resuspend in 0.1 mL of HENS buffer per mg of protein in the initial sample.
4. Add 1/3 vol of biotin-HPDP solution and 1/100 vol of 100 mM L-ascorbic acid (*see* **Notes 8** and **9**). After this stage, it is not necessary to protect samples from light.
5. Incubate 1 h at room temperature.
6. Precipitate with acetone as in **step 2**.
7. Resuspend in HENS buffer as in **step 3**.

At this point, the biotin-switch treatment is complete, and previously *S*-nitrosylated proteins are now biotinylated. An aliquot may be analyzed in order to determine the degree of biotinylation by Western blot, detecting the biotinylated proteins with avidin or anti-biotin antibody (*see* **Note 10**).

3.4. Avidin Capture

1. Add 2 v of neutralization buffer.
2. At this point it is critical to be sure that the pellet from the acetone precipitation has been completely resuspended. If not, proteins that were not resuspended can be found in the elution fraction, even if they were not biotinylated. More neutralization buffer can be added, and the samples must be centrifuged at high speed (15,000g) for 1 min, discarding the pellet.
3. Add the avidin resin previously washed in equilibration buffer. Add 15 μL of resin per mg of protein.
4. Incubate for 1 h at room temperature with agitation.
5. Centrifuge at 400g for 2 min. Discard the supernatant containing the unbound proteins.
6. Wash the resin five times with washing buffer, centrifuging as in **step 5**.
7. Add 1 v of elution buffer and incubate for 20 min at 37°C. We mix the suspension by pipetting up and down before, in the middle, and after the incubation (*see* **Note 11**).
8. Centrifuge for 1 min at high speed and recover the supernatant, containing the eluted proteins that were formerly biotinylated.
9. A second elution can be performed to increase the recovery, and it can be added to the first one.
10. Add sample buffer for SDS-polyacrylamide gel electrophoresis, including reducing agent (2-mercaptoethanol or dithiothreitol), and boil as usual.

At this point, the sample contains the proteins that were previously *S*-nitrosylated in the original extract. In principle, any proteomic technique can be applied, and the proteins can be treated with reducing agents, because the biotin label has been removed. We have performed SDS-polyacrilamide gel electrophoresis with Coomassie staining to separate the proteins (because we did not obtain a high number of proteins), and in-gel digestion and MALDI-time-of-flight peptide mass fingerprinting and electrospray ionization-in time-tandem mass spectrometry to identify the proteins *(3)*.

4. Notes

1. Prepare it using 9 vol of HEN buffer, 1 vol of 25% SDS and 1/100 vol of 2 *M* MMTS in DMF (prepare fresh MMTS solution each time you perform the protocol).
2. Nitrosothiols are photosensitive. From now on, protect the tube from light to diminish destruction of the nitrosothiol bond by light. Take a similar caution with cell extracts, and even with cell cultures after they are treated with S-nitroso-L-cysteine, and until S-nitrosothiols are reduced (addition of L-ascorbate).
3. Thaw aliquots just once, as S-nitroso-L-cysteine is very labile. For that reason, we usually store aliquots of 500 µL and we prepare 4 mL of solution, although the aliquot size and the total quantity can be scaled as necessary.
4. We obtain S-nitroso-L-cysteine solutions between 30 and 40 m*M*, which represents a yield between 60 and 80%.
5. Remember: protect the samples from the light, because it decomposes the S-nitrosothiols.
6. We use the BCA assay from Pierce, with the samples diluted 1/4 in water, and the standard curve prepared in the same buffer dilution. Jaffrey et al. *(6)* diluted the extracts to 0.8 mg/mL. We try to use a more diluted concentration to ensure a complete blocking; however, this can result in a worse acetone precipitation, if volume sample is small. In this case, we suggest using the higher concentration, 0.8 mg/mL.
7. We obtain approx 1 mg from a 100-mm dish of cells. Thus, we recommend starting with about five 100-mm dishes or the equivalent number of cells from bigger dishes.
8. A control for determining endogenously biotinylated proteins, or proteins that are basally purified or detected, is treating half of the sample with just the vehicle of the biotinylating reagent, DMF.
9. It has been recently proposed that ascorbate reduction is limited in these conditions, and it can be improved by raising its concentration up to 30 m*M* and performing the incuation for 3 h *(11)*. It has just been published that increasing ascorbate concentration higher than 1 m*M* induces false-positive signals, because it provokes the reduction of disulfide bridges *(17)*. Our own observations (with the collaboration of Carlos Tarín) corroborate this fact. Thus, care should be taken on reduction conditions by using the correct ascorbate concentration and assessing the possibility of nonspecific reduction.
10. Special care must be taken to use nonreducing buffer and electrophoresis conditions, because the biotin labeling is incorporated via a disulfide bond and is thus reversed by reduction. Indeed, it is better not to boil the samples with loading

buffer, to avoid spurious reactions of remaining biotin-HPDP (6). Endogenously biotinylated proteins will be also detected and they can be identified as the bands that are seen in the control without biotin-HPDP.

11. Elution of proteins bound by biotin–avidin interactions is usually performed in harsh conditions, because it is a very strong interaction and very difficult to reverse. In this case, the incorporation of the biotin via the disulfide bond is an advantage, because it allows an easy elution by incubation with a reducing agent, allowing also the elimination of the endogenously biotinylated proteins, which remain bound to the avidin. However, this has the disadvantage of losing the label from the protein.

12. When performing biotin detection, because this is fairly sensible, a smearing of high-signal bands can be obtained. The control without biotin-HPDP can help to discriminate if it is a failure in the blocking step (if the signal is only in biotin-HPDP-treated samples), or if it is an unspecific detection. If the blocking step is inefficient, try to reduce the protein concentration, be sure that MMTS is used fresh every time, or even reduce the scaling of the assay. If there is an unspecific detection, be careful to eliminate all the biotin-HPDP reagent, perform the electrophoresis just after the biotin-switch protocol, and separate during the electrophoresis, even in different gels, the samples that were not treated with biotin-HPDP, as this reagent may diffuse.

13. If no signal is obtained in the biotin detection, you can include a positive control treating the extract, without blocking, with biotin-HPDP, because this will biotinylate the free thiol cysteines, which are much more abundant than the S-nitrosylated cysteines.

14. Finally, as a rule of thumb, always try to be more specific than sensitive. For example, when separating two phases, avoid maintaining traces from the discarded fraction, even if you lose part of your sample.

Acknowledgments

This research has been supported by grants from the Plan Nacional de I+D+I, Ministerio de Ciencia y Tecnología (SAF 2000-0149 and SAF 2003-01039), Comunidad Autónoma de Madrid (08.4/0030.1/2003), and Ministerio de Sanidad y Consumo (RECAVA network and CP03/00025), all in Spain.

References

1. Stamler, J. S., Lamas, S., and Fang, F. C. (2001) Nitrosylation: the prototypic redox-based signaling mechanism. *Cell* **106,** 675–683.

2. Martínez-Ruiz, A. and Lamas, S. (2004) S-nitrosylation: a potential new paradigm in signal transduction. *Cardiovasc. Res.* **62,** 43–52.

3. Martínez-Ruiz, A. and Lamas, S. (2004) Detection and proteomic identification of S-nitrosylated proteins in endothelial cells. *Arch. Biochem. Biophys.* **423,** 192–199.

4. Yang, Y. and Loscalzo, J. (2005) S-nitrosoprotein formation and localization in endothelial cells. *Proc. Natl. Acad. Sci. USA* **102,** 117–122.

5. Kaneko, R. and Wada, Y. (2003) Decomposition of protein nitrosothiols in matrix-assisted laser desorption/ionization and electrospray ionization mass spectrometry. *J. Mass Spectrom.* **38,** 526–530.

6. Jaffrey, S. R., Erdjument-Bromage, H., Ferris, C. D., Tempst, P., and Snyder, S. H. (2001) Protein S-nitrosylation: a physiological signal for neuronal nitric oxide. *Nat. Cell Biol.* **3,** 193–197.

7. Kuncewicz, T., Sheta, E. A., Goldknopf, I. L., and Kone, B. C. (2003) Proteomic analysis of S-nitrosylated proteins in mesangial cells. *Mol. Cell. Proteomics* **2,** 156–163.

8. Foster, M. W. and Stamler, J. S. (2004) New insights into protein S-nitrosylation: mitochondria as a model system. *J. Biol. Chem.* **279,** 25,891–25,897.

9. Rhee, K. Y., Erdjument-Bromage, H., Tempst, P., and Nathan, C. F. (2005) S-nitroso proteome of *Mycobacterium tuberculosis*: enzymes of intermediary metabolism and antioxidant defense. *Proc. Nat. Acad. Sci. USA* **102,** 467–472.

10. Gao, C., Guo, H., Wei, J., Mi, Z., Wai, P. Y., and Kuo, P. C. (2005) Identification of S-nitrosylated proteins in endotoxin-stimulated RAW264.7 murine macrophages. *Nitric Oxide* **12,** 121–126.

11. Zhang, Y., Keszler, A., Broniowska, K. A., and Hogg, N. (2005) Characterization and application of the biotin-switch assay for the identification of S-nitrosated proteins. *Free Radical Biol. Med.* **38,** 874–881.

12. Zhang, Y. and Hogg, N. (2004) Formation and stability of S-nitrosothiols in RAW 264.7 cells. *Am. J. Physiol. Lung Cell Mol. Physiol.* **287,** L467–474.

13. Zhang, Y. and Hogg, N. (2004) The mechanism of transmembrane S-nitrosothiol transport. *Proc. Natl. Acad. Sci. USA* **101,** 7891–7896.

14. Jourd'heuil, D., Gray, L., and Grisham, M. B. (2000) S-nitrosothiol formation in blood of lipopolysaccharide-treated rats. *Biochem. Biophys. Res. Commun.* **273,** 22–26.

15. DeMaster, E. G., Quast, B. J., Redfern, B., and Nagasawa, H. T. (1995) Reaction of nitric oxide with the free sulfhydryl group of human serum albumin yields a sulfenic acid and nitrous oxide. *Biochemistry* **34,** 11,494–11,499.

16. Brown, K. A., Vora, A., Biggerstaff, J., et al. (1993) Application of an immortalized human endothelial cell line to the leucocyte: endothelial adherence assay. *J. Immunol. Methods* **163,** 13–22.

17. Landino, L. M. (2005) *Free Rad. Biol. Chem.* **340,** 347–352.

18

The Proteome and Secretome
of Human Arterial Smooth Muscle Cell

Annabelle Dupont and Florence Pinet

Summary

Smooth muscle cells (SMCs) play a crucial role in cardiovascular diseases. Proteomic analysis using two-dimensional gel electrophoresis (2DE) associated with mass spectrometry allows characterization of the proteome and secretome of human smooth muscle. The presence of a distinct SMC population in the arterial wall implies that under normal conditions, SMCs are phenotypically heterogeneous. Intracellular and secreted proteins from a primary culture of SMCs obtained from patients undergoing coronary bypass surgery were analyzed using 2DE in order to determine their specific features. The 2D reference maps show that SMCs are involved in a wide range of biological functions. They could constitute a useful tool for a wide range of investigators involved in vascular biology, allowing them to investigate SMC protein changes associated with cardiovascular disorders or environmental stimuli.

Key Words: Arteria proteome; arteria secretome; two-dimensional gel electrophoresis; mass spectrometry.

1. Introduction

Smooth muscle cells (SMCs) are important actors in the pathogenesis of atherosclerosis and of restenosis after angioplasty or stent application. In both phenomena, one of the characteristic changes is the accumulation of SMCs within the intima. The combined action of growth factors, proteolytic agents, and of extracellular matrix proteins, produced by a dysfunctional endothelium and/or inflammatory cells, induces migration of SMCs from the media and their proliferation *(1)*. Moreover, during this processes, SMCs switch from a contractile to a synthetic phenotype *(2,3)*.

The SMCs play a predominant role in various physiological events such as maintaining vessel wall tone, repair of wound healing, and development *(4,5)*. As a result, numerous studies have been performed over the past two decades

From: *Methods in Molecular Biology, vol. 357: Cardiovascular Proteomics: Methods and Protocols*
Edited by: F. Vivanco © Humana Press Inc., Totowa, NJ

to elucidate the molecular events in arterial SMC (ASMC) that are associated with the pathogenesis and progression of diseases affecting these processes. Despite these studies, the mechanisms involved and especially their patho-physiological significance are still poorly understood. Proteomic analysis can be used to elucidate these complex cellular processes. This is an innovative approach for determining the coordinated changes in protein levels in tissues and cells that can provide a comprehensive view of biological phenomena *(6)*. Two-dimensional gel electrophoresis (2DE), a widely used proteomic tool, is a powerful technique for mapping thousands of known and unknown polypeptides simultaneously *(7)*. ASMC cultures have been extensively used to explore the regulation of differentiated SMC function and structure during atheroma development and restenosis formation *(8)*. ASMCs derived from different sources display distinct, stable differences in gene expression that can be maintained in cell culture. Therefore, they constituted an appropriate model for a proteomic approach to explore the mechanisms that control the changes in these cells during disease, and to evaluate the interindividual variability of their biological activities in relation to development of vascular lesions. We proposed studying ASMCs from human internal mammary arteries obtained from patients displaying symptoms of coronary disease. This artery is widely used for aorto-coronary bypass grafting and could be easily used for rescues *(9)*. To the best of our knowledge, no detailed 2D reference map of human ASMCs is yet available to the scientific community. The main objective of this study was the construction of two 2D reference maps of human ASMCs: one of the proteome (the intracellular proteins) and the other of the secretome (the proteins released into the cell culture medum) *(10)*. The identified proteins are involved in protein synthesis as well being components of the cytoskeleton, confirming the synthetic phenotype of ASMCs observed in culture *(11)*. These two 2D reference maps of human ASMCs provide a basis for vascular biology investigations, and open the way to investigations of ASMC protein changes associated with pathological states of exterior stimuli, for the purpose of developing diagnostic markers and detecting novel drug targets.

2. Materials

2.1. Cell Culture

1. HAM-F10 Medium (Gibco-BRL, Bethesda, MD) supplemented with 30% fetal calf serum and 10% horse serum (Biowest).
2. Collagenase type 1 at 235 U/mg (Gibco-BRL); Elastase at 3.73 U/mg (Worthington, Lakewood, NJ); soybean trypsin inhibitor (Sigma, St Louis, MO).
3. Solution of trypsin (0.25%) and 1 mM, EDTA from Gibco-BRL.
4. Solution of proteinase-inhibitor. One tablet of Complete™ (Roche Diagnostics, Meylan, France) should be dissolved in 10 mL of appropriate buffer.

5. Phosphate buffered saline (10X; Gibco-BRL) was diluted 1:10 in double-distilled (dd) H_2O.

2.2. Two-Dimensional Gel Electrophoresis

1. Loading buffer for immobilized pH gradient (IPG) strips was purchased from Genomic Solutions (Steinheim, Germany).
2. Equilibration buffer: 37.5 mM Tris-HCl, pH 8.8, 6 M urea, 2% (w/v) sodium dodecyl sulfate (SDS), 30% glycerol, and 2% (w/v) dithiothreitol (DTT). For the next step, in the same buffer 2.5% (w/v) iodoacetamide was added.
3. 30% Duracryl™ (Genomic Solutions) and TEMED (Bio-Rad, Hercule, CA) (*see* **Note 1**).
4. Ammonium persulfate: prepare 10% solution in water (*see* **Note 2**) and store for 1 wk at 4°C.
5. Running buffer: 25 mM Tris, 192 mM glycine containing 0.5% (w/v) SDS.
6. Silver staining:
   a. Fixation buffer: 30% (v/v) ethanol, 5% (v/v) acetic acid.
   b. Sensitization buffer: 0.02% (w/v) sodium thiosulfate, H_2O washes, 0.2% silver nitrate.
   c. Developing solution: 0.028% (v/v) formalin, 0.0125% (w/v) sodium thiosulfate, and 2.4% (w/v) sodium carbonate.
7. Blue Coomassie staining:
   a. Fixation buffer: 50% (v/v) ethanol with 2% orthophosphoric acid.
   b. Staining solution: 34% (v/v) methanol containing 17% ammonium sulfate, 2% (w/v) orthophosphoric acid, and 1 g CBB-G-250.
8. *See* **Note 3** for precautions when using the technique.

2.3. Mass Spectrometry

1. Destaining of gel: 50 mM NH_4HCO_3 buffer, pH 8.8, containing 50% acetonitrile.
2. Digestion of proteins in the gel: 50 mM NH_4HCO_3, pH 8.0, containing 20 μg/mL of trypsin (Promega, Madison, WI).
3. Matrix solution: dihydrobenzoic acid (DHB) matrix solution (10 mg DHB in 2 mL of 50% methanol/50% ddH_2O).

3. Methods

The isolation of SMCs should be carried quickly after receiving the surgery piece (internal mammary). Extraction of intracellular and secreted proteins should be prepared at 4°C to avoid degradation of proteins. The two steps of 2DE could be performed separately in the time. In that case, the IPGs can be stored at −20°C before the equilibration steps.

3.1. Preparation of Samples for 2DE

1. Primary cultures of human ASMCs were prepared according to the technique patented by Dr. J. B. Michel (INSERM Patent no. 00.09055) *(12)*. ASMCs were isolated

from a residual segment of human internal mammary arteries obtained from patients undergoing coronary bypass grafting. The media was stripped from the underlying adventitia and then finely minced and digested for 45 min at 37°C in 5 mL of HAM-F10 medium containing 3 mg of collagenase, 7 mg of elastase, and 5 mg of soybean trypsin inhibitor (*see* **Note 4**). The enzymatic reaction was stopped by adding 30% fetal calf serum. Medium was changed every 3 d. Cells were trypsinized at confluence and reseeded at a 1:2 ratio. Confluent cells at passage 2 were washed three times with PBS 1X and then incubated for 24 h in serum-free HAM-F10 medium before the proteins were extracted.

2. SMCs were washed three times with 25 mmol/L Tris-HCl, pH 7.4, and scraped in buffer containing 50 mmol/L Tris-HCl, pH 8.6, 10 mmol/L EDTA, 65 mmol/L DTT, proteinase-inhibitor-coktail tablet, 2000 U/mL DNase I, and 2.5 mg/mL RNase A. Cells were then lysed in ice using a mixer suitable for 1.5-mL Eppendorf tubes. The proteins in the medium were precipitated overnight at −20°C by adding 5 v of precooled acetone (*see* **Note 5**). The protein pellet was collected by centrifuging at 2200g for 20 min at 4°C, and washed three times in ice-cold 95% ethanol. The pellet was dried, and then solubilized in the same buffer as the intracellular proteins using a mixer. Protein concentration was determined by commercial Bradford reagent (Bio-Rad). Aliquots of the proteins (100 µL) were prepared, supplemented with 2 M thiourea and 7 M urea, and stored at −20°C until use (*see* **Note 6**).

3.2. Two-Dimensional Gel Electrophoresis

1. Protein samples (100 µg for analytical gels or 500 µg for preparative gels) were mixed with loading buffer for IPG strips (Genomic solutions) in 200 µL final volume and 200 µL of urea solubilization/rehydratation buffer for IPG strips (Genomic solutions) to obtain a final volume of 400 µL. The mixture was applied in gel for reswelling with a dry IPG 180 mm, pH 3.0–10.0 linear gradient (Immobiline Drystrip; GE Healthcare, Uppsala, Sweden) on a Protean isoelectric focusing cell system (Bio-Rad). Complete sample uptake into the strips was achieved after 9 h at 20°C without applying any current. Focusing was performed at 50 V for 9 h, at 200 V for 1 h, and at 1000 V for 1 h followed by a slow ramping to 10,000 V for 6 h, and was completed at 10,000 V for 4.5 h. The current was limited to 50 µA per strip, and the temperature maintained at 20°C for all isoelectric focusing steps. For SDS-polyacrilamide gel electrophoresis, the IPG strips were incubated in equilibration buffer containing 37.5 mM Tris-HCl, pH 8.8, 6 M urea, 2% (w/v) SDS, 30% (v/v) glycerol, and 2% (w/v) DTT for 15 min, and then incubated for 15 min in equilibration buffer supplemented with 2.5% (w/v) iodoacetamide. The equilibrated IPG strips were transferred for the second dimension (SDS-polyacrilamide gel electrophoresis) onto 12% Duracryl™ gels (170 × 200 × 1.5 mm). Electrophoresis was carried out at 10°C using a Protean II xi 2D system (Bio-Rad) with running buffer at 15 mA per gel for 16 h (*see* **Note 7**).

2. The 2D gels were silver-stained according to the protocol previously described by Shevchenko et al. *(13)* with minor modifications. Gels were fixed overnight, then rinsed four times for 10 min in dd water. Gels were sensitized for 1-min, followed

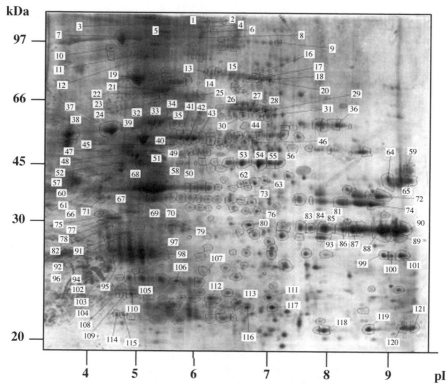

Fig. 1. Representative 2D gels images of intracellular proteins extracted from human smooth muscle cells (SMC). Proteins (100 µg) were separated on a linear pH gradient 3.0–10.0 immobilized pH gradient strip, followed by a 12% Duracryl™ sodium dodecyl sulfate polyacrilamide gel electrophoresis. The gel was then silver-stained. Approximately 1500 silver-stained polypeptide spots were detected by Progenesis software program. Identification of the proteins was performed by matrix-assisted laser desorption/ionization–time-of-flight mass spectrometry and proteins identified in both human macrophages and SMCs are presented in **Table 1**.

by two 1 min ddH$_2$O washes and then incubated for silver staining for 30 min before being washed again for 10 s in ddH$_2$O. Proteins were then visualized using developing solution (*see* **Note 8**), then the reaction was stopped by adding 1% (v/v) acetic acid. An example of silver-stained 2D gels of SMCs is shown in **Fig. 1**.

3. The preparative gels for mass spectrometry are stained using Coomassie blue according to Neuhoff et al. (*14*). Gels were fixed for at least 2 h, and rinsed three times in ddH$_2$O. Gels were then incubated for 1 h in staining solution.

3.3. 2D Image Analysis

Silver-stained 2D gels were digitized at 200 density per inch resolution using an Imagescanner® scanner (GE Healthcare). A calibration filter using different shades of gray was applied to transform pixel intensities into optical density (OD) units. The images were exported into Progenesis V2003-01 2D gel image analysis software (Nonlinear Dynamics, Newcastle upon Tyne, UK) for analysis. Briefly, after automatic detection, the background was removed from each gel and the images were edited manually, for example, adding, splitting, and removing spots. One gel was chosen as the master gel, and used for automatic matching of spots in the other 2D gels. In order to perform automatic determination of the observed isoelectric point (pI) and Mr values with Progenesis, the digitized images were first calibrated. The Mr and pI values of the proteins were assigned after calibrating the 2D gels by running standard protein markers, covering the 20–200 kDa and 3.5–9.3 pI range according to the information provided by the manufacturer (*see* **Note 9**).

3.4. Matrix-Assisted Laser Desorption/Ionization Mass Spectrometry

1. To identify the protein spots on the gel pieces, preparative 2D gels were excised, cut into 1–2 mm^2 pieces, and destained at room temperature for 1–2 h. After washing, the gel pieces were dehydrated and dried thoroughly in a vacuum centrifuge. The dried gel pieces were then digested with 20 μL trypsin solution at 37°C overnight. The samples were then dried in a vacuum centrifuge, and 10 μL of dd water was added before matrix-assisted laser desorption/ionization mass spectrometry (MALDI-MS) analysis.

2. For acquisition of the mass spectrometric peptide maps of the proteins, 1 μL of the generated cleavage products was mixed with 1 μL of DHB matrix. The mixture was air-dried at room temperature prior to the acquisition of the mass spectra. MALDI-MS was performed using a Voyager DE STR mass spectrometer (PerSeptive Biosystems, Framingham, MA) equipped with a 337.1-nm nitrogen laser and with the delayed extraction facility. All spectra were acquired in a positive ion reflector mode (*see* **Note 10**). The spectra are internally calibrated (using the Data Explorer TM software) using three peptides arising from trypsin autoproteolysis ([M+H]$^+$ 842.5100; [M+H]$^+$ 1045.5642; [M+H]$^+$ 2211.1046). Tryptic monoisotopic peptide masses were searched for the National Center for Biotechnology Information using Mascot software (http://www.matrixscience.com) with the following parameters: human species, one missed cleavage site, and a mass tolerance setting of 50 ppm. Chemical partial modification such as oxidation of methionine and carbamidomethylation of cysteine were taken into consideration for the queries. The criteria used to accept identifications include the extent of sequence coverage, the number of peptides matched (minimum of 4), the score of probability (minimum of 70 for the Mowse Score), the mass accuracy, and also when a protein appeared as the top candidates in the first pass of search where no restriction was applied to the species. An example of proteins identified in common between human macrophages and SMCs is presented in **Table 1**. Data were provided from two publications (*10, 15*). Data are also available on the following website: http://www.pasteur-lille.fr/en/research/u5082dgels/.

Table 1
Proteins in Common Between Human
Arterial SMC and Macrophages[a]

Biological functions	Protein name	Accession number
Carbohydrate metabolism	α-Enolase	P06733
	Fructose-biphosphate aldolase A	P04075
	Glucose-6-phosphate 1-deshydrogenase	Q8IUA6
	Glyceraldehyde 3-phosphate dehydrogenase	P04406
	Phosphoglycerate kinase 1	Q8NI87
	Pyruvate kinase M1 isozyme	P14618
	Triosephosphate isomerase	P60174
Structural	Actin	P02571
	Gelsolin	P06396
	Heat shock protein 27 kDa	P04792
	Lamin A	P02545
	Moesin	P26038
	Rho GDP-dissociation inhibitor 1	P52565
	Tropomyosin α 3 chain	P06753
	Vimentin	P08670
Cell death/defense	Cathepsin D	P07339
	Endoplasmin	P14625
	Glutathione transferase omega 1	P09211
	Heat shock cognate 71 kDa protein	P11142
	Heat shock protein 90-β	P08238
	Peroxiredoxin 1	Q06830
	Peroxiredoxin 6	P30041
Protein metabolism	78 kDa glucose-related protein	P11021
	Elongation factor Tu	P49411
	Protein disulfide isomerase	P07237
Energy generation	ATP synthase β chain	P06576
	Vacuolar ATP synthase catalytic subunit A	P38606
Membrane budding	Annexin A1	P04083
	Annexin A2	P07355
	Annexin A5	P08758
Channel	Voltage-dependent anion-selective channel protein 1	P21796
Cell cycle	Prohibitin	P35232
Lipid metabolism	Apolipoprotein A-1	P02647
Miscellaneous	Serotransferrin	P02787
	Serum albumin	P02768
	Transferrin receptor 1	P02786

[a]Protein names and functions have been assigned according to SWISS-PROT and PUBMED numbers.

4. Notes

1. TEMED is best stored at room temperature in a dessicator. Buy small bottles because it may decline in quality (gels will take longer to polymerize) after opening.
2. Unless stated otherwise, all the solutions should be prepared in water that has a resistance of 18.2M Ω-cm and total organic content less than five parts per billion. This standard is referred to as "water" in this text.
3. To avoid contamination with keratin, work in a closed space, wearing gloves.
4. If the residue of internal mammary arteries is less then 5 mm, do not spend time isolating the cells. The amount will be too low for the primary culture cells to grow.
5. Usually, the proteins are precipitated using trichloracetic acid but the concentrations of proteins needed should be >5 µg/µL. In the case of a lower concentration of proteins (<1 µg/µL), it is better to use acetone for the precipitation.
6. Aliquots of 100 µg of proteins are prepared in order to use an aliquot thawed only once for 2DE.
7. The electrophoresis is achieved when the bromophenol blue had reached the bottom of the gel.
8. The time for visualization is variable. The scientist should decide to stop the reaction when an appropriate level of staining is achieved.
9. At that time we were using (Progenesis® software); since then, a new generation of software dedicated to 2D analysis has become available.
10. Typically, 200 laser shots were recorded per sample

Acknowledgments

This work received fundings from "Fondation de France" and Leducq Foundation.

References

1. Ross, R. (1999) Atherosclerosis: an inflammatory disease. *N. Engl. J. Med.* **340,** 115–126.
2. Campbell, G. and Campbell, J. (1990) The phenotypes of smooth muscle cell expressed in human athroma. *Ann. NY Acad. Sci.* **598,** 143–158.
3. Thyberg, J., Blomgren, K., Hedin, U., and Dryski, M. (1995) Phenotypic modulation of smooth muscle cells during the formation of neointimal thickenings in the rat carotid artery after ballon injury: an electron microscopic and stereological study. *Cell Tissue Res.* **281,** 421–433.
4. Schwartz, S. M. (1997) Smooth muscle migration in vascular development and pathogenesis. *Transpl. Immunol.* **5,** 255–260.
5. Katsuyama, H., Wang, C. L., and Morgan, K. G. (1992) Regulation of vascular smooth muscle tone by caldesmon. *J. Biol. Chem.* **267,** 14,555–14,558.
6. Hanash, S. (2003) Disease proteomics. *Nature* **422,** 226–232.
7. Ong, S. E. and Pandey, A. (2001) An evaluation of the use of two-dimensional gel electrophoresis in proteomics. *Biomol. Eng.* **18,** 195–205.

8. Thyberg, J., Nilsson, J., Palmberg, L., and Sjolund, M. (1985) Adult human arterial smooth muscle cells in primary culture: modulation from contractile to synthetic phenotype. *Cell Tissue Res.* **239,** 69–74.
9. Rosenfeldt, F. L. and Wong, J. (1993) Current expectations for survival and complications in coronary artery bypass grafting. *Curr. Opin. Cardiol.* **8,** 910–918.
10. Dupont, A., Corseaux, D., Dekeyzer, O., et al. (2005) The proteome and secretome of human arterial smooth muscle cells *Proteomics* **5,** 585–596.
11. Thyberg, J. (1996) Differentiated properties and proliferation of arterial smooth muscle cells in culture. *Int. Rev. Cytol.* **169,** 183–265.
12. Battle, T., Arnal, J. F., Challah, M., and Michel, J. B. (1994) Selective isolation of rat aortic wall layer and their cell types in culture-application to converting enzyme activity measurement. *Tissue Cell.* **26,** 943–955.
13. Shevchenko, A., Wilm, M., Vorm, O., and Mann, M. (1996) Mass spectrometry sequencing of proteins silver-stained polyacrylamide gels. *Anal. Chem.* **68,** 850–858.
14. Neuhoff, V., Arold, N., Taube, D., and Ehrardt, W. (1988) Improved staining of proteins in polyacrylamide gels including isoelectric focusing gels with clear background at nanogram sensitivity using Coomassie Brilliant Blue G-250 and R-250. *Electrophoresis* **9,** 255–292.
15. Dupont, A., Tokarski, C., Dekeyzer, O., et al. (2004) Two-dimensional maps and databases of the human macrophage proteome and secretome. *Proteomics* **4,** 1761–1778.

19

Real-Time In Vivo Proteomic Identification of Novel Kinase Substrates in Smooth Muscle

Anne A. Wooldridge and Timothy A. Haystead

Summary

Relaxation of smooth muscle can occur through agonists (such as nitric oxide) that activate guanylyl cyclase and stimulate the production of cGMP, activating its target, cGMP-dependent protein kinase (PKG). This kinase can raise the Ca^{2+} threshold for contraction, thus causing Ca^{2+} desensitization, but the mechanism for this event is not completely understood. Ca^{2+} sensitization/desensitization pathways are essential for maintenance of normal smooth muscle tone, and abnormalities in these pathways have been shown to be key components in the pathogenesis of diseases such as hypertension and asthma in humans. Our laboratory has devised a proteomic method to specifically address the question of what proteins are early phosphorylation targets in calcium desensitization. Using ileum smooth muscle, we metabolically labeled the muscle with (^{32}P)-orthophosphate, permeabilized the muscle, established constant calcium concentrations, and stimulated with 8-bromo-cGMP, which activates PKG. Proteins whose phosphorylation state changed in response to cGMP at constant levels of calcium were separated with two-dimensional gel electrophoresis, identified by autoradiography, and sequenced with nanospray mass spectrometry. Using this technique, we identified a previously uncharacterized PKG phosphoprotein, which we have termed CHASM (Calponin Homology Smooth Muscle protein). Using physiological muscle bath contraction studies, we have validated CHASM as a component of calcium desensitization pathways in smooth muscle.

Key Words: cGMP; PKG; calcium desensitization; smooth muscle; CHASM; metabolic labeling; protein phosphorylation.

1. Introduction

The ability to identify novel signaling proteins in cardiovascular smooth muscle could lead to better understanding of cardiovascular physiology and eventually, new therapies. Techniques in proteomics provide excellent tools for the identification of components of signaling pathways, and these techniques are

From: *Methods in Molecular Biology, vol. 357: Cardiovascular Proteomics: Methods and Protocols*
Edited by: F. Vivanco © Humana Press Inc., Totowa, NJ

especially powerful if very specific events are addressed. Validation of new protein targets after identification is an essential step in the process. In our laboratory, we use real-time proteomic techniques combined with physiological validation to study calcium sensitization and calcium desensitization signaling pathways in smooth muscle.

Contraction or relaxation of smooth muscle is primarily determined by the level of phosphorylation of the myosin light chain (MLC)-20. To initiate contraction, an action potential or binding of a contractile agonist causes an increase in intracellular Ca^{2+}, which activates myosin light-chain kinase (MLCK), a Ca^{2+}/calmodulin-dependent enzyme. MLCK phosphorylates MLC-20 on serine 19, resulting in contraction of smooth muscle through increases in myosin ATPase activity and cross-bridge cycling *(1,2)*. Smooth muscle myosin phosphatase (SMPP)-1M, a heterotrimeric enzyme consisting of a large myosin targeting subunit (MYPT1), a PP1 catalytic subunit, and small subunit of unknown function, dephosphorylates MLC-20 resulting in relaxation of smooth muscle.

Contraction can also occur in response to certain signals in the absence of changes in intracellular Ca^{2+}, a phenomenon called Ca^{2+} sensitization. The reverse, calcium desensitization, is relaxation of smooth muscle in the absence of changes in Ca^{2+}. Regulation of SMPP-1M is thought to be a primary mechanism for explaining Ca^{2+} sensitization/desensitization in smooth muscle *(3)*. Ca^{2+} sensitization induced by activation of G protein-coupled receptors acting through RhoA involves phosphorylation of threonine 696 of MYPT1, thus inhibiting SMPP-1M activity. Several kinases have been shown to phosphorylate this site including Rho-kinase *(4)* and MYPT1 associated kinase (MYPT1K or ZIPK) *(5)*. MYPT-1K was identified as the endogenous SMPP-1M-associated kinase by our laboratory using metabolic labeling of porcine bladder smooth muscle with (^{32}P)-orthophosphate, protein fractionation, in-gel kinase assays, and de novo protein sequencing *(6)*. This kinase was also confirmed by another group to be the endogenous SMPP-1M-associated kinase *(7)*.

Relaxation of smooth muscle is achieved by either removal of the contractile agonist (passive relaxation) or through agonists that activate guanylyl cyclase and stimulate the production of cGMP, thus activating its target, cGMP-dependent protein kinase (PKG) *(2)*. cGMP/PKG lowers intracellular Ca^{2+} through multiple mechanisms, but it also raises the Ca^{2+} threshold for contraction, thus causing Ca^{2+} desensitization *(3,8)*. The mechanism by which cGMP/PKG signaling results in Ca^{2+} desensitization is not completely understood. One hypothesis for cGMP-induced calcium desensitization is activation of SMPP-1M. We used real-time smooth muscle physiology, metabolic labeling, and phosphorylation site mapping to determine that PKG phosphorylates MYPT1 and reduces phosphorylation of the inhibitory threonine site, thus indirectly activating SMPP-

1M *(9)*. PKG is likely the sole target for cGMP's relaxant effects in smooth muscle because disruption of the *PKG1* gene in mice eliminates the nitric oxide (NO)/cGMP-mediated relaxation in smooth muscle, but does not affect cAMP/PKA signaling *(10)*.

We sought to identify targets of PKG phosphorylation involved in Ca^{2+} desensitization pathways in smooth muscle using permeabilized ileum smooth muscle at constant submaximal calcium levels, stimulated with cGMP, and metabolically labeled with $[\gamma\text{-}^{32}P]$-ATP. Smooth muscle lysates were resolved with two-dimensional gel electrophoresis (2DE). Proteins that changed phosphorylation state in response to cGMP stimulation were determined using autoradiography. These proteins were cut from the gel and subjected to *de novo* sequencing using nanospray mass spectrometry (MS). We identified a novel phosphoprotein, which we have termed CHASM (<u>C</u>alponin <u>H</u>omology <u>A</u>ssociated <u>S</u>mooth <u>M</u>uscle protein), and we validated this protein as a component of PKG-mediated Ca^{2+} desensitization using physiological smooth muscle contraction studies *(11)*. The techniques used in the discovery and validation of this protein are described herein, and could be applied to other questions in the study of regulation of smooth muscle contraction and relaxation.

2. Materials

2.1. Smooth Muscle Tissue Harvest and Metabolic Labeling

1. Silicone elastomer poured in 2-cm tissue-culture plates. (Sylgard elastomer, Dow Corning, Midland, MI). Using the Sylgard manufacturer's directions, pour a layer of silicone elastomer into the bottom of as many 2-cm tissue-culture plates as are needed for experiments (*see* **Note 1**). The plates need to dry at least 24 h before use.
2. *Staphylococcus aureus* α-toxin (List Biological Laboratories, Campbell, CA). Reconstitute 0.25 mg of α-toxin with 500 μL water (*see* **Note 2**). This product is a hazardous bacterial toxin, so barrier precautions (gloves, eye protection, etc.) should be used when handling.

2.2. Stock Solutions for Muscle Studies

1. $CaMs_2$: 100 m*M* $CaCO_3$, 200 m*M* methane sulfonic acid (Ms). Add Ms slowly to $CaCO_3$ that is stirring. Combine initially in 1/2 vol. Stir for approx 20 min and then quantity sufficient (q.s.) to 100 mL with water. Store at 4°C.
2. $MgMs_2$: 100 m*M* MgO, 200 m*M* Ms. Combine initially in 1/2 vol. Add MgO (Sigma) to water and add Ms slowly while stirring. Q.S. with water after MgO is dissolved. Do not pH. Store at 4°C.
3. KMs: 1 *M* Ms, 1 *M* KOH, add Ms to 1/2 vol of water. Add KOH and allow to dissolve, q.s. to volume. Adjust pH to 7.0 using 1 *N* KOH. Store at 4°C (*see* **Note 3**).
4. 100 m*M* K_2EGTA: 100 m*M* KOH, 100 m*M* EGTA. Q.S. to desired total volume with water. Do not pH. Store at 4°C.

5. 100 mM CaEGTA: 100 mM CaCO$_3$, 100 mM EGTA, 195 mM KOH. Dissolve EGTA and CaCO$_3$ in 80% final volume water. Stir for 30 min and add KOH until the solution clears. Let stand overnight at room temperature and then q.s. to final volume. Do not pH. Store at 4°C.
6. 0.2 M PIPES in water, store at 4°C.
7. 0.2 M Creatine phosphate (Na$_2$CP; Calbiochem). Dissolve in water. Store in 5 mL aliquots at −20°C.
8. 0.1 M Adenosine triphosphate (Na$_2$ATP). Dissolve in water, pH to 7.0 using NaOH. Store in 5-mL aliquots at −20°C.

2.3. Working Solutions for Muscle Studies

1. HEPES-buffered normal Kreb's solution: 137.4 mM NaCl, 5.9 mM KCl, 1.2 mM CaCl$_2$, 1.2 mM MgCl$_2$, 11.6 mM HEPES, 11.5 mM dextrose, pH to 7.3 with NaOH. Store at 4°C for 1 wk.
2. NES: 150 mM NaCl, 4 mM KCl, 2 mM CaMs$_2$, 1 mM MgMs$_2$, 5 mM HEPES, pH to 7.4 at room temperature with Tris-base. Add glucose at 1 g/L (5.6 mM). Store at 4°C for 1 wk.
3. KES: 150 mM KMs, 4 mM KCl, 2 mM CaMs$_2$, 1 mM MgMs$_2$, 5 mM HEPES, pH to 7.4 at room temperature with Tris-base, add glucose at 1 g/L (5.6 mM). Store at 4°C for 1 wk.

2.4. Intracellular Solutions for Muscle Studies

For all solutions (except pCa6.3), make up to 100 mL with water and pH to 7.1 with KOH. Make 10-mL aliquots and store at −20°C. Aliquots can be used for 1 wk after thawing, but keep at 4°C. Warm aliquots to room temperature immediately before use.

1. Maximum calcium solution (CaG): 30 mM PIPES, 10 mM Na$_2$CP, 5.14 mM Na$_2$ATP, 7.3 mM MgMs$_2$, 47 mM KMS, 10 mM CaEGTA.
2. Calcium-free solution (G1): 30 mM PIPES, 10 mM Na$_2$CP, 5.16 mM Na$_2$ATP, 7.3 mM MgMs$_2$, 74 mM KMs, 1 mM K$_2$EGTA.
3. Calcium-free, high-EGTA solution (G10): 30 mM PIPES, 10 mM Na$_2$CP, 5.16 mM Na$_2$ATP, 7.3 mM MgMs$_2$, 74 mM KMs, 10 mM K$_2$EGTA.
4. Labeling calcium-free solution (G10'): 30 mM PIPES, 10 mM Na$_2$CP, 0.5 mM Na$_2$ATP, 7.3 mM MgMs$_2$, 74 mM KMs, 10 mM K$_2$EGTA.
5. Submaximal calcium (pCa6.3): For 1 mL: mix 645 µL of CaG with 355 µL G10. Make fresh before use.
6. 100 mM 8-bromo-cGMP (Calbiochem), make 50-µL aliquots in water and and store at −20°C. Make a working solution of 5 mM at the time of use by diluting the 100 mM stock with submaximal calcium (pCa6.3).
7. Trichloroacetic acid 10% (v/v) in acetone, make 20 mL. Fill Eppendorf tubes (1 tube/sample) with 1 mL, poke holes in the tubes with a 20G needle, and store in an insulated freezer container at −80°C.

2.5. 2DE and Autoradiography

1. Muscle lysis buffer: 5 M urea, 4% CHAPS, 1 mM dithiothreitol, 10 nM microcystin (Alexis Biologicals). Use precaution when handling microcystin, a known carcinogen. Prepare 5 mL fresh and keep at 4°C until use (*see* **Note 4**).
2. 10% Triton X-100 (v/v) incubated with amberlite MB-150 (Sigma) mixed bed exchanger (*see* **Note 5**).
3. 30% Acrylamide, 0.8% *bis* (w/v) in water (neurotoxin when unpolymerized so use precautions).
4. TEMED. Store at room temperature. Ammonium persulfate, prepare 1 mL of 10% solution in water. Freeze in 100-μL single-use aliquots at −20°C.
5. Basic upper chamber buffer: two pellets of NaOH to 250 mL water, stir under vacuum.
6. Acid lower chamber buffer: 680 μL O-phosphoric acid to 1 L of water, stir under vacuum.
7. Separation buffer (4X): 1.5 M Tris-HCl, pH 8.8, store at room temperature
8. Running buffer (10X): For 4 L, dissolve 120 g Tris base, 560 g glycine, and 40 g sodium dodecyl sulfate (SDS) in water. Mix 1:10 with water for working solution.
9. 4X Sample loading buffer with β-mercaptoethanol: 200 mM Tris-HCl, pH 6.8, 44% glycerol (v/v), 8% SDS (w/v), 0.08% bromophenol blue (w/v), 10% β-mercaptoethanol (v/v). Dissolve Tris in 1/2 vol water. Titrate to pH 6.8 (right at edge of titration curve, so go slowly). Dissolve SDS in solution. Add bromophenol blue. Add solution to glycerol and mix well. Bring to final desired volume with water. Store at 4°C.

2.6. Silver Stain

1. Fixing solution: 50% (v/v) methanol, 10% (v/v) acetic acid, store at room temperature.
2. Sodium thiosulfate, 0.2 g/500 mL water, prepare fresh before each use.
3. Silver nitrate solution: 0.9 g AgNO$_3$ in 500 mL water. Can be reused several times. Do not dispose of silver nitrate down sinks. Store at room temperature.
4. Developer: 10 g Potassium carbonate, 20 mL sodium thiosulfate solution from previous step, 250 μL of 40% formaldehyde in 500 mL water. Prepare fresh before each use.

2.7. Preparation of Tryptic Peptides for MS

1. 100 mM sodium thiosulfate, make 50 mL and store at 4°C.
2. 30 mM potassium ferrocyanate, make 50 mL and store at 4°C, protect from light.
3. 20 mM ammonium bicarbonate, store at room temperature.
4. 50 mM ammonium bicarbonate, store at 4°C.
5. 50 mM ammonium bicarbonate, 50% (v/v) acetonitrile (ACN), store at room temperature. 50% (v/v) ACN/5% (v/v) formic acid (make fresh in glass bottles, keep at room temperature).
6. Sequencing-grade porcine trypsin (Promega), 20 μg: add 40 μL of ice-cold trypsin resuspension buffer (packaged with the trypsin) to 20 μg of trypsin. Freeze 5-μL aliquots at −20°C.

2.8. Expression of Recombinant Protein

1. IPTG (Roche). Prepare 1 M solution in water and freeze 300-µL aliquots.
2. Ampicillin, prepare 10 mL of a 100 mg/mL stock in water, store 1-mL aliquots at −20°C.
3. STE buffer: 150 mM NaCl, 50 mM Tris-HCl, 1 mM EDTA, 1 µg/mL leupeptin (Roche), aprotinin (Roche), pefabloc(Roche). Store at 4°C.
4. High-salt STE buffer: 0.5 M, 50 mM Tris-HCl, 1 mM EDTA, 1 µg/mL leupeptin, aprotinin, pefabloc. Store at 4°C.
5. Protein addition buffer: 30 mM PIPES, pH 7.1, 165 mM KMs, 5 mM MgMs$_2$. Store at 4°C (*see* **Note 6**).

2.9. Muscle Tension Measurements

1. β-Escin (Sigma, St. Louis, MO), make single-use 20-µL aliquots of 5 mM β-escin in water and store at −20°C.
2. A23187 (Calbiochem). Make a 10 mM stock in dimethyl sulfoxide. Store at 4°C. Make a 1 mM working solution in water and store at 4°C (*see* **Note 7**).
3. Calmodulin, bovine brain (Calbiochem): dilute in protein addition buffer to 5 mg/mL. Store in 25-µL single-use aliquots at −20°C.

3. Methods

3.1. Smooth Muscle Tissue Harvest and Metabolic Labeling

1. Bubble Kreb's solution with a mixture of 95% O_2, 5% CO_2 and warm to 37°C before beginning the experiment (*see* **Note 8**).
2. Euthanize a 3- to 4-kg adult white rabbit according to an approved Institutional Animal Care and Use Committee method for your program (*see* **Note 9**).
3. Immediately after euthanasia, place 70% ethanol on the fur of the abdomen and open the peritoneal cavity with scalpel and scissors. Remove the ileum, trimming the mesentery as close to the serosal surface as possible so that the ileum straightens into a 20-cm-long section. Flush as much food material from the lumen using a 10 cc syringe filled with Kreb's solution. Cut 1-cm tubular pieces from the intact ileum using and place the pieces into Kreb's (*see* **Note 10**).
4. Using curved forceps, thread a 1-cm tubular section of ileum over a 10-cc pipet (approx 0.7 cm diameter). Using a razor blade, gently score the ileum longitudinally on both sides of the mesenteric band. Using curved forceps, grasp the superficial layer (which is the smooth muscle) and peel around the pipet until the smooth muscle layer is freed from the underlying mucosa and submucosa (*see* **Fig. 1**). Place the isolated smooth muscle layers into fresh Kreb's and allow to rest at 37°C for 15 min with O_2/CO_2 bubbling (*see* **Note 11**).
5. Choose six sheets of ileum with the best integrity (*see* **Note 12**); three sheets will be used as control and three sheets will be stimulated with 8-bromo-cGMP. Using minutien dissection pins (Fine Science Tools [FST]) and Dumont no. 5 straight forceps (FST), pin the each sheet loosely on all sides in individual 2-cm silicone-

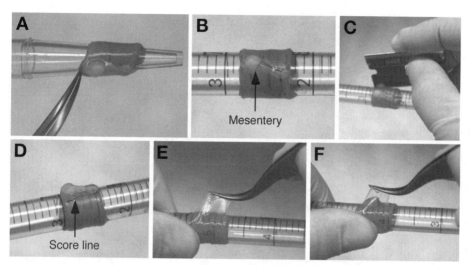

Fig. 1. Smooth muscle tissue harvest from rabbit ileum. (**A**) One-centimeter tubular section of ileum is gently pulled onto a 10-mL pipet from the tapered end. Note that the forceps are grasping at the edge of the tissue. (**B**) The single ileal mesenteric band is identified. (**C**) Using a razor blade, the ileum is gently scored on either side of the mesenteric band. (**D**) The top layer (smooth muscle) is cut, leaving the mucosal layers intact. (**E**) The smooth muscle layer is grasped with curved forceps and gently pulled as the pipet is turned. (**F**) The layer is peeled all the way around until the second score line on the other side of the mesenteric band is reached.

bottom dish filled with Kreb's. Allow tissues to equilibrate at 37°C for 15 min. After equilibration, stretch the tissue gently in all directions, but take care that the circular fibers (these are the most prominent) are oriented parallel to each other so that all stretching is even. Add more pins so that the tissue can be trimmed to 1 cm by 0.5 cm without losing tension.

The next steps should be performed at room temperature on an orbital shaker at low-speed shaking. All buffer volumes on the sheets of muscle are 2 mL unless otherwise stated. An aspirator works well for removing buffers between washes, but note that after labeling with ^{32}P, buffers will be radioactive. Appropriate steps must be used to reduce exposure and environmental contamination.

6. Add NES and allow to equilibrate for 30 min.
7. Permeabilize the muscles with 20 µg/mL α-toxin in calcium-free solution with 1 m*M* EGTA (G1) for 40 min at room temperature (*see* **Note 13**).
8. Wash the muscles 3 × 5 min in G1.
9. Contract the muscles in submaximal calcium solution, pCa6.3, for 10 min.

From this point through **Subheading 3.3.**, appropriate shielding, gloves, protective eyewear, and proper waste disposal for working with radioactive ^{32}P must be utilized.

10. Wash in G10' for 5 min. Add [γ-^{32}P]-ATP (1 mCi/muscle sheet) to an appropriate volume of G10' (2 mL/sheet) in a conical tube. Add 2 mL of this mixture to each muscle sheet in the silicone dishes and incubate for 5 min.

11. Stimulate tissues with 8-bromo-cGMP (100 μM) or vehicle (equal volume of water) for 5 min. Add the agonist or vehicle directly to the muscles that are incubating in the labeling buffer from **step 10**.

12. Remove all buffer and snap-freeze muscle in one of two ways:
    a. Quickly unpin the muscle and place in an Eppendorf tube (poke a hole in the top of the tube with a 20G needle first) and drop into liquid nitrogen (*see* **Note 14**). Store tubes at −80°C until use.
    b. Pour liquid nitrogen-cooled freon on the muscle in the dish. Break the frozen muscle into pieces and drop into Eppendorf tubes (with holes from a 20G needle in the top) that are filled with 1 mL 10% trichloracetic acid (v/v) in acetone and that have been prechilled to −80°C. Immediately place tubes with muscles at −80°C for storage (*see* **Note 15**).

3.2. 2DE and Autoradiography

3.2.1. Muscle Lysates

1. For liquid nitrogen frozen tissues: grind muscles into a fine powder using a mortar and pestle resting inside a styrofoam container with enough liquid nitrogen to cover the bottom of the mortar. Dilute the ground muscle 3:1 (v/w) in 2D muscle lysis buffer and homogenize in 4-mL volume glass homogenizer (Duall 20, Kontes glass).

2. For freon frozen tissues: remove from −80°C, place at −20°C for 1 h, 4°C for 1 h, and finally room temperature for 1 h. Wash the muscles three times in 100% acetone using an aspirator or pipet to remove the acetone. Allow the muscles to air-dry after the final wash. Homogenize the muscles 3:1 (v/w) in 2D muscle lysis buffer (*see* **Note 16**).

3. Centrifuge lysates at more than 13,000g for 15 min. Remove and retain supernatant.

4. In a small side-arm flask, mix the following: 5.5 g urea, 2 mL H_2O, 2 mL 10% Triton-X, 1.33 mL 30% acrylamide/0.8% *bis*. Stir in a warm water bath until urea dissolves. Add 100 μL 3.0–10.0 ampholytes, 200 μL 5–7 ampholytes, and 200 μL 6–8 ampholytes (BioLyte, Bio-Rad). Stir under a vacuum for 5 min (*see* **Note 17**).

5. Place as many nontreated capillary tubes as are needed (1 per sample and some extra) into a casting chamber with the blue side of the capillary tube up. Add 10 μL 10% ammonium persulfate and 10 μL TEMED to gel/ampholyte solution and drizzle down the side wall of the casting chamber until the gel level is just below the blue line on the capillary tube (*see* **Note 18**).

6. After tube gels are polymerized, remove from casting chamber, rinse with water, and dry by rolling on a paper towel (*see* **Note 19**). Affix tube gels into Bio-Rad

IEF chamber. Make sure all sample wells and plugs are sealed. Add acid buffer to lower chamber and carefully add base to upper chamber (*see* **Note 20**).

7. Clean out sample wells with 200 µL of base solution. Add up to 100 µL of muscle lysate supernatant (which has been lysed in 2D sample buffer). Load the same amount of lysate for both vehicle and treated tissues.
8. Run first dimension at 300 V (constant voltage) overnight (*see* **Note 21**).
9. Clean enough glass plates (Bio-Rad minigel system) for 1-mm gels with a rinsable detergent (e.g., Alconox, New York, NY), rinse with 70% ethanol, and air-dry.
10. Prepare 12% separating gels. (Prepare enough solution for as many 1-mm gels as are needed.) For two 1-mm minigels: 2.5 mL of separating buffer, 4 mL 30% acrylamide/0.8% *bis*, 3.35 mL water, 100 µL 10% SDS, 10 µL TEMED, and 100 µL 10% APS. Pour the gel and leave enough space for a stacking gel. Place a layer of isopropanol on top of the gel after it is poured so that a level surface will be created as the gel polymerizes, which will take about 20 min.
11. Rinse the gel with water twice to remove isopropanol after polymerization.
12. Prepare 4% stacking gels: for 10 mL of stacker solution: 6.1 mL water, 2.5 mL stacking buffer, 1.3 mL 30% acrylamide/0.8% *bis*, 100 µL 10% SDS, 20 µL TEMED, 100 µL 10% APS. Do not place a comb in the stacking gel, just allow it to polymerize.
13. Set up the gel apparatus and fill with the appropriate amount of running buffer.
14. Blow the first dimension tube gels out of the capillary tube using a water-filled 3-mL syringe with a plastic adapter fittted to the capillary tube into a small plastic weigh boat filled with 1 mL of 4X protein sample loading buffer (*see* **Note 22**).
15. Lay the tube gel between the glass plates of the second-dimension gel on top of the stacking gel. Using a flattened metal forcep, gently push the tube gel down so that it lies flat across the stacking gel (*see* **Note 23**).
16. Run the second dimension at 100 V until the dye front runs into the separating gel and then continue at 200 V until the dye front just runs off the gel (*see* **Note 24**).

3.2.2. Silver Staining and Autoradiography

1. Remove the gel from the gel apparatus and place in small glass dish containing fixing solution. Gently agitate for at least 20 min on an orbital shaker. Rinse the gel 4 × 5 min in water.
2. Add sodium thiosulfate solution for exactly 90 s.
3. Rinse the gel three times quickly in water.
4. Add silver nitrate solution for 10–20 min. Gel should turn golden brown during this step.
5. Dispose of silver nitrate solution in appropriate waste and rinse the gel three times quickly in water.
6. Add developer solution. Leave the gel in this solution until brown protein staining begins to appear. Stop the developing reaction when it appears to be complete (time varies from experiment to experiment) by pouring off developer and adding fixing solution for 20 min.

7. Transfer the gel to water for 30 min. Place the gel between two pieces of cellophane, taking care to keep out bubbles. Dry gel completely in a gel dryer.

8. Place the dried gel in a film cassette between two intensifying screens. Add a piece of luminescent tape so that the gels are easy to align with the film after exposure. Add film and place at −80°C (*see* **Note 25**).

3.3. Preparation of Tryptic Peptides for MS

1. Align the gels and the autoradiography film and identify spots with increased phosphorylation in response to cGMP (*see* **Note 26**). Using a no. 15 surgical blade, cut out the spots on the silver-stained gel that correspond to increased density on the autoradiography. Place the excised spots into Eppendorf tubes.

2. Wash the spots in water 3 × 10 min to rehydrate gel and remove SDS.

3. Remove water and add 200 μL of a 1:1 mixture of 100 mM sodium thiosulfate and 30 mM potassium ferricyanide. Make the 1:1 mixture fresh before each use. Incubate 10 min at room temperature.

4. Wash the gel chunks 3 × 15 min in water. The yellow color will disappear after these washes and the gel chunk should appear clear.

5. Add 500 μL of 50% ACN/50% 100 mM ammonium bicarbonate. Wash 2 × 15 min in this mixture.

6. Add 500 μL 100% ACN for 15 min. Gel will appear white after this step. Aspirate away ACN.

7. Using the pre-prepared 5-μL aliquots of trypsin, add 120 μL of ice-cold 50 mM ammonium bicarbonate. 30 μL of this dilution will be needed per spot to be sequenced, so prepare enough trypsin for all samples. This dilution gives a final concentration of 0.02 μg/μL. Add 30 μL of this dilution to each sample and store on ice for 1 h. The final amount of trypsin per sample is 0.6 μg/sample (*see* **Note 27**).

8. Discard the liquid from each sample. Add 30 μL of 50 mM ammonium bicarbonate and incubate overnight at 37°C.

9. Quick spin tubes, remove supernatant, and place into new tube (*see* **Note 28**).

10. Add 30 μL of 20 mM ammonium bicarbonate to the gel piece. Incubate for 15 min and then combine with supernatant from **step 9**.

11. Add 30 μL 50% ACN/5% formic acid and incubate for 15 min. Combine with previous supernatants. Repeat this step.

12. Dry down samples in a speedvac until volume is approx 20 μL. Further sample processing will be determined by the mass spectrometer that is used.

13. We used a QSTAR Pulsar hybrid mass spectrometer (Applied Biosystems) to derive *de novo* peptide sequences from the 2D gel spots (*see* **Fig. 2**). Peptide sequences were searched against protein and DNA databases using the FASTS algorithm *(12)*. We identified an acidic, 45 kd, previously uncharacterized protein from the RIKEN database that corresponded RIKEN cDNA putative gene product 1110030K22 (NM_024230) (**Fig. 3**). The sequence of this gene product showed the presence of a calponin homology domain in the carboxy terminus. We designated the protein CHASM *(11)*.

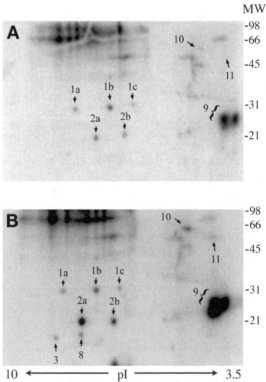

Fig. 2. Autoradiograph of **(A)** vehicle and **(B)** 100 µ*M* 8-Br-cGMP-stimulated rabbit ileum. Sheets of rabbit ileum smooth muscle were permeabilized with *Staphylococcal aureus* α-toxin, contracted with submaximal calcium, and incubated with [γ-^{32}P]ATP for 5 min before agonist stimulation. Muscle lysates were processed for two-dimensional gel electrophoresis, and phosphorylated proteins visualized by autoradiography. Phosphorylated proteins were identified as follows: (1a–c) heat shock protein (HSP)27, (2a,b) HSP20, (3) ubiquitin conjugating enzyme, (9) telokin, and (10) RIKEN putative gene product. Proteins labeled 8 and 11 were below detection limits. (From **ref. *11*** with permission from Elsevier Science.)

3.4. Expression of Recombinant Protein

In order to perform experiments to verify that a protein is actually involved in the signaling pathway of interest, the ability to express recombinant protein is essential. We obtained an I.M.A.G.E. cDNA clone 3593616 (Research Genetics) corresponding to the sequence of CHASM. We polymerase chain reaction-amplified the open reading frame for CHASM and inserted it into the bacterial expression vector pGEX 6-P-1. (Amersham Biosciences). This expression vector is particularly useful for the muscle bath studies that will be described in

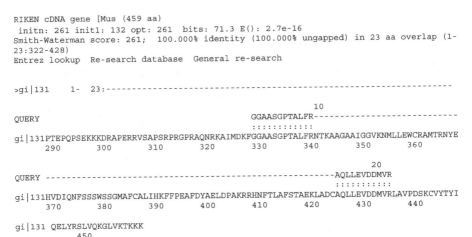

```
RIKEN cDNA gene [Mus (459 aa)
  initn: 261 init1: 132 opt: 261  bits: 71.3 E(): 2.7e-16
Smith-Waterman score: 261;  100.000% identity (100.000% ungapped) in 23 aa overlap (1-
23:322-428)
Entrez lookup  Re-search database  General re-search


>gi|131    1- 23:-----------------------------------------------------------------

                                              10
QUERY                                  GGAASGPTALFR-------------------------
                                       ::::::::::::
gi|131PTEPQPSEKKKDRAPERRVSAPSRPRGPRAQNRKAIMDKFGGAASGPTALFRNTKAAGAAIGGVKNMLLEWCRAMTRNYE
    290       300       310       320       330       340       350       360

                                                                       20
QUERY -----------------------------------------------------AQLLEVDDMVR
                                                          ::::::::::::
gi|131HVDIQNFSSSWSSGMAFCALIHKFFPEAFDYAELDPAKRRHNFTLAFSTAEKLADCAQLLEVDDMVRLAVPDSKCVYTYI
    370       380       390       400       410       420       430       440

gi|131 QELYRSLVQKGLVKTKKK
    450
```

Fig. 3. Alignment of peptides sequenced by mass spectrometry, using the FASTS algorithm. The identity of protein 10 (from **Fig. 2**) was determined. (From **ref. 11** with permission from Elsevier Science.)

Subheadng 3.5. because the glutathione-*S*-transferase (GST) tag can be removed with a protease. PreScission™ Protease (Amersham Biosciences) has a GST-tag, so it is removed simultaneously with the GST portion of the fusion protein. Thus there is no contamination of the protein in the muscle bath with the protease. The rest of this method will describe expression and purification of recombinant CHASM for muscle bath studies.

1. Pick a single colony of plasmid-containing bacteria and grow in 5 mL of Luria-Bertani broth with 50 μg/mL ampicillin overnight at 37°C (*see* **Note 29**).
2. Transfer the 5 mL overnight cultures to 1 L of Luna-Bertani broth containing 50 μg/mL ampicillin. Incubate at 37°C until the OD is 0.6–0.8 at 600 nm.
3. Induce bacterial production of protein by adding 300 μ*M* IPTG for 2 h at 37°C (*see* **Note 30**). Spin down 1-L cultures at 3635*g* for 20 m at 4°C. Immediately remove supernatant and freeze pellet at −80°C overnight (the pellet can also be snap-frozen in liquid nitrogen for immediate use, if desired).
4. Thaw pellets on ice and resuspend pellets in 40 mL of sodium chloride–Tris–EDTA (STE) buffer with 1 μg/mL leupeptin, aprotinin, and pefobloc.
5. Add 800 μL of 50 mg/mL lysozyme (Sigma; made fresh before use) per pellet and incubate on ice for 10 min.
6. Add 400 μL of 10% NP40 detergent per pellet and mix. At this stage, if the lysate is the consistency of egg whites, you will need to sonicate the lysate until it has a thin consistency. After sonication, centrifuge at 11,265*g* for 30 min at 4°C.
7. While the lysates are spinning, wash glutathione Sepharose™ beads (Amersham Pharmacia Biosciences; 1.3 mL of original bead slurry per 1 L of bacterial culture) three times in 10-column volumes of STE buffer.

8. Transfer lysates to 250-mL disposable conical tubes and add the washed beads. Rock gently at 4°C for 2 h to allow protein binding to the glutathione Sepharose.
9. Spin the conical tubes at 528g for 20 min and carefully remove supernatant. Transfer the beads to a column. The beads can be diluted in high-salt STE if it makes transfer easier. Wash by gravity flow with 50 mL of high-salt STE at 1 mL/min. Wash by gravity flow with 50 mL of STE at 1 mL/min.
10. Use 80 U of Precission™ protease per mL of bead bed volume. Dilute protease in STE buffer, mix with beads using a large bore pipet and leave overnight at 4°C.
11. Elute the protease fraction and wash the column twice with 1 bed volume of STE buffer. Check the fractions with a Bradford protein assay to confirm the presence of protein. A portion of each fraction should be run on a 12% polyacrylamide gel and silver-stained to confirm the correct protein molecular weight and the purity.
12. For use in muscle bath studies: buffer exchange the protein in protein addition buffer using 10,000 molecular-weight cutoff spin columns (Amicon ultra, Millipore). Combine the protein-containing fractions in a 4 mL capacity spin column and spin at 1885g for 10 min (do not allow the column to spin dry). Add 3 mL of protein addition buffer and spin again. The volume left in the column should be 100 µL. Repeat three times. After buffer exchange, continue to concentrate the protein until the concentration of the protein is appropriate. Check the final concentration of protein with a Bradford protein assay. Store protein at 4°C until use (*see* **Note 31**).

3.5. Muscle Bath Contraction Studies

These experiments are performed in a bubble chamber muscle bath system adapted from the apparatus used by the Somlyo lab at the University of Virginia. The muscles are hung on a hook attached to a SensorOne AE801 force transducer and to a fixed hook. The transducer is interfaced to a computer physiograph (AD instruments, Macintosh Powerlab, Chart software). The volume of the muscle bath is 140 µL.

1. Collect and pin rabbit ileum in the same manner as described in **Subheading 3.1., steps 1–5** (the tissue does not have to be trimmed to 0.5 × 1 cm).
2. Using a razor blade holder (WPI), break a piece of safety razor blade 5 mm long. Orienting the cuts so that longitudinal sections are obtained, cut a "ladder" in the muscle so that each strip is 250 µm wide and 5 mm long. Cut many more strips than are needed because muscles may be pressed or stretched during the tying process (**Fig. 4A,B**).
3. Cut a 1 cm piece of 2-0 surgical braided silk (Ethicon). Unweave the silk threads and remove one thread. Under a dissecting microscope, unweave this thread and remove one single fiber. These fibers are used to tie loops on each end of the cut muscle strips (**Fig. 4C**).
4. Still working under the dissecting scope, use no. 5 Dumont fine-point forceps (FST) and tie surgical knots and small loops on each end of the muscle. Tie the loops 1 mm from each end so that 3 mm of muscle strip are between the loops

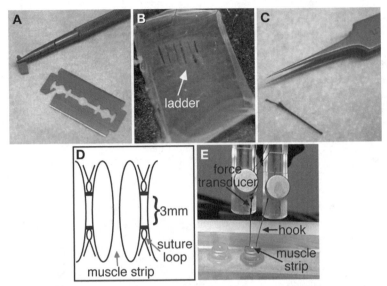

Fig. 4. Preparation of muscle strips for the bubble chambers. (**A**) A razor blade holder with a 5-mm piece of safety razor blade is used to create the ladder in the ileum. (**B**) 5 mm × 250 μ*M* strips are cut in a longitudinal orientation in pinned, stretched ileum smooth muscle. (**C**) A 1-cm piece of surgical braided silk is cut and a single strand removed. That single strand will be unwoven under a dissecting microscope to generate another single strand, which is used to tie loops on the muscle strips. (**D**) Schematic drawing of the muscle strips with loops tied at each end. (**E**) Set up of the "bubble" chamber.

(**Fig. 4D**). Leave long tags when the suture is cut after tying the loops so that they can be handled by these tags when hanging in the baths.

5. Cut away the muscle strips from the sheet of ileum using superfine spring action scissors (WPI), taking care to only handle the muscle on the ends outside the loops.
6. Add 140 μL of NES to the bubble chambers.
7. Working under a dissecting microscope, hang the one loop over the fixed end of the muscle bath. Hang the second loop over the hook that is glued to the force transducer. Allow the muscles to equilibrate in NES for 30 min (**Fig. 4E**).
8. Set the resting tension to 1.3 times the resting muscle length. Equilibrate in NES for 30 min. After equilibration, change the buffer to 140 μL KES and look for rapid contractile response, indicating that the muscles are viable. Move the muscles back to NES.
9. Permeabilize the muscles in pCa6.3 with 50 μ*M* β-escin and for 40 min. Successful permeabilization will be evident in pCa6.3 because the muscle will steadily contract as the membrane becomes permeable to calcium (*see* **Note 32**).

A   **B**

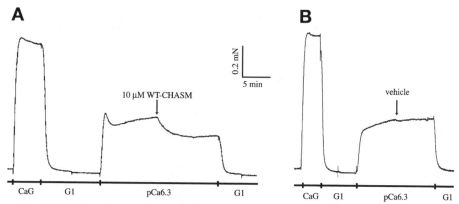

Fig. 5. Calcium desensitization of rabbit ileum by recombinant CHASM. β-Escin-permeabilized rabbit ileum smooth muscle strips were maximally contracted in CaG, washed in calcium-free solution (G1), contracted with submaximal calcium (pCa 6.3), and treated with either **(A)** 5 μ*M* recombinant CHASM or **(B)** vehicle at the plateau of contraction. (From **ref.** *11* with permission from Elsevier Science.)

10. All subsequent steps will require 1 μ*M* CaM in the buffers because the membrane has been permeabilized.
11. Add 10 μ*M* A23187 for 10 min. This empties the sarcoplasmic reticulum of calcium stores.
12. Wash three times for 5 min in G10.
13. Maximally contract the muscle in CaG for 5 min.
14. Wash three times for 5 min in G10. Change buffer to pCa6.3. After the calcium-induced contraction plateaus (usually about 10 m), add 5 μ*M* recombinant CHASM or equal volume of vehicle. Because calcium levels are kept constant, any relaxation effect that CHASM has is owing to an increase in the threshold for calcium induced contraction (calcium desensitization; *see* **Fig. 5**).
15. Wash in G10 three times for 5 min.
16. Maximally contract with CaG for 5 min.
17. Wash three times for 5 min in G10.

4. Notes

1. Do not pour the silicone layer so thick that the plate lids will not go on with dissection pins present. Make several extra plates so that the radioactive plates can be discarded or allowed to decay.
2. All solutions are to be prepared in water that has a resistance of 18.6 *M*Ω and a total organic content of less than 5 parts per billion. This standard is referred to as "water" in this text.
3. This pH step is very close to the edge of the titration curve so go very slowly.

4. A small amount of bromophenol blue can be added at this step so that the samples are light blue in color and easier to see to load.

5. The mixed bed exchanger removes cations and anions from the triton solution because ionic content will interfere with the first-dimensional gel.

6. Protein addition buffer ensures that the pH and the ionic strength of the muscle bath are not changed when recombinant proteins are added.

7. Stock solutions of compounds that will be added to the muscle baths are formulated such that the addition to the bath is a 1:100 dilution. This reduces the volume added so that effects on electrolyte and calcium concentrations in the bath is minimal.

8. Kreb's solution should be maintained at 37°C and bubbled continuously with 95% O_2/5% CO_2 at every step where "Kreb's" is mentioned.

9. We euthanize our rabbits by anesthetizing with halothane and then performing jugular exsanguination. Halothane diffuses quickly from tissues and has little effect on ex vivo muscle contractility.

10. Extreme care needs to be taken to only handle the ileum from one end, only flush with Kreb's solution, and to not stretch the ileum longitudinally. Overstretching will severely compromise the viability of the tissue in the contractility studies. Avoid overdistending with buffer as the fecal material is flushed out.

11. Again, take care not to put pressure on any portion of the ileum other than the original end that was grasped with the forceps for peeling and do not overstretch.

12. Tissues that wrap up tightly after peeling and placing back into Kreb's seems to perform better in the muscle studies, so they are good tissues to choose for pinning. Also avoid tissues that have overstretched portions that occurred during the peeling process.

13. To establish constant calcium concentrations, the smooth muscle must be permeabilized. To maintain most of the integrity of the membrane but allow Ca^{2+} ions and nucleotides to flow freely, use *S. aureus* α-toxin for permeabilization to produce small pores that are 1–1.5 nm. After permeabilization, muscles require ATP and Na_2CP in the buffers.

14. Unpinning the muscle before placing in liquid nitrogen is easier to perform, but does not maintain tension on the muscle, adding another variable to the experiment. Changing tension on the muscle can change the phosphorylation state of contractile proteins.

15. Freezing within the dish with freon maintains tension on the muscle up to the point of freezing, but the tissues have to be "popped" off of the silicone layer, which breaks them into difficult to handle pieces and damages the silicone.

16. 2D samples have to be lysed and run in buffers with no salt and no SDS because the proteins are being separated according to charge.

17. Stirring under vacuum helps polymerization later on.

18. Move the capillary tubes around with a needle to avoid bubble formation.

19. Use a p1000 disposable pipet tip to push the block of tubes and polymerized acrylamide out of casting chamber. Discard tubes with bubbles.

20. It is essential to make sure all wells are plugged so that acid and base do not mix at all or the pH gradient will not run properly.

21. To run the first dimension in a shorter period of time, run for 1 h at 200 V, and then 5 h at 750 V. Proteins run to their isoelectric point and stop. The current should be very low (0.2 mA).
22. The acid side of the tube gel will turn yellow in the 4X sample buffer, which aids in orientation.
23. Be consistent about which way the tube gel is laid across the stacking gel, e.g., base on the left, acid on the right. Many basic proteins will not enter the gel from a concentrated protein lysate so a "base line" will appear after silver staining, which also helps orient the gel.
24. Unincorporated ^{32}P will run in the dye front, so it either needs to be cut off or run off the gel into the buffer and disposed of as liquid waste with the running buffer before doing the autoradiograph.
25. The amount of time the gel stays on film depends on the amount of radioactivity in the gel. This can vary from 1 h to 1 wk.
26. Often there is "hot spot" on the film with no detectable silver-stained protein associated with it. These proteins are usually very low abundance and are rarely able to be sequenced.
27. Trypsin is activated by warming and putting in neutral or basic pH buffer. Keeping the trypsin and the gel on ice for 1 h allows the trypsin to enter the gel, but it does not have catalytic activity. The excess trypsin is then removed and the gel is placed at 37°C, thus activating the proteolytic activity. This procedure reduces the level of digested trypsin peptides that appear on MS and dampen the signal of the peptides of interest.
28. Use the same pipet tip for each individual sample for every elution step so that maximum recovery of peptides is obtained. Some peptides will be retained in the plastic tip.
29. The amount of protein produced from this protocol varies between proteins and between experiments. Five mL will be needed per liter of large culture. The number of liters required will have to be adjusted as needed.
30. This step can be adjusted with different concentrations of IPTG, different induction temperatures, and different incubation lengths depending on the protein.
31. For muscle bath studies, no more than 10 μL can be added to a single bath without changing the ion concentrations because the bath volume is only 140 μL. The typical final concentrations of recombinant protein in the baths are 5–15 μ*M*. The final concentration of the protein to be added will depend on the molecular weight of the protein so that no more than 10 μL is added to a single bath.
32. β-Escin permeabilization creates a larger pore size (>17,000 KD) than α-toxin, so calmodulin is permeable, but it does not uncouple membrane surface receptors, so normal receptor physiology is still intact. Larger recombinant proteins can also be introduced into the muscle using this type of permeabilization.

Acknowledgments

The authors would like to thank Elizabeth Snyder and Nicole Kwiek for assistance in figure and manuscript preparation and Everett McCook for technical assistance. This work was supported by PO1 HL19242-27.

References

1. Horowitz, A., Menice, C. B., Laporter, R., et al. (1996) Mechanisms of smooth muscle contraction. *Physiol. Rev.* **76,** 967–1003.
2. Woodrum, D. A. and Brophy, C. M. (2001) The paradox of smooth muscle physiology. *Mol. Cell Endocrinol.* **177,** 135–143.
3. Somlyo, A. P. and Somlyo, A. V. (2003) Ca2+ sensitivity of smooth muscle and nonmuscle myosin II: modulated by G proteins, kinases, and myosin phosphatase. *Physiol. Rev.* **83,** 1325–1358.
4. Feng, J., Ito, M., Ichikawa, K., et al. (1999) Inhibitory phosphorylation site for Rho-associated kinase on smooth muscle myosin phosphatase. *J. Biol. Chem.* **274,** 37,385–37,390.
5. MacDonald, J. A., Borman, M. A., Muranyi, A., et al. (2001) Identification of the endogenous smooth muscle myosin phosphatase-associated kinase. *Proc. Natl. Acad. Sci. USA* **98,** 2419–2424.
6. Borman, M. A., MacDonald, J. A., Muranyi, A., et al. (2002) Smooth muscle myosin phosphatase-associated kinase induces Ca2+ sensitization via myosin phosphatase inhibition. *J. Biol. Chem.* **277,** 23,441–23,446.
7. Endo, A., Surks, H. K., Mochizuki, S., et al. (2004) Identification and characterization of zipper-interacting protein kinase as the unique vascular smooth muscle myosin phosphatase-associated kinase. *J. Biol. Chem.* **279,** 42,055–42,061.
8. Lincoln, T. M., Dey, N., and Sellak, H. (2001) Invited review: cGMP-dependent protein kinase signaling mechanisms in smooth muscle: from the regulation of tone to gene expression. *J. Appl. Physiol.* **91,** 1421–1430.
9. Wooldridge, A. A., MacDonald, J. A., Erdodi, F., et al. (2004) Smooth muscle phosphatase is regulated in vivo by exclusion of phosphorylation of threonine 696 of MYPT1 by phosphorylation of Serine 695 in response to cyclic nucleotides. *J. Biol. Chem.* **279,** 34,496–34,504.
10. Pfeifer, A., Klatt, P., Massberg, S., et al. (1998) Defective smooth muscle regulation in cGMP kinase I-deficient mice. *EMBO J.* **17,** 3045–3051.
11. Borman, M. A., MacDonald, J. A., and Haystead, T. A. (2004) Modulation of smooth muscle contractility by CHASM, a novel member of the smoothelin family of proteins. *FEBS Lett.* **573,** 207–213.
12. Mackey, A. J., Haystead, T. A., and Pearson, W. R. (2002) Getting more from less: algorithms for rapid protein identification with multiple short peptide sequences. *Mol. Cell. Proteomics* **1,** 139–147.

20

Proteomic Analysis of Vascular Smooth Muscle Cells Treated With Ouabain

Alexey V. Pshezhetsky

Summary

Apoptosis of vascular smooth muscle cells (VSMC) plays an important role in remodeling the vessel walls, one of the major determinants of long-term blood pressure elevation and an independent risk factor for cardiovascular morbidity and mortality. Apoptosis in VSMC can be inhibited by inversion of the intracellular $[Na^+]/[K^+]$ ratio after the sustained blockage of the Na^+,K^+-ATPase by ouabain. Using two-dimensional gel electrophoresis followed by tandem mass spectroscopy, we compared proteomes of control VSMC and of those with ouabain-inhibited Na^+,K^+-ATPase and found that ouabain treatment led to overexpression of numerous soluble and membrane-bound proteins. Among proteins, which showed the highest level of ouabain-induced expression, we identified mortalin (also known as GRP75 or PBP-74), a member of the heat shock protein 70 superfamily and a marker for cellular mortal and immortal phenotypes. Further experiments showed that mortalin RNA and protein levels are induced in ouabain-treated VSMC, and that transient transfection of cells with mortalin cDNA inhibited serum deprivation-induced apoptosis via inactivation of the tumor suppressor gene, *p53*.

Key Words: Vascular smooth muscle cells; apoptosis; Na^+,K^+-ATPase; ion transport; p53; ouabain; proteomics; mass spectroscopy.

1. Introduction

Remodeling of the blood vessels plays an important role in a variety of human vascular disorders, including hypertension *(1–3)*, atherosclerosis, *(4)* arterial injury, and restenosis after angioplasty *(5–7)*. Apoptosis (programmed cell death) of vascular smooth muscle cells (VSMC) has recently been identified as the main factor contributing to the regulation of their number during remodeling *(8–12)*, which inspired numerous studies of the mechanisms of the induction and progression of VSMC apoptosis. As in the other cell types, the execution phase of apoptosis in VSMC is triggered by activation of the caspase cascade

From: *Methods in Molecular Biology, vol. 357: Cardiovascular Proteomics: Methods and Protocols*
Edited by: F. Vivanco © Humana Press Inc., Totowa, NJ

leading to cleavage of intracellular proteins and final disintegration of the cell. In contrast, the induction phase is specific for different subtypes of remodeling and involves the integration of multiple pro- and anti-apoptotic signals, including the expression of death receptors, proto-oncogenes, and tumor suppressor genes *(13–19)*.

In VSMC, the inhibition of Na^+,K^+-pump with ouabain or in K^+-free medium blocks apoptosis triggered by a number of factors, including serum deprivation *(20)*. The effect is specific for VSMC and is caused by the inversion of $[Na^+]_i/[K^+]_i$ ratio *(20–24)*. We have analyzed the VSMC proteome using a combination of two-dimensional gel electrophoresis (2DE) and tandem mass spectrometry (MS/MS) to identify proteins induced after the inhibition of Na^+,K^+-pump. Among identified gene products, a heat shock protein (HSP)70 family protein, mortalin, was shown to inhibit VSMC apoptosis via inactivation of pro-apoptotic tumor suppressor gene, p53 *(25)*.

2DE MS/MS still remains a most broadly used proteomic technique owing to its ability to display both changes in abundance and posttranslational modification of proteins. However, at present it is recognized that this method has significant limitations. First, the combination of limited sample capacity and low detection sensitivity of 2DE makes the detection of low-abundance proteins impossible *(26)*. Because proteins expressed at a low level may represent a large portion of a given proteome, it is apparent that 2DE does not provide a true representation of all expressed proteins. Second, 2DE cannot be applied to separate transmembrane proteins, proteins with high or low isoelectric point (p*I*) or with high *Mr*. Third, the majority of spots on 2D gels contains more than one protein and differentially modified or processed forms of a protein migrate to different positions in the gel, thus complicating quantification. Therefore, in addition to the 2DE MS/MS method provided in this chapter, a description of a complementary technique where cell proteins are separated on one-dimensional sodium dodecyl sulfate polyacrylamide gel electrophoresis (SDS-PAGE), identified by MS/MS and quantified by comparing the number and the intensity of the peptide spectra.

2. Materials

2.1. Cell Culturing and Fractionation

1. VSMC can be obtained from the aorta of 10–13-wk-old male Brown Norway rats (BN.lx) as described *(27)*. Primary cells can be immortalized and sensitized to serum deprivation-induced apoptosis through stable transfection with adenoviral protein E1A *(28)*.
2. Dulbecco's modified Eagle's medium (DMEM) with 10% calf serum (CS), 100 U/mL penicillin and 100 µg/mL streptomycin (all from Invitrogen, Burlington, ON).

E1A-transfected cells (VSMC–E1A) are cultured in the same medium with the addition of 500 μg/mL Geneticin (Sigma, Oakville, ON).

3. Minimum essential medium (MEM) supplemented with L-glutamine and 0.2% CS but without L-cysteine, L-methionine, and cystine (all from MP Biomedicals, Irvine, CA).

4. A mixture of ^{35}S-cysteine and ^{35}S-methionine (Trans ^{35}S-label, MP Biomedicals) is dissolved in MEM to a concentration of 50 μCi/mL.

5. Ouabain (Sigma) is dissolved in water (*see* **Note 1**) at a final concentration of 10 m*M* and stored in aliquots at −80°C.

6. Buffers for sequential extraction of cell proteins (3-fraction method). Buffer 1: 2.5 m*M* aminopropanol (98% pure, Sigma), pH 9.5 containing a complete protease inhibitor cocktail (Roche Diagnostics, Laval, QC). Buffer 2: 40 m*M* Tris base (99.9% pure, Sigma), pH 9.5, containing 8 *M* urea (electrophoresis-grade; Bio-Rad, Hercules, CA), 4% (w/v) CHAPS (Sigma), and 2 m*M* tributylphosphine (TBP; Aldrich Chemical Company, Milwaukee, WI), (*see* **Note 2** for the details of preparation). Buffer 3: 40 m*M* Tris base, pH 9.5 containing 5 *M* urea, 2 *M* thiourea (MP Biochemicals), 2% (w/v) CHAPS, 3% (w/v) sulfobetain (SB) Z 310 detergent (Sigma), and 2 m*M* TBP.

7. Homogenization buffer for subcellular fractionation (4-fraction method): 0.25 m*M* sucrose, 10 m*M* triethanolamine (TEA), 10 m*M* acetic acid, 25 m*M* KCl, 1 m*M* EDTA (both Sigma), and a complete protease inhibitor cocktail (Roche Diagnostics). Adjust pH to 7.8 with acetic acid or TEA.

2.2. Two-Dimensional Polyacrylamide Gel Electrophoresis

1. Sample buffer for 2D: 8 m*M* urea, 20 m*M* TBP, 4% (w/v) CHAPS, 2% (v/v) carrier ampholites (3.0–10.0 biolytes, Bio-Rad) (*see* **Note 3**), 40 m*M* Tris base, 50 m*M* thiourea, and 0.001% Orange G dye.

2. Immobilized pH gradient (IPG) strips (18 or 17 cm, pH range 3.0–10.0) are from Bio-Rad (17 cm) or Amersham Biosciences (18 cm).

3. Solutions for equilibrating IPG strips before the second dimension. Solution 1: 100 m*M* Tris-HCl, pH 6.8, 6 *M* urea, 30% (v/v) glycerol, 4% (w/v) SDS, and 2% (w/v) dithiothreitol (DTT). Solution 2: 100 m*M* Tris-HCl, pH 6.8, 6 *M* urea, 30% (v/v) glycerol, 4% (w/v) sodium dodecyl sulfate (SDS), and 2.5% (w/v) iodoacetamide.

4. Gel buffer for second dimension: 1.5 *M* Tris-HCl, pH 8.8.

5. Acrylamide stock (30.8%T/0.8%C): 30% (w/v) acrylamide, 0.8% (w/v) piperazine diacrylamide (*see* **Note 4** for the details of preparation).

6. 10% (w/v) ammonium persulfate solution (APS) in water; prepare fresh each time.

7. 10% (v/v) TEMED; prepare fresh each time.

8. 100 m*M* sodium thiosulfate in water; prepare fresh.

9. Displacing solution: mix 50 mL of 1.5 *M* Tris-HCl buffer, pH 8.8, 100 mL of glycerol, 2 mg of bromphenol blue, and 50 mL of water. Prepare fresh; stored solution may develop bacterial growth.

10. Water-saturated *N*-butanol: mix 40 mL of *N*-butanol and 4 mL of water and shake; use the upper phase. Can be stored at room temperature indefinitely.

11. Running buffer: mix 60.5 g of Tris base (final concentration 25 m*M*), 288 g of glycine (final concentration 192 m*M*), and 20 g of SDS (final concentration 0.1% [w/v]). Add water to 20 L; do not adjust pH.

12. Agarose solution: 1% (w/v) agarose, 0.1% (w/v) SDS, 0.25 m*M* Tris-HCl, pH 6.8. Prepare 10 mL, add 1 mg of Bromophenol blue.

13. Coomassie staining solutions. Staining: 2 g of Comassie R250, 500 mL of methanol, 100 mL of acetic acid, 400 mL of water. Solution has to be filtered after preparation. Destaining: 500 mL of methanol, 100 mL of acetic acid, 400 mL of water.

2.3. SDS-Polyacrylamide Gel Electrophoresis

1. Sample buffer (4X SDS reducing buffer): mix 3 mL of water, 1 mL of 0.5 *M* Tris-HCl, pH 6.8, 1.6 mL of glycerol, 1.6 mL of 10% (w/v) SDS solution in water, 0.4 mL of β-mercaptoethanol, and 0.4 mL of 0.5% (w/v) bromophenol blue solution in water.

2. Separating buffer (4X): 1.5 *M* Tris-HCl, pH 8.8, 0.4% (w/v) SDS.

3. Stacking buffer (4X): 0.5 *M* Tris-HCl, pH 6.8, 0.4% (w/v) SDS.

4. Acrylamide stock (30%T/1%C): 29% (w/v) acrylamide, 1% (w/v) bisacrylamide.

5. 10% (w/v) APS in water; prepare fresh.

6. Running buffer (5X): 125 m*M* Tris base, 960 m*M* glycine, 0.5% (w/v) SDS. Do not adjust pH.

2.4. Protein Identification

1. Gel destaining solutions: 200 m*M* ammonium bicarbonate, pH 8.0; 50% (v/v) methanol in 10% (v/v) acetic acid; 40% (v/v) ethanol. All solvents should be high-performance liquid chromatography (HPLC)-grade.

2. Trypsin solution: 25 m*M* ammonium bicarbonate buffer, pH 8.0, containing 0.01 μg/μL of trypsin (sequencing-grade, Promega). Solution is prepared fresh each time.

3. Solutions for extraction and HPLC of peptides: 50% (v/v) solution of methanol in 5% (v/v) acetic acid; 10% and 90% (v/v) solutions of acetonitrile (ACN) in 0.1% (v/v) formic acid. All solvents and water should be HPLC-grade.

3. Methods

Major obstacle for the comparative proteomics is the high dynamic range of cellular proteins (10^7 and more), which makes the detection of low-abundance proteins challenging even after loading milligram quantities of samples on one-dimensional and 2D gels *(26)*. Therefore a pre-fractionation of the sample in order to concentrate on a subproteome of interest becomes an important step to obtain reliable results. With fresh tissues (especially soft tissues such as liver) pre-fractionation is easily achieved through purification of individual organelles; however, low yields normally do not allow applying this technique to cultured cells. Methodically simple but still effective pre-fractionation can be achieved by sequential extraction of VMSC proteins first with high-pH buffer solution to remove soluble proteins (~40–50% of total cell protein) and then with CHAPS detergent and urea to solubilize membrane proteins (~40% of protein) *(29)*. The remaining pellet (~20% of protein) is dissolved in a buffer containing SB Z 310 and thiourea. More complicated but still applicable to cultured cells (10^6 cells or more are required) is a combination of lysis and centrifugation to separate cells into four fractions: nuclear fraction, particulate, cytosol, and combined membranes. For better reproducibility, the fractionation of the control and ouabain-treated samples should be performed simultaneously rather then on different occasions.

For comparison of proteomes, the proteins from subcellular fractions are further separated by either 2D or by SDS-polyacrilamide gel electrophoresis (PAGE), quantified and identified by MS/MS. 2DE approach provides an advantage of separating protein isoforms (such as proteins having different posttranslational modifications), as well as of direct quantification of proteins on the stained gels by computer-based image analysis. Transmembrane hydrophobic proteins are better detected after separation by SDS-PAGE, but in this case additional techniques such as labeling with stable isotopes *(30)* or counting the number of peptides detected by mass spectrometer for each particular protein *(31)* must be used.

3.1. Culturing of Cells and Induction of Apoptosis

1. VSMC or VSMC-E1A are cultured in DMEM with 10% CS and antibiotics and passaged when reaching confluence with trypsin/EDTA. If 3×10^6 of wild-type cells and 0.75×10^6 of E1A-transfected cells are seeded in T-150 culture flasks, the experimental cultures will reach confluence in 48–72 h. The primary cells are normally used between the 10th and 16th passages. At 80% confluence, the primary cells are starved for 24 h in DMEM supplemented with 0.2% CS.
2. For the induction of apoptosis by starvation, the cells are washed at 100% confluence with DMEM and incubated in DMEM for 24 h (primary cells) or 6–12 h (E1A-transfected cells). During this time 6–10% of primary cells and 30–50% of E1A-transected cells usually develop apoptosis.

3.2. Ouabain Treatment

1. Ouabain is added to the culture medium in a final concentration of 1 m*M* and the cells are incubated for a period of 1–16 h. Fifteen million cells (one confluent T-150 flask) are required for each experimental data point.
2. The medium is removed by aspiration, each flask is placed on ice, and the cells are washed twice by ice-cold phosphate buffered saline (PBS).
3. Ten mL of ice-cold PBS is added to each flask and the cells are removed by gentle scraping with a polyethylene cell scraper (Nalge Nunc International, Rochester, NY). The cell suspension is transferred into a conical 15-mL polyethylene tube and centrifuged for 10 min at 1000*g* and 4°C. At this point the cell pellet can be frozen if necessary and kept at −80°C until the time of analysis.

3.3. Metabolic Labeling of the Cells (Optional)

1. The cells are pre-incubated for 2 h in a MEM supplemented with L-glutamine and 0.2% CS but without L-cysteine, L-methionine, and cystine.
2. A mixture of ^{35}S-cysteine and ^{35}S-methionine is added in a concentration of 50 μCi/mL of medium and the cells are incubated for 3–6 h with or without 1 mM ouabain.
3. The medium is removed by aspiration; cells are placed on ice, washed twice with ice-cold PBS, and harvested as described in **Subheading 3.2.3.**

3.4. Cell Fractionation (Three-Fraction Method)

Frozen cell pellet is mixed with 1 mL of 2.5 mM aminopropanol buffer, pH 9.5, containing a complete protease inhibitor cocktail, vortexed for 30 s, and sonicated three times for 5 s at 60 W. During the sonication, the tube is chilled with a plastic bag containing ice. The homogenate is centrifuged for 1 h at 100,000g. The supernatant is quickly frozen in liquid nitrogen, lyophilized, re-dissolved in 0.5 mL of the 2D sample buffer, and kept at room temperature (extract 1). The pellet (pellet 1) is extracted with a 0.5 mL of 40 mM Tris base buffer, pH 9.5, containing 8 M urea, 4% CHAPS, and 2 mM TBP and centrifuged for 1 h at 100,000g. The supernatant (extract 2) is kept at room temperature and the pellet (pellet 2) is dissolved in 0.5 mL of 40 mM Tris base buffer, pH 9.5, containing 5 M urea, 2 M thiourea, 2% CHAPS, 3% SB Z 310, and 2 mM TBP (extract 3). Before 2D-PAGE the extract 2 and the extract 3 are supplemented with 10 μL of concentrated biolytes 3–10 (final concentration 2% [v/v]) and 5 μL of 0.1% Orange G dye (final concentration 0.001% w/v). The extract 2 is also supplemented with 5 μL of 5 M thiourea (final concentration 50 mM).

3.5. Cell Fractionation (Four-Fraction Method)

1. All steps have to be done on ice. Mix cell pellet from 10^6–10^8 cells with 1 mL of homogenization buffer and homogenize using 25 strokes in a Potter glass-glass tissue grinder.
2. Transfer the homogenate into a 1.5-mL Eppendorf tube. Centrifuge at 1000g for 10 min. Transfer the supernatant into a new Eppendorf tube and keep on ice.
3. Re-suspend the pellet in 0.5 mL of the homogenization buffer, transfer to a tissue grinder, homogenize with additional 10 strokes, and centrifuge at 1000g for 10 min. The pellet represents the nuclear fraction (~35% of total cell protein).
4. Combine the first and the second supernatants and centrifuge at 17,000g for 15 min. The pellet (~10% of total protein) represents the particulate fraction.
5. The supernatant is centrifuged at 100,000g for 1 h. The pellet (~20% of total protein) represents the combined membrane fraction, and the supernatant (~30% of total protein), the cytosol fraction.
6. Quality of separation can be tested by assaying the DNA content and activities of marker enzymes in the obtained fractions (*see* **Note 5**).

3.6. 2D (First Dimension)

1. The instructions are provided for the use of either a Multiphor II (Amersham Pharmacia Biotech) or a Protean isoelectric focusing (IEF) cell (Bio-Rad) for the separation in the first dimension (isoelectric focusing) and an Ettan Dalt system (Amersham Pharmacia Biotech) for the separation in the second dimension; however, they can be adopted to other formats.

2. From 0.5–2 mg of protein dissolved in 500 µL of the sample, 2D buffer can be loaded on each 18-cm IPG strip. It is crucial that the same amount of protein from the control and ouabain-treated cells is loaded on the gel. Therefore the protein concentration is measured using RC DC Protein assay kit (Bio-Rad) and samples are diluted to equal concentration with 2D sample buffer. The IPG strips are individually rehydrated in the sample solutions. As the strips rehydrate, the proteins from the samples are absorbed and distributed over the entire length of the strip (*see* **Note 6**). If a Protean IEF cell is used, rehydration can be performed directly in the IEF tray. In this case, IPG strips are placed gel-side down into the grooves of the tray filled with the sample and re-hydration is performed overnight at the constant voltage of 50 V. Alternatively, each IPG trip is placed in a 2-mL plastic pipet cut to the correct length, the protein solution is introduced using an autopipet, and the ends of the plastic pipet are sealed with Para-Film. Ensure that the solution is dispensed on the gel side of the IPG. The tube is left on a flat surface allowing the rehydration to proceed. If necessary the pipet is inverted a number of times to redistribute the remaining liquid. After 24 h, the IPG is completely rehydrated and no sample solution should be left outside the strip.

3. The rehydrated IPGs are placed in the grooves of the tray on the cooling plate (*see* **Note 7**) and covered with a light mineral oil. IEF is performed at 20°C. (If Multiphor II is used, the thermostatic circulator should be turned on 15 min before starting the experiment.)

4. The voltage is linearly increased from 300 to 3500 V for 15 h, followed by 30–50 h at 3500 V (Multiphor II) or 30 h at 7500 V (Protean cell) up to a total of 200-300 kVh. If necessary, after this step strips can be stored at −80°C (*see* **Note 8**).

5. After separation in the first dimension, the strips must be equilibrated in a solution containing 100 mM Tris-HCl, pH 6.8, 6 M urea, 30% (v/v) glycerol, 4% (w/v) SDS, and 2% (w/v) DTT for 12–15 min at 80°C. Then the strips are equilibrated for another 5 min in the same solution containing 2.5% (w/v) iodoacetamide instead of DTT and a trace of bromophenol blue.

3.7. 2D (Second Dimension)

In the second dimension, a vertical gradient (8–16% of acrylamide) slab gels and the discontinuous Laemmli system must be used (*see* **Note 9**). Gels should be poured in a multigel casting chamber to ensure that they are uniform in composition. This is especially important for gradient gels. The casting procedure varies depending on the type of apparatus, so we recommend following the manufacturer's instructions. However, the following general rules have to be respected.

1. Measure the volume of the casting chamber when it is fully loaded with gel plates to calculate exactly the volumes of lower limit solution and higher limit solution to be prepared. For Ettan Dalt system, the formulation in **Table 1** for twelve 1-mm thick gels has to be used.

% T	1.5 M Tris-HCl, pH 8.8 (mL)	Acrylamide solution (mL)	Water (mL)	Glycerol
8	100	107	188	0
16	100	213	57	27

Mix the solutions thoroughly and then remove the dissolved air from the solutions under vacuum (place the gel solutions into the vacuum dessicator until both solutions stop bubbling).

2. Prior to assembling the casting chamber, the glass plates should be carefully cleaned with a detergent, wiped with a methanol-soaked tissue, and dried. The entire gel-casting system should also be clean and dry.

3. Add the initiators to both heavy and light solutions. For 8% gel solution: add 4 mL of 10% APS, 0.86 mL TEMED, and 80 μL of 100 mM sodium thiosulphate. For 16% gel solution: add 2 mL of 10% APS, 0.14 mL TEMED, and 80 μL of 100 mM sodium thiosulphate (*see* **Note 10**).

4. Check that both gradient maker valves are closed. Pour the light solution into the mixing chamber and the heavy solution into the reservoir chamber. Start the magnetic stirrer in the mixing chamber, open the outlet valve on the gradient maker, and turn on the pump. Allow the tubing to fill with the gel solution and then open the valve between the chambers.

5. Allow the gel solution to fill the casting chamber until there is a gap of 4 cm at the top of the glass plates. This should require nearly all the solution from the gradient maker. Stop the pump to prevent air bubbles from entering the casting chamber. Shut the clamp on the inlet hose of the casting chamber. Disconnect the pump tubing from the casting chamber inlet, flush the gradient maker and pump tubing with water.

6. Overlay the gel solution in each cassette with 0.75 mL of water-saturated N-butanol. Cover the casting unit with a plastic wrap and allow the gels to polymerize overnight. Do not move the unit until polymerization is complete. Then remove the N-butanol overlay and replace it with 1.5 M Tris-HCl, pH 8.8. Gels can be stored at 4°C for up to 2 d.

7. At least 2 h before the run fill the tank with SDS running buffer, connect circulating thermostat and set it at 10°C.

8. Place the IPG strips on the top of the gels.

9. Apply 10 μL of Mr markers solution in SDS-PAGE Laemmli buffer to a piece of filter paper (approx 4 × 4 mm) and place it on the acidic side of the strip. Pre-stained markers will allow monitoring of the progress of the electrophoresis run.

10. Melt agarose in a microwave and pour it on the strip.

11. Place the gels into the Ettan Dalt multicell, add running buffer to the upper chambers, and run gels at a constant current of 5 mA/gel for 2 h and then at 15 mA/gel overnight.

3.8. SDS-PAGE

1. To achieve better detection of hydrophobic and low-abundance proteins SDS-PAGE should be used instead of 2D. In this case cellular fractions (75–100 µg of total protein in a total volume of 50 µL) are mixed with an equal volume of 2X Laemmli sample buffer and incubated in a boiling water for 5 min. Because multiple samples (up to 20) can be analyzed on the same gel, there is no need to use multigel running chambers. Therefore we provide protocols for a Protean IIxi system (Bio-Rad) containing 2 gels of 1.5 mm thickness.

2. As in the case of 2D, it is critical that the glass plates for the gels, spacers, and combs are washed with a detergent, rinsed with distilled water, and dried.

3. Prepare a 1.5-mm thick, 10% separating gel by mixing 12.5 mL of 4X separating buffer with 17 mL acrylamide/*bis* solution, 20.5 mL distilled water, 250 µL 10% (v/w) APS, and 25 µL TEMED. Pour the gel, leaving space for a stacking gel, and overlay with water-saturated n-butanol. The gel should polymerize in about 30 min.

4. Remove the *N*-butanol solution.

5. Prepare 4% stacking gel by mixing 5 mL of 4X stacking buffer with 2.7 mL acrylamide/*bis* solution, 12 mL water, 100 µL 10% (v/w) APS, and 20 µL TEMED. Pour the gel mix and insert the comb. The stacking gel should polymerize within 30 min.

6. Prepare the running buffer by diluting 200 mL of the 5X running buffer with 800 mL of water.

7. Once the stacking gel has set, carefully remove the comb and assemble the gel in the electrophoresis unit. Add the running buffer to the upper and lower chambers of the unit and load 100 µL of each sample (between 100 and 200 µg of protein) into each well. Use one well for molecular-weight markers.

8. Complete the assembly of the gel unit and connect to a power supply. The gel can be run overnight at 15 mA/gel and 10°C.

9. Gradient gels for SDS-PAGE can be prepared using the instructions to prepare gradient second-dimension gels for 2D (*see* **Subheading 3.7.** for details) and the formulation in **Table 2** for one Protean II 1.5-mm-thick gel:

% T	1.5 M Tris-HCl, pH 8.8 (mL)	Acrylamide solution (mL)	Water (mL)	Glycerol (mL)	APS 10% (mL)	TEMED (mL)
8	17.5	18.7	33.5	0	0.5	0.05
16	17.5	37.3	7.4	7.4	0.5	0.05

3.9. Protein Detection and Image Analysis

1. Gels containing 0.5–2 mg of protein should be stained with Coomassie Brilliant Blue. The gel is stained with Coomassie staining solution for 30 min and then destained

until the bands are visible and the background is nearly clear. If less protein is applied to the gel more sensitive stains should be used such as: silver stain compatible with mass spectrometric protein analysis *(32)* or Sypro Ruby (Molecular Probes).

2. Gels containing ^{35}S-labeled proteins are detected by autoradiography using a PhosphorImager (Molecular Dynamics) or similar equipment.

3. To quantify and compare proteomes digitized gel images are analyzed using imaging software such as PDQuest (Bio-Rad) or Image Master 2D (Amersham Biosciences). In all programs, algorithms are applied to remove background noise, gel artifacts, and horizontal or vertical streaking from the image. Then programs detect protein spot positions and boundaries, quantify spot intensities, assign the reference numbers to each spot, and calculate isoelectric points and apparent molecular mass values. To create the reference map a set of gels (up to 100 images) of the similar samples are aligned and automatically matched using several "landmark" spots. Then one reference gel (in the case of PDQuest, it is a synthetic gel produced by merging the spot information from several gels in a set) is selected. The intensities of protein spots in the series of gels are compared with the matching spots on the reference gels and the protein spots subjected to up- or downregulation by ouabain are marked and excised from gels (*see* **Fig. 1**).

4. The protein band patterns on one-dimensional SDS gels are normally not analyzed. Instead the entire lanes are cut into 50–80 gel slices approx 2–3 mm long and 5 mm wide; it is important that the bands containing ouabain-treated and control samples are cut in the same manner (*see* **Fig. 2A**).

3.10. Protein Characterization and Identification

1. Gel pieces containing protein spots of interest are placed in 1.5-mL Eppendorf tubes and destained with sequential washes (at least 1 h each) using 500 µL of 200 m*M* ammonium bicarbonate, 500 µL of 50% (v/v) methanol in 10% (v/v) acetic acid, and 500 µL of 40% ethanol (v/v). Gels stained with silver are destained by washing them in a 5 m*M* solution of potassium ferricyanide and 5 m*M* solution of sodium thiosulfate.

2. Gel pieces are then rinsed three times with water, soaked in ACN for 30 min and reswelled with 10–20 µL of trypsin (Promega) solution (0.01 µg/µL in 50 m*M* ammonium bicarbonate) prior to overnight incubation at 37°C. Gel pieces have to stay wet during the digest.

3. Digestion solutions containing peptide fragments are put aside and the gel pieces are extracted two to three times with 30–50 µL of 60% (v/v) ACN in 0.1% (v/v) formic acid. All extracts and digestion solutions are combined and evaporated

Fig. 1. *(Opposite page)* Composite 2DE maps of soluble and membrane vascular smooth muscle cells (VSMC) proteins showing changes in protein expression after the ouabain treatment. VSMC treated with 1 m*M* ouabain for 3 h were extracted with **(A)** buffer and then with **(B)** CHAPS detergent. Protein extracts were analyzed by two-dimensional polyacrylamide gel electrophoresis (electrofocusing in the pH range 3.0–10.0

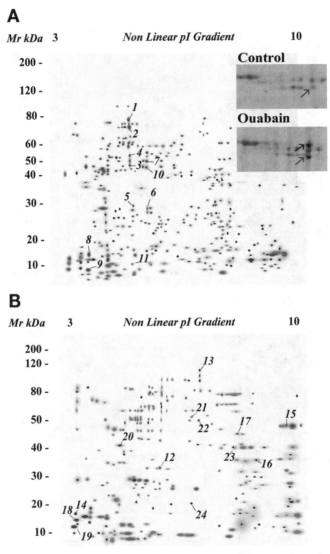

followed by sodium dodecyl sulfate polyacrylamide gel electrophoresis on gradient 8–18% gels); the gels were stained with Coomassie blue and compared using PDQuest software. Highlighted are protein spots that were changed twofold and more on all gels. Red spots represent proteins induced by ouabain; blue spots, proteins suppressed by ouabain; green spots, proteins expressed at the same level in control and ouabain-treated cells. Numbers indicate protein spots identified by tandem mass spectrometry. Inset: enlarged fragments of two-dimensional gels showing the induction of mortalin (red arrow, spot 2 on the composite gel) and GRP78 (blue arrow, spot 1 on the composite gel) in ouabain-treated VSMC. From **ref. 25** with permission. (*See* ebook for color version of this figure.)

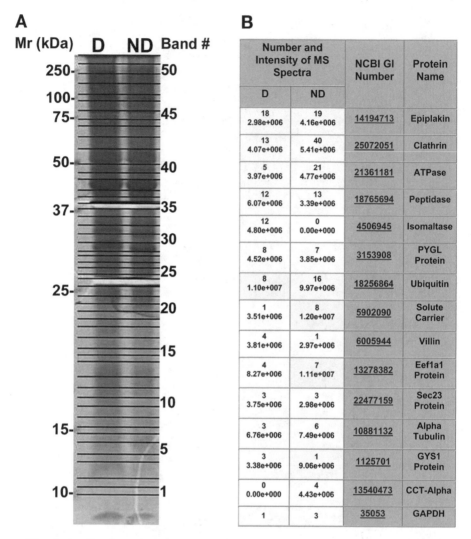

Fig. 2. Application of sodium dodecyl sulfate polyacrylamide gel electrophoresis tandem mass spectrometry method for the comparative analysis of cellular proteome. Plasma membranes of the differentiated (D) and nondifferentiated (ND) Caco 2 cells were solubilized in reducing Laemmi sample buffer, and alkylated with iodacetamide. **(A)** 200 µg of total protein extracts were analyzed by SDS-PAGE on gradient 8–18% gel. Gel was stained with Coomassie blue and entire lanes were cut into 51 gel bands. Position of *Mr* markers are shown on the left and numbers of the gel bands, on the right side of the gel. **(B)** Example of the data output table for the gel band 35, showing the NCBI gene identifications (GI) numbers, and names of the identified proteins as well as the total number and mean intensity of MS spectra of the tryptic peptides detected for each protein in the samples from the differentiated (D) and nondifferentiated (ND) cells.

using a Speedvac concentrator. Dry pellets are dissolved in 30 µL of 2% (v/v) ACN in 0.1% (v/v) formic acid.

4. Conditions for LC-MS/MS analysis of tryptic peptides are selected depending on the type of the instrument used for the analysis. In our laboratory, all experiments were conducted using a system consisting of a nanoflow liquid chromatographer and an ion trap 1100 series LC MSD mass spectrometer (NanoFlow Proteomics Solution, Agilent Technologies, Santa Clara, CA). Peptides are typically separated by reversed-phase HPLC on a Zorbax 300SB-C18 column (Agilent) with a gradient of 3–90% (v/v) ACN in 0.1% (v/v) formic acid at a flow rate of 300 nL/min. The column eluent is sprayed directly into the orthogonal nanospray source mounted on the mass spectrometer. Liquid chromatography-MS acquisitions are obtained with alternating survey MS scans (1 s/scan) and MS/MS precursor selection (2 MS/MS of 3 s each per MS/MS scan). Mass spectral resolution (50% full width at half maximum definition) is typically 0.35 u. Collision activation is performed using helium as a collision gas. Data are automatically acquired and processed using a Spectrum Mill software (Agilent Spectra are searched against NCBInr database (NCBI, Bethesda, MD).

5. Data generated by the Spectrum Mill consist of information related to protein identity (GI number) and information that can be used to estimate protein quantity in the sample (number of peptide spectra detected for the particular protein; average intensity of these sprectra). An example of data output table demonstrating the upregulation for a number of cell proteins is shown on **Fig. 2B**.

4. Notes

1. All solutions are prepared using a Milli Q water or HPLC-grade water when indicated.

2. Preparation of 200 mM TBP stock. Mix 1 mL of TBP and add 19 mL of isopropanol. All the procedures for preparing TBP solutions should be done under a chemical hood. Flush all bottles containing TBP with oxygen-free nitrogen. Remove spills with a wet paper towel. Store the TBP stock in the dark at 4°C for less than 2 wk.

3. The concentration of carrier ampholines in the 2D sample buffer may vary depending on the type of apparatus used for IEF. We use 2% (v/v) with Multiphor and 0.5% (v/v) with Protean IEF cell.

4. Use only high-grade acrylamide such as electrophoresis grade from Bio-Rad. Monomeric acrylamide is toxic; wear mask. Add 5 g of Bio-Rad deionizing resin (AG 501 X8 [D]) for every 2 L of solution and stir for 10 min. If the resin de-colorizes, add additional 5 g and repeat until resin is not de-colorized. Filter the solution through Whatman No. 1 paper using a Buchner funnel. The acrylamide stock should be stored in a fridge for less than 2 wk.

5. Quality of cell fractionation can be assayed using the following markers: total DNA for nuclear fraction *(33)*; glutamate dehydrogenase *(34)*, LAMP-2 (Western blot), β-hexosaminidase, and β-galactosidase for particulate *(35)*; GPDH for cytosol *(36)*; Na$^+$,K$^+$-ATPase (Western blot), and carboxylesterase *(37)* for membrane fraction. Inexpensive monoclonal antibodies against human LAMP-2 and Na$^+$,K$^+$-

ATPase are available from Developmental Studies Hybridoma Bank (University of Iowa, Iowa City, IA).

6. The rehydration method allows more complete application of protein then the cup loading because the sample does not precipitate. In addition, larger amounts (up to 3 mg) of sample can be loaded using this method.

7. Ensure that the IPG tray, strip aligner sheet, sample cup bar (if required), and electrodes are clean and dry prior to assembling the apparatus. The cleaning should be done in a solution of laboratory detergent. This removes mineral oil and protein from the previous IEF runs. After cleaning the tray, strip aligner, sample cup bar, and electrode should be rinsed thoroughly in Milli Q water and dried.

8. Because the pH gradient is fixed in the gel, the focused proteins are stable at their p*I*, allowing IPGs to be stored at −80°C indefinitely without having a detrimental effect on the final 2D pattern. The IPGs are bound to a plastic sheet, so gel cracking resulting from expansion and contraction, associated with freezing and thawing, is avoided and the IPGs retain their original dimensions after thawing. It is convenient to store the IPGs in plastic tubes with screw caps, which can then be used for the equilibration step.

9. Gradient gels (with increasing %T and usually constant %C) have two advantages over homogeneous gels with constant %T and %C: they allow proteins with a wide range of molecular weights to be analyzed simultaneously and the decreasing pore size sharpens the bands, improving resolution. When separating components in a narrow molecular weight range, homogeneous gels generally give better separation. However, if selected %T is too low, there will be insufficient retardation, whereas, if it is too high, the molecules will not penetrate the gel sufficiently. Typically gradient 8–16% gels are better for crude samples, such as cell lysates.

10. Addition of sodium thiosulphate helps to reduce background when using silver stain.

Acknowledgment

The author thanks Dr. Sebastien Taurin, Mrs. Karine Landry, Mrs. Linda Cote, and Dr. Mila Ashmarina for helpful advice and Mrs. Carmen Movila for help in preparing the manuscript. This work was supported in part by operating grants from Canadian Institutes of Health Research (FRN 15079 and MT-38107), and by an equipment grant from Canadian Foundation for Innovation.

References

1. Folkow, B. (1982) Physiological aspects of primary hypertension. *Physiol. Rev.* **62,** 347–504.
2. Heagerty, A. M., Aalkjaer, C., Bund, S. J., Korsgaard, N., and Mulvany, M. J. (1993) Small artery structure in hypertension. Dual processes of remodeling and growth. *Hypertension* **21,** 391–397.
3. Mulvany, M. J., Baandrup, U., and Gundersen, H. J. (1985) Evidence for hyperplasia in mesenteric resistance vessels of spontaneously hypertensive rats using a three-dimensional disector. *Circ. Res.* **57,** 794–800.

4. Glagov, S., Weisenberg, E., Zarins, C. K., Stankunavicius, R., and Kolettis, G. J. (1987) Compensatory enlargement of human atherosclerotic coronary arteries. *N. Engl. J. Med.* **316,** 1371–1375.
5. Clarkson, T. B., Prichard, R. W., Morgan, T. M., Petrick, G. S., and Klein, K. P. (1994) Remodeling of coronary arteries in human and nonhuman primates. *JAMA* **271,** 289–294.
6. Post, M. J., Borst, C., and Kuntz, R. E. (1994) The relative importance of arterial remodeling compared with intimal hyperplasia in lumen renarrowing after balloon angioplasty. A study in the normal rabbit and the hypercholesterolemic Yucatan micropig. *Circulation* **89,** 2816–2821.
7. Geary, R. L., Williams, J. K., Golden, D., Brown, D. G., Benjamin, M. E., and Adams, M. R. (1996) Time course of cellular proliferation, intimal hyperplasia, and remodeling following angioplasty in monkeys with established atherosclerosis. A nonhuman primate model of restenosis. *Arterioscler. Thromb. Vasc. Biol.* **16,** 34–43.
8. Cho, A., Courtman, D. W., and Langille, B. L. (1995) Apoptosis (programmed cell death) in arteries of the neonatal lamb. *Circ. Res.* **76,** 168–175.
9. Cho, A., Mitchell, L., Koopmans, D., and Langille, B. L. (1997) Effects of changes in blood flow rate on cell death and cell proliferation in carotid arteries of immature rabbits. *Circ. Res.* **81,** 328–337.
10. Kumar, A. and Lindner, V. (1997) Remodeling with neointima formation in the mouse carotid artery after cessation of blood flow. *Arterioscler. Thromb. Vasc. Biol.* **17,** 2238–2244.
11. Hamet, P., Richard, L., Dam, T. V., et al. (1995) Apoptosis in target organs of hypertension. *Hypertension* **26,** 642–648.
12. Hamet, P., Thorin-Trescases, N., Moreau, P., et al. (2001) Workshop: excess growth and apoptosis: is hypertension a case of accelerated aging of cardiovascular cells? *Hypertension* **37,** 760–766.
13. Fukuo, K., Hata, S., Suhara, T., et al. (1996) Nitric oxide induces upregulation of Fas and apoptosis in vascular smooth muscle. *Hypertension* **27,** 823–826.
14. Fukuo, K., Nakahashi, T., Nomura, S., et al. (1997) Possible participation of Fas-mediated apoptosis in the mechanism of atherosclerosis. *Gerontology* **43 (Suppl 1),** 35–42.
15. Sandau, K., Pfeilschifter, J., and Brune, B. (1997) Nitric oxide and superoxide induced p53 and Bax accumulation during mesangial cell apoptosis. *Kidney Int.* **52,** 378–386.
16. Messmer, U. K., Ankarcrona, M., Nicotera, P., and Brune, B. (1994) p53 expression in nitric oxide-induced apoptosis. *FEBS Lett.* **355,** 23–26.
17. Zhao, Z., Francis, C. E., Welch, G., Loscalzo, J., and Ravid, K. (1997) Reduced glutathione prevents nitric oxide-induced apoptosis in vascular smooth muscle cells. *Biochim. Biophys. Acta* **1359,** 143–152.
18. Geng, Y. J., Henderson, L. E., Levesque, E. B., Muszynski, M., and Libby, P. (1997) Fas is expressed in human atherosclerotic intima and promotes apoptosis of cytokine-primed human vascular smooth muscle cells. *Arterioscler. Thromb. Vasc. Biol.* **17,** 2200–2208.

19. Xie, K., Wang, Y., Huang, S., et al. (1997) Nitric oxide-mediated apoptosis of K-1735 melanoma cells is associated with downregulation of Bcl-2. *Oncogene* **15**, 771–779.

20. Orlov, S. N., Thorin-Trescases, N., Kotelevtsev, S. V., Tremblay, J., and Hamet, P. (1999) Inversion of the intracellular Na+/K+ ratio blocks apoptosis in vascular smooth muscle at a site upstream of caspase-3. *J. Biol. Chem.* **274**, 16,545–16,552.

21. Isaev, N. K., Stelmashook, E. V., Halle, A., et al. (2000) Inhibition of Na(+),K(+)-ATPase activity in cultured rat cerebellar granule cells prevents the onset of apoptosis induced by low potassium. *Neurosci. Lett.* **283**, 41–44.

22. Zhou, X., Jiang, G., Zhao, A., Bondeva, T., Hirszel, P., and Balla, T. (2001) Inhibition of Na,K-ATPase activates PI3 kinase and inhibits apoptosis in LLC-PK1 cells. *Biochem. Biophys. Res. Commun.* **285**, 46–51.

23. Penning, L. C., Denecker, G., Vercammen, D., Declercq, W., Schipper, R. G., and Vandenabeele, P. (2000) A role for potassium in TNF-induced apoptosis and gene-induction in human and rodent tumour cell lines. *Cytokine* **12**, 747–750.

24. McConkey, D. J., Lin, Y., Nutt, L. K., Ozel, H. Z., and Newman, R. A. (2000) Cardiac glycosides stimulate Ca2+ increases and apoptosis in androgen-independent, metastatic human prostate adenocarcinoma cells. *Cancer Res.* **60**, 3807–3812.

25. Taurin, S., Seyrantepe, V., Orlov, S. N., et al. (2002) Proteome analysis and functional expression identify mortalin as an antiapoptotic gene induced by elevation of [Na+]i/[K+]i ratio in cultured vascular smooth muscle cells. *Circ. Res.* **91**, 915–922.

26. Gygi, S. P., Corthals, G. L., Zhang, Y., Rochon, Y., and Aebersold, R. (2000) Evaluation of two-dimensional gel electrophoresis-based proteome analysis technology. *Proc. Natl. Acad. Sci. USA* **97**, 9390–9395.

27. Hadrava, V., Tremblay, J., and Hamet, P. (1989) Abnormalities in growth characteristics of aortic smooth muscle cells in spontaneously hypertensive rats. *Hypertension* **13**, 589–597.

28. Bennett, M. R., Evan, G. I., and Schwartz, S. M. (1995) Apoptosis of rat vascular smooth muscle cells is regulated by p53-dependent and -independent pathways. *Circ. Res.* **77**, 266–273.

29. Molloy, M. P., Herbert, B. R., Walsh, B. J., et al. (1998) Extraction of membrane proteins by differential solubilization for separation using two-dimensional gel electrophoresis. *Electrophoresis* **19**, 837–844.

30. Han, D. K., Eng, J., Zhou, H., and Aebersold, R. (2001) Quantitative profiling of differentiation-induced microsomal proteins using isotope-coded affinity tags and mass spectrometry. *Nat. Biotechnol.* **19**, 946–951.

31. Wiener, M. C., Sachs, J. R., Deyanova, E. G., and Yates, N. A. (2004) Differential mass spectrometry: a label-free LC-MS method for finding significant differences in complex peptide and protein mixtures. *Anal. Chem.* **76**, 6085–6096.

32. Rabilloud, T., Carpentier, G., and Tarroux, P. (1988) Improvement and simplification of low-background silver staining of proteins by using sodium dithionite. *Electrophoresis* **9**, 288–291.

33. Graham, J. and Ford, T. C. (1984) Marker enzymes and chemical assays for the analysis of subcellular fractions, in *Centrifugation: A Practical Approach*, *2nd ed.* (Rickwood, D., ed.), IRL Press, Washington, DC.
34. Schmidt, E. (1974) Glutamate dehydrogenase, in *Methods of Enzymatic Analysis* (Bergmeyer, H. V., ed.), Verlag-Chemie, Weinheim, Germany, pp. 650–656.
35. Rome, L. H., Garvin, A. J., Allietta, M. M., and Neufeld, E. F. (1979) Two species of lysosomal organelles in cultured human fibroblasts. *Cell* **17,** 143–153.
36. Ryazanov, A. G., Ashmarina, L. I., and Muronetz, V. I. (1988) Association of glyceraldehyde-3-phosphate dehydrogenase with mono- and polyribosomes of rabbit reticulocytes. *Eur. J. Biochem.* **171,** 301–305.
37. Burdette, R. A. and Quinn, D. M. (1986) Interfacial reaction dynamics and acyl-enzyme mechanism for lipoprotein lipase-catalyzed hydrolysis of lipid p-nitro-phenyl esters. *J. Biol. Chem.* **261,** 12,016–12,021.

21

Isolation and Culture of Adult Mouse Cardiac Myocytes

Timothy D. O'Connell, Manoj C. Rodrigo, and Paul C. Simpson

Summary

Cardiac myocytes are activated by hormonal and mechanical signals and respond in a variety of ways, from altering contractile function to inducing cardio-protection and growth responses. The use of genetic mouse models allows one to examine the role of cardiac-specific and other genes in cardiac function, hypertrophy, cardio-protection, and diseases such as ischemia and heart failure. However, studies at the cellular level have been hampered by a lack of suitable techniques for isolating and culturing calcium-tolerant, adult mouse cardiac myocytes. We have developed a straightforward, reproducible protocol for isolating and culturing large numbers of adult mouse cardiac myocytes. This protocol is based on the traditional approach of retrograde perfusion of collagenase through the coronary arteries to digest the extracellular matrix of the heart and release rod-shaped myocytes. However, we have made modifications that are essential for isolating calcium-tolerant, rod-shaped adult mouse cardiac myocytes and maintaining them in culture. This protocol yields freshly isolated adult mouse myocytes that are suitable for biochemical assays and for measuring contractile function and calcium transients, and cultured myocytes that are suitable for most biochemical and signaling assays, as well as gene transduction using adenovirus.

Key Words: Cardiac myocyte; cell isolation; cell culture; hypertrophy; cell signaling; apoptosis; adenovirus; β-adrenergic.

1. Introduction

Cardiac myocytes respond to a variety of hormonal, neural, mechanical, and electrical stimuli by altering their force and rate of contraction (*1*). In addition, cardiac myocytes hypertrophy as a compensatory response to either physiological stimuli, such as exercise, or pathological stimuli, such as hypertension (*2*). The signaling pathways regulating cardiac myocyte function, hypertrophy, and death encompass a wide array of signaling molecules and pathways, including G protein-coupled receptors, cytokine receptors, and a plethora of intracellular signaling molecules (*3–6*), and these processes have important clinical relevance (*7*).

From: *Methods in Molecular Biology, vol. 357: Cardiovascular Proteomics: Methods and Protocols*
Edited by: F. Vivanco © Humana Press Inc., Totowa, NJ

The advent of transgenic and knockout mouse technology has advanced our ability to examine the role of many cardiac-specific and other genes in cardiac function, hypertrophy, cardio-protection, ischemia, and heart failure. However, it would be extremely beneficial to study these genetic manipulations at the myocyte level. Although techniques for isolating and culturing myocytes in other animals, such as rat or rabbit, have been refined, the isolation and culture of adult mouse cardiac myocytes remains challenging.

This chapter describes a straightforward protocol for the isolation and the short-term (24 h) and long-term (72 h) culture of adult mouse cardiac myocytes. The protocol is an adaptation of procedures used to isolate myocytes from other animals, such as rats and rabbits, and prior work in the mouse (8–10). Briefly, the heart is excised rapidly and mounted on a perfusion apparatus. The heart is then perfused with a calcium-free buffer to arrest contraction, followed by a collagenase-based enzyme solution to digest the extracellular matrix of the heart. After digestion, myocytes are dispersed into a single-cell suspension and calcium is reintroduced. Once the myocytes are equilibrated (calcium-tolerant), they are plated in culture dishes, allowed to attach, and finally washed and cultured overnight.

Using this protocol, we routinely isolate roughly $1–1.5 \times 10^6$ rod-shaped myocytes per heart (11), which is approximately two- to threefold higher than other published reports in the mouse heart (8–10). The goal of the protocol is rod-shaped myocytes, to mimic cells in the intact heart, rather than the round myocytes ("meatballs") that are seen when the myocytes hypercontract. Freshly isolated myocytes are highly useful for biochemical and other assays, because this preparation eliminates "contamination" by the many nonmyocytes assayed when the intact heart is used. The short-term (24 h) culture protocol allows the myocytes to recover and stabilize after the isolation procedure, and is suitable for signaling assays and early growth and death end points. The long-term (72 h) culture is a platform for late growth responses and genetic manipulation by adeno-virus, or by other approaches. We have used the freshly isolated myocytes to measure contractile function and calcium transients. We have used the short-term cultured myocytes in a wide variety of applications, including measuring production of second messengers like cAMP, activation of signaling proteins using phospho-specific antibodies, and changes in gene expression (11,12). Finally, we have used the long-term cultured cells for high-efficiency gene transduction with adenovirus (11). The following describes the protocol and illustrates these various applications.

2. Materials

See details of solutions in **Tables 1–6**. All solutions should be made with culture-grade 18.2 MΩ H_2O.

Table 1
Perfusion Buffer (1X and 10X Recipes) [a]

Compound	Mol. wt (g/mol)	Final conc. (mM)	1X add g/L	10X add g/L
NaCl	58.4	120.4	7.03	70.3
KCl	74.6	14.7	1.1	11
KH$_2$PO$_4$	136.1	0.6	0.082	0.82
Na$_2$HPO$_4$	142	0.6	0.085	0.85
MgSO$_4$-7H$_2$O	246.5	1.2	0.30	3
Na-HEPES [b]	1 M	10	10 mL	100 mL
NaHCO$_3$	84	4.6	0.39	
Taurine	125.1	30	3.75	
BDM	101.1	10	1	
Glucose	180.2	5.5	1	

[a]All reagents are from Sigma Chemical (St. Louis, MO), except Na-HEPES (Gibco-BRL, Bethesda, MD). For 1X perfusion buffer made fresh from powder at the time of use, dissolve ingredients in 990 mL of 18.2 MΩ H$_2$O and filter-sterilize. Adjust the pH to 7.0 with sterile HCl as needed. For 10X perfusion buffer, dissolve in 900 mL of 18.2 MΩ H$_2$O, add 100 mL of Na-HEPES, and store at 4°C for up to 1 wk. To make 1X perfusion buffer, add NaHCO$_3$, taurine, BDM, and glucose fresh on the day of use, and filter-sterilize. Adjust the pH to 7.0 with sterile HCl as needed.
[b]Na-HEPES comes as a 1 M liquid.

Table 2
Myocyte Digestion Buffer (50 mL/Heart) [a]

	Final conc.	Amount
Perfusion buffer		50 mL
Collagenase II	2.4 mg/mL	120 mg

[a]Prepare fresh for each heart and add to the reservoir just prior to digestion, to prevent heat inactivation of the enzyme. The absolute amount of enzyme will vary from lot to lot and from different manufacturers (*see* **Note 7**).

2.1. Myocyte Isolation

1. Perfusion buffer (*see* **Table 1**): reagents are from Sigma Chemical (St. Louis, MO), except Na-HEPES (Gibco-BRL, Bethesda, MD), and buffer is optimally made fresh each day, using 18.2 MΩ H$_2$O, and sterile-filtered with a 0.22-μm filter. The buffer pH should be 6.9–7.0; if not, adjust the pH to 7.0 with sterile HCl after filtration. For simplicity, if many preparations are to be done in a single week, a 10X buffer without NaHCO$_3$, taurine, 2,3-butanedione monoxime (BDM), and glucose can be made and stored at 4°C, then when ready for use these four reagents are added fresh, the solution is filtered, and the pH is adjusted to 7.0.

Table 3
Myocyte Stopping Buffer (20 mL/Heart)

	Final conc.	Volume
Perfusion buffer		18 mL
Calf serum	10%	2 mL
100 mM CaCl$_2$	12.5 µM	2.5 µL

Table 4
Myocyte Plating Medium

	Final conc.	Volume (mL)
MEM (Eagle's w/HBSS)		42.5
Calf serum	10%	5
BDM	10 mM	1
Penicillin	100 U/mL	0.5
Glutamine	2 mM	0.5
ATP	2 mM	0.5

Table 5
Myocyte Culture Medium[a]

	Final conc.	Volume (mL)
MEM (Eagle's w/HBSS)		48.5
BSA	1 mg/mL (0.1%)	0.5
Penicillin	100 U/mL	0.5
Glutamine	2 mM	0.5

[a]This myocyte culture medium is for short-term culture (24 h). For long-term culture (72 h), the myocyte culture medium is also supplemented with 10 mM BDM and ITS medium supplement (*see* **Table 6** and **Note 17**). Even in the short-term culture, addition of a low concentration of BDM (1 mM, 100 µL of BDM stock) can improve myocyte morphology and survival, without any detectable effect on the assays shown in this chapter. BSA in the short-term culture can be reduced to 0.1 mg/mL (0.01%) without detectable negative effects.

2. Myocyte digestion buffer (*see* **Table 2**): add to perfusion buffer 120 mg per 50 mL crude Collagenase Type II for a final 2.4 mg/mL (Worthington Biochemical, Lakewood, NJ). Note that the amount of collagenase will vary with the particular lot being used. After adding the collagenase, this buffer should be sterile-filtered again with a 0.22-µm filter to remove any undissolved particulate.
3. Myocyte stopping buffer (*see* **Table 3**): add to perfusion buffer 10% (v/v) calf serum (CS) (HyClone, Logan, UT) and 12.5 µM CaCl$_2$ (Sigma Chemical). CS is not heat-inactivated.

Table 6
Stock Solutions[a]

Calf serum: (HyClone, cat. no. SH30073). Store the CS in 25-mL aliquots in sterile 50-mL tubes at −20°C.

Bovine serum albumin (endotoxin and lipid-free): (Sigma, cat. no. A-8806). Prepare 100 mg/mL stock in H_2O (100X, 10%, 5g in 50-mL) and sterile-filter. Store BSA in 5-mL aliquots at −20°C.

$CaCl_2$ (1 M): (Sigma, cat. no. C-7902). Prepare 1 M by adding 14.7 g/100 mL H_2O. Filter with 150-mL filter unit (Fisher, cat. no. 09-740-28E) and store at room temperature. Make 1:10 dilution to get 100 mM stock for the stop buffer and calcium reintroduction.

BDM: 2,3-Butanedione monoxime: (Sigma, cat. no. B-0753). Prepare 500 mM stock in H_2O (50X, 2.25 g/50 mL H_2O). Warm the solution to dissolve BDM and sterile-filter into a 100-mL filter bottle. Store in 5-mL aliquots at 4°C. Warm before use to dissolve any precipitate.

Penicillin-G: (Sigma, cat. no. P-7794, powder). Penicillin powder is 1600 U/mg (6.25 mg/10,000 U). Prepare 10,000 U/mL stock in H_2O (100X, 312.5 mg/50 mL H_2O) and sterile-filter. Store in 5-mL aliquots at −20°C.

L-Glutamine: (Gibco, cat. no. 25030-81). This is 200 mM (100X) stock supplied in a 50-mL bottle. Store in 5-mL aliquots at −20°C.

Laminin coating solution: (BD Bioscience, cat. no. 354232). The stock solution is 2 mg/mL, and is stored in 100-µL aliquots at −70°C. Dilute 1:200 in PBS (Ca/Mg-free) for working solution of 10 µg laminin/mL PBS. Do not re-use diluted laminin. Unused coated dishes can be saved overnight, but this is not advised.

Heparin: (ICN/MP Biomedicals, cat. no. 101932). The stock solution is 1000 IU/mL, and is stored at room temperature. Dilute 1:10 in PBS (Ca/Mg-free) for working solution of 100 IU/mL. Do not re-use diluted heparin.

Na-ATP: (Sigma, cat. no. A-6419). Prepare 200 mM stock (100X, 1 g/9 mL H_2O), sterile-filter, store 0.5-mL aliquots at −20°C. ATP is not stable and is added fresh to the medium just prior to use, 0.5 mL in 50 mL MEM.

ITS medium supplement (insulin, transferrin, selenium): (Sigma, cat. no. I-1884). Make a 100X stock solution from the lyophilized powder (which contains 25 mg insulin, 25 mg transferring, and 25 µg Na selenite) by adding 50 mL H_2O. Sterile-filter the ITS, and store 1-mL aliquots at −20°C. Add 1 mL to 100 mL MEM for long-term myocyte culture, for final 5 µg/mL insulin, 5 µg/mL transferrin, and 5 ng/mL selenium (*see* **Note 17**).

[a]Note that all water is culture-grade 18.2 MΩ H_2O.

4. Myocyte plating medium (*see* **Table 4**): plating medium is minimum essential medium (MEM) with Hank's Balanced Salt Solution (HBSS) (Gibco-BRL), containing 10% (v/v) CS, 10 mM BDM, 100 U/mL penicillin, 2 mM glutamine, and 2 mM Na-ATP. ATP is not stable and is added fresh to the medium prior to use from a 200 mM sterile stock solution that is stored at −20°C.

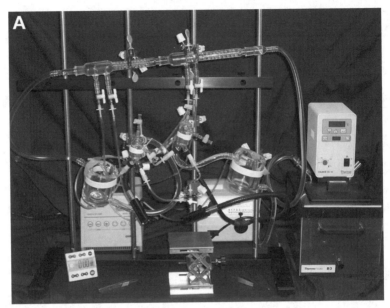

Fig. 1. Perfusion system. The perfusion system has two halves, one for the perfusion buffer and one for the digestion buffer. Each half contains a water-jacketed reservoir for warming solutions to 37°C, from which fluid is pumped through a peristaltic pump, through a heat exchanger to maintain solution temperature, through a bubble trap, then finally out through the cannula. Fluid flow is switched from perfusion buffer to digestion buffer by clamping the appropriate tube at the pump, and turning the three-way valve above the cannula. To maintain the perfusate a 37°C, the water-jacketed circulation system is maintained at 42°C (but this must be empirically determined). **(A)** Overview of the perfusion system. The peristaltic pump and pump tubing are from Rainin Instruments (Oakland, CA); the circulating water bath is from Haake (VWR Instruments); the illuminator, light pipes and focusing lens, and the thermocouple thermometer and probe are from Cole Parmer Instruments; and finally the ring stand is from Radnotti Glass.

5. Myocyte culture medium (*see* **Table 5**): culture medium is MEM with HBSS (Gibco-BRL), containing 1 mg/mL bovine serum albumin (BSA) (endotoxin- and lipid-free), 100 U/mL penicillin, and 2 mM glutamine.
6. Laminin-coated dishes: laminin (*see* **Table 6**) (BD Biosciences, San Jose, CA) is stored at −70°C as 100-µL aliquots of a 2 mg/mL stock. Thaw the laminin in the refrigerator, and add 20 mL of ice-cold phosphate-buffered saline (PBS; CaCl$_2$/MgCl$_2$-free, Gibco-BRL) to the laminin stock, for a final concentration of 10 µg/mL. For each heart to be isolated, coat 15 to 20 35-mm dishes or 5 to 7 60-mm dishes. To coat 35-mm dishes, cover the bottom with 1 mL of laminin coating solution; to coat 60-mm dishes, add 3 mL. Incubate at room temperature for at least 1.5–2 h with gentle shaking on a rocker platform (do not use an orbital shaker).

B

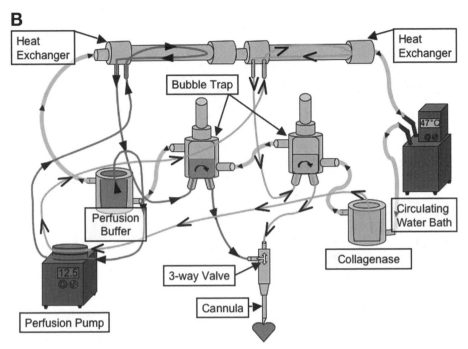

Fig. 1. (*continued*) **(B)** Schematic diagram of the perfusion system and fluid flow. Perfusion buffer or myocyte digestion buffer are loaded in the water-jacketed reservoir and pumped through the peristaltic pump, through the water-jacketed heat exchanger, through a water-jacketed bubble trap, through a three-way valve that controls flow from each half of the system and out through the cannula. The water-jacketed reservoir and heat exchanger are from Ace Glass (Vineland, NJ); the bubble trap, ring clamps, tubing, and tubing adaptors are from Radnotti; and the one-way and three-way valves are from Cole Parmer.

Remove the laminin coating solution just prior to plating the myocytes. It is highly preferable to use the dishes on the day they are made. However, if the plates are not used on the day they are prepared, wrap the dishes individually in parafilm, and store in the refrigerator overnight. Do not refreeze laminin or reuse the coating solution.

7. Perfusion system (*see* **Fig. 1**): the perfusion system is shown in **Fig. 1**. The perfusion system has two halves, one for the perfusion buffer and one for the digestion buffer. Each half contains a water-jacketed reservoir for warming solutions to 37°C, from which fluid is pumped through a peristaltic pump, through a heat exchanger to maintain solution temperature, through a bubble trap, then finally out through the cannula. Fluid flow is switched by changing the pump head attachment and diverting flow from one side to the other, at the three-way valve above the cannula. To maintain the perfusate at 37°C at the outflow of the cannula, the

Table 7
Equipment and Supplies

Culture incubator and metal trays, laminar flow culture hood with vacuum aspirator, centrifuge, phase contrast microscope, light microscope.

Peristaltic pump and pump tubing (Rainin Instruments, Oakland, CA); circulating water bath (Haake, VWR Instruments); iluminator, light pipes, focusing lens, thermocouple thermometer, and probe (Cole Parmer Instruments); ring stand (Radnotti Glass); water-jacketed reservoir and heat exchanger (Ace Glass); bubble trap, ring clamps, tubing, and tubing adaptors (Radnotti); one-way and three-way valves (Cole Parmer); Digi-Sense thermometer and insulated wire probe (Cole Parmer); cannula made of 20G needle, with the point removed and filed flat, and notches 1 and 2 mm above the tip as reference points.

Isoflurane and oxygen; atomizer and induction chamber (Vetland), nose cone made from a small funnel.

Surgical tools: clip, scissors, retractor, forceps, and Dumont microsurgery fine-tip forceps with microblunted, atraumatic tips angled at 45° (Fine Science Tools) (autoclaved before use); 6-0 surgical silk (about 15 cm); sterile 1-mL syringes with needles (for heparin injection); 70% ethanol

Sterile 60-mm bacterial culture dishes (Valmark).

Sterile culture plastics: transfer pipettes with 2-, 1.5-, and 1-mm openings (Fisher Scientific); 5- and 10-mL pipets; culture dishes, 35- and 60-mm (Fisher Scientific, Falcon 35-3001 and 35-3004); conical polypropylene tunes, 15-mL and 50-mL.

Hemocytometer (VWR, cat. no. 15170-079).

water-jacketed circulation system is maintained at 42°C to 48°C (which must be empirically determined). A Digi-Sense thermometer and insulated wire probe (Cole Parmer, Vernon Hills, IL) are used to measure perfusate temperature. *See* **Subheading 3.10.** for cleaning the perfusion system.

8. Cannula: the cannula for the perfusion system is a 20-gage (G) needle, with the point removed and filed flat. To aid in the cannulation, we make notches 1 and 2 mm above the tip as reference points.

2.2. Removal and Cannulation of the Heart

See **Table 7** for list of special materials and equipment.

1. Heparin: 1000 IU/mL (CN/MPBiomedicals), prepare at 100 U/mL in PBS (Ca^{2+}/ Mg^{2+}-free; Gibco-BRL).
2. Surgical tools: all of our surgical tools and a small clip used to hold the aorta to the cannula are from Fine Science Tools (FST, Foster City, CA). For the fine-tip forceps, we recommend the Dumont microsurgery forceps with micro-blunted, atraumatic tips angled at 45° (FST).
3. Isoflurane atomizer, including the induction chamber (Vetland, Louisville, KY).

2.3. Heart Perfusion and Enzyme Digestion

1. Calcium chloride: 100 mM CaCl$_2$ in H$_2$O, sterile filtered with a 0.22-μm filter.

2.4. Myocyte Dissociation

1. Valmark 60-mm dish: this is a sterile bacterial culture dish made by Valmark (Midwest Scientific, St. Louis, MO) to which myocytes will not adhere. Although we assume other nonculture treated dishes would work, we have not tried other vendors.
2. Sterile transfer pipets with 2-, 1.5-, and 1-mm openings (Fisher Scientific, Pittsburgh, PA).

2.5. Calcium Reintroduction

1. ATP: 200 mM Na-ATP (Sigma) made in H$_2$O, stored frozen in aliquots at −20°C, thawed immediately prior to use.
2. Calcium chloride: 100 mM CaCl$_2$ in H$_2$O, sterile-filtered with a 0.22-μm filter.

2.6. Plating Myocytes and Culture

1. Incubators: our incubators are kept at 37°C with 2% CO$_2$. When MEM with HBSS, which contains 0.35 g/L NaHCO$_3$, is placed in a 2% CO$_2$ incubator, the resultant pH of the medium is about 6.9–7.0.
2. Falcon culture dishes, 35-mm (15–20 per heart) or 60-mm (5–7 per heart), coated with laminin as above.
3. Myocyte culture medium (*see* **Table 5**).
4. Hemocytometer.

2.7. Long-Term Culture (72-h) of Adult Mouse Myocytes

1. Long-term myocyte culture medium: myocyte culture medium (*see* **Table 5**), with additional 10 mM BDM, and insulin, transferrin, and selenium (ITS Medium Supplement, Sigma) (*see* **Table 6** and **Note 17**).
2. Falcon culture dishes, 35- or 60-mm, coated with laminin.

3. Methods

The heart perfusion is done on an open lab bench, but maintaining clean technique and using sterile solutions. The work is transferred to a laminar-flow culture hood at **Subheading 3.4., step 1**.

3.1. Preparation for Myocyte Isolation From One to Two Hearts

This section describes the initial preparation for myocyte isolation. Briefly, all the buffers are prepared, culture dishes are coated with laminin, and the perfusion system is primed.

1. Prepare 500 mL of perfusion buffer for each heart (50 mL will be used to make the myocyte digestion buffer).

2. Prepare 50 mL of myocyte digestion buffer for each heart, and store each 50 mL in a separate 50-mL tube. Do not warm the enzyme for each heart to 37°C until ready to isolate myocytes; exposure to prolonged high temperature might inactivate the enzyme.
3. Prepare myocyte stopping buffer, used to inactivate the collagenase and other proteases in the crude mixture.
4. Prepare myocyte plating medium and myocyte culture medium. Equilibrate both at 37°C in a 2% CO_2 incubator for at least 2 h to adjust temperature and pH.
5. Prepare laminin-coated culture dishes.
6. Prepare the perfusion apparatus (*see* **Fig. 1**). Set the circulating water bath so that the outflow from the tip of the cannula is 37°C, as measured with a digital thermometer. Check the flow rate of the pump and adjust to 4 mL/min (*see* **Note 1**).
7. Run 100 mL of purified water through the perfusion system.
8. Add perfusion buffer (50 mL) and myocyte digestion buffer (50 mL) to the correct reservoirs, prime the perfusion system with buffers (run perfusion buffer and myocyte digestion buffer through system for 5 min), eliminate air bubbles, and allow time to warm to 37°C (about 10 min).
9. Add 10 mL of room temperature perfusion buffer to a 60-mm culture dish for heart collection. Add 10 mL of room temperature perfusion buffer to another 60-mm culture dish for heart cannulation, and place on an adjustable stage under the perfusion apparatus.
10. Position the cannula with the tip close to the surface of the perfusion buffer in the 60-mm dish.
11. Cut a small piece of 6-0 surgical silk (about 15 cm), knot loosely, and place on the adjustable stage (this will be used to secure the aorta to the cannula).

3.2. Removal and Cannulation of the Heart

This section describes the removal and the subsequent cannulation of the heart on the perfusion system. Briefly, the mouse is injected with heparin to prevent coagulation of blood in the coronary arteries. The mouse is then anesthetized, the chest opened and the heart rapidly removed and cannulated. Perfusion with the calcium-free perfusion buffer is started immediately and blood should rapidly clear from the coronary arteries, indicating proper cannulation and good perfusion.

1. Anesthetize the mouse with isoflurane and 100% O_2. Set the isoflurane atomizer dial to 3% (scale 1 to 5% of total flow), turn the O_2 valve to 0.5 L/min, and place animal inside the induction chamber. When the mouse is anesthetized, it will lose consciousness and roll over on its side. Check with a toe pinch to ensure that the mouse is fully anesthetized. Transfer the mouse to the surgery/perfusion area and place under a nose cone connected to the anesthesia system (*see* **Note 2**).
2. Once the mouse is anesthetized, inject the mouse intraperitoneally with 0.5 mL heparin, diluted in PBS to 100 IU/mL. Injecting after anesthesia reduces stress to the animal and improves the isolation.

3. Wait a few minutes for the heparin to circulate. Wipe the chest with 70% ethanol. Adjust the isoflurane (usually 1.5%) as necessary to ensure proper level of anesthesia (movement indicates that the anesthesia is too shallow, whereas irregular respiration indicates that it is too deep). Check with a toe pinch to ensure that the mouse is fully anesthetized.

4. Open the peritoneal cavity and chest with small scissors and use forceps to peel back the rib cage to expose the heart. Lift the heart gently using forceps. Identify and cut the pulmonary vessels, which will make it easier to identify and cut the aorta. Cut the transverse aorta between the carotid arteries, cut the right carotid artery at the same time, and immediately place the heart in a 60-mm dish containing 10 mL of perfusion buffer at room temperature (*see* **Note 3**).

5. Remove extraneous tissues (thymus and lungs), if necessary, and transfer heart to the second 60-mm dish with perfusion buffer at room temperature.

6. Working under magnification (*see* **Note 4**), cannulate the heart using fine-tip forceps to slide the aorta onto the cannula so that the tip of the cannula is just above the aortic valve (check the 1-mm notch on the cannula to ensure proper cannulation; *see* description of cannula in **Subheading 2.1., step 8**). Attach a small clip to the end of the aorta on the cannula to prevent the heart from falling. Start the perfusion immediately (4 mL/min). Tie the aorta to the cannula with 6-0 silk thread. Total time to cannulate the heart should be less than 1 min (*see* **Note 5**).

3.3. Heart Perfusion and Enzyme Digestion

This section describes the enzymatic digestion of the heart. Briefly, the heart is perfused with a collagenase solution to digest the extracellular matrix. During the digestion, the heart will initially become very hard to the touch, owing to a large increase in vascular resistance upon introduction of enzyme. As the perfusion continues the heart will become pale, swollen, and flaccid, indicating a good digestion.

1. Once cannulated, perfuse the heart with perfusion buffer for 4 min at 4 mL/min, to flush blood from the vasculature and remove extracellular calcium to stop contractions (*see* **Note 6**). Measure the temperature of the heart with an insulated wire probe attached to a digital thermometer, placing the temperature probe into ventricle to ensure that the temperature is 37°C. This does not need to be done each time, but should be done periodically to ensure reproducibility.

2. After 4 min, switch to myocyte digestion buffer and perfuse for 3 min at 4 mL/min. Collect the myocyte digestion buffer and discard (*see* **Notes 7** and **8**).

3. After 2–3 min, add 15 µL of 100 mM $CaCl_2$ to the myocyte digestion buffer in the reservoir and continue to digest for 8 min at 4 mL/min, although digestion times can vary from heart to heart. At this point, the calcium concentration of the myocyte digestion buffer is roughly 40 µM (15 µL of 100 mm $CaCl_2$ in roughly 35 mL of enzyme buffer). From this point on, the myocyte digestion buffer can be collected and added back to the reservoir for reuse until the digestion is completed. The total digestion time is usually about 11 min. If the heart is well-perfused during

the enzyme digestion, the heart will become swollen and turn slightly pale, and separation of muscle fibers on the surface of the heart might become apparent. Digestion should be terminated if the heart feels spongy when gently pinched.

3.4. Myocyte Dissociation

This section describes the dissociation of myocytes following enzymatic digestion of the heart. Briefly, the heart is removed from the cannula and gently teased apart. A buffer containing serum is added at this point to stop the enzyme digestion and prevent overdigestion. Finally, an initial count of the isolated cells is made to evaluate the digestion of the heart.

1. Once enzyme digestion of a heart is complete (heart appears swollen, pale, and flaccid), cut the heart from the cannula just below the atria using sterile, fine scissors. Place the ventricles in a sterile 60-mm Valmark dish containing 2.5-mL of myocyte digestion buffer. From this point forward, all subsequent steps are performed under a laminar flow culture hood using sterile technique (*see* **Note 9**).
2. Tease the ventricles into 10–12 small pieces with fine-tip forceps. Add 5 mL room temperature myocyte stopping buffer to the dish. Pipet gently several times with a sterile plastic transfer pipet (2-mm opening). This process takes 60–90 s. The tissue should be very flaccid, almost falling apart on its own, and require very little force to dissociate, which will indicate a good digestion (*see* **Note 10**).
3. Transfer the cell suspension to a 15-mL polypropylene conical tube. Rinse the plate with 2.5 mL of myocyte stopping buffer, and combine with the cell suspension for a final volume of 10 mL. Myocyte stopping buffer contains serum to inactivate proteases; the final CS concentration is 5%.
4. Continue to dissociate the heart tissue gently, using sterile plastic transfer pipets with different sized openings (1.5- and 1-mm diameters), until all the large pieces of heart tissue are dispersed in the cell suspension. Avoid vigorous agitation to minimize shearing of the cells (*see* **Note 10**). This process should take 3–5 min. Bring the final volume of cell suspension to 10 mL with myocyte stopping buffer.
5. Transfer 80 µL cell suspension to a microcentrifuge tube, and use duplicate 10 µL aliquots to count rod-shaped and round myocytes in a hemocytometer (VWR, cat. no. 15170-079). Calculate the total number of myocytes, the number of rod-shaped myocytes, and the percent of rod-shaped myocytes. Record these values as the initial number of cells obtained (*see* **Note 11**). For example, 10 µL of the cell suspension is loaded onto each side of the hemocytometer, five grids (center, 4 corners) are counted on each side, and the two sides are averaged. If 60 rod-shaped myocytes and 30 round myocytes are counted, this would give a yield of 1.8 million total myocytes, with 1.2 million rod-shaped (67%):
   a. $[(60 \text{ rod} + 30 \text{ round})/5 \text{ grids}] \times 10^4$ (a constant) $= 1.8 \times 10^5$ cells/mL or 1.8×10^6 myocytes in the original 10 mL
   b. Rod-shaped myocytes are 1.2×10^6 (67% rods) by the same formula
6. The important numbers for determining the quality of the isolation are the total number of myocytes (rod and round) and the total number of rod-shaped myocytes

(or the percentage of rod-shaped myocytes). If at this point, the total myocyte yield from a heart is low (<1 million), or the percent of rod-shaped myocytes is low (<50%), this would indicate a less then optimal isolation and a decision as to whether to continue can be made.

7. While counting, allow the remaining myocytes to sediment by gravity for a few minutes at room temperature in the 15-mL tube.
8. Centrifuge for 3 min at $20g$ (400 rpm on Precision Durafuge 100 benchtop centrifuge). Gently resuspend the pellet in 10 mL myocyte stopping buffer (final calcium concentration 12.5 μM).

3.5. Isolation of Myocyte From Multiple Hearts

Often, one might wish to isolate myocytes from more than one heart, sometimes with different genotypes. We have isolated myocytes sequentially from up to four hearts of the same genotype or two hearts with different genotypes.

1. If two hearts of different genotypes are to be digested sequentially, resuspend the combined pellet from the first heart in 10 mL of myocyte stopping buffer, and leave at room temperature while the second heart is digested. Once the second heart is in 10 mL of myocyte stopping buffer, proceed to the calcium reintroduction with both preparations.
2. If two hearts of the same genotype are to be digested sequentially, resuspend the combined pellet of the first heart in 5 mL myocyte stopping buffer, and leave at room temperature while the second heart is digested. Resuspend the combined pellet of the second heart in 5 mL myocyte stopping buffer and combine with cells from the first heart so that the total volume of myocyte stopping buffer is 10 mL. Conduct calcium reintroduction procedures on the combined preparations.

3.6. Calcium Reintroduction

This section describes the gradual reintroduction of calcium to produce calcium-tolerant myocytes. Briefly, the myocytes are incubated in buffer with increasing concentrations of calcium, finally achieving a concentration of 1.2 mM Ca^{2+} as in our culture medium.

1. Prior to the calcium reintroduction, add 100 μL of 200 mM ATP to the tube, so that the final concentration of ATP in the buffer is 2 mM (*see* **Note 12**).
2. At this point, the calcium concentration is 12.5 μM. A three-step calcium reintroduction is done at room temperature to bring the calcium concentration to 1.2 mM.
3. Prepare three 15-mL tubes containing 10 mL myocyte stopping buffer:
   a. 100 μM calcium: 10 μL of 100 mM $CaCl_2$ in 10 mL myocyte stopping buffer.
   b. 400 μM calcium: 40 μL of 100 mM $CaCl_2$ in 10 mL myocyte stopping buffer.
   c. 900 μM calcium: 90 μL of 100 mM $CaCl_2$ in 10 mL myocyte stopping buffer.
4. Centrifuge the myocytes for 3 min at $20g$. Remove the supernatant, which contains nonmyocytes and some round myocytes, and to the pellet, carefully introduce the contents of tube 1 by mixing very gently with a 1.5-mm plastic pipet.

5. Let the myocytes stand for 2 min, then repeat **step 4**; continue in this fashion through tube 3.
6. Resuspend the final pellet in 5 mL of myocyte plating medium at 37°C for a final calcium concentration of 1.2 mM.
7. Count rod-shaped and round myocytes using a hemacytometer. Count both sides of a hemocytometer (10 grids, take mean of 5 grids), with duplicate cell aliquots. Calculate the total number of myocytes, the number of rod-shaped myocytes, and the percent of rod-shaped myocytes (*see* **Subheading 3.4., step 5** for a sample calculation). Record these values as the number of myocytes for plating. The most important number is the total number of rod-shaped myocytes, and the goal is at least 1 million. If the total myocyte number is low (total cells less than 1 million), or if myocyte quality is poor (less than 60% rod-shaped myocytes), a decision may be made as to whether to continue with the preparation or to start over, depending on the judgment of the technician.
8. For electrophysiological studies or measurements of contraction and calcium transients, myocyte may be suspended in any suitable buffer, as in **Subheading 3.6., step 6**.

3.7. Plating Myocytes, Myocyte Attachment, and Culture

This section describes myocyte plating and short-term culture (24 h). Briefly, the myocytes are plated on laminin-coated dishes and allowed to attach. After attachment, the nonattached cells, mostly round myocytes and a few rod-shaped myocytes, are washed away. Typically, we get more than 70% plating efficiency; that is, more than 70% of the rod-shaped myocytes initially plated actually attach. After washing, the cultures are about 90% rod-shaped in the dish, and we retain roughly 90% of these rod-shaped myocytes overnight in culture *(11)*.

1. Myocyte plating medium, which is MEM containing calf serum, BDM (a contraction inhibitor), glutamine, ATP, and penicillin (*see* **Table 4**), should be equilibrated for 2–3 h at 37°C in a 2% CO_2 incubator. Myocyte culture medium should also be equilibrated for the same time. Calculate the total number of rod-shaped myocytes and determine the volume of myocyte plating medium required to adjust the concentration of rod-shaped myocytes to 25,000 rod-shaped myocytes/mL in a 50-mL tube. Make sure that the myocytes are resuspended well by gently pipetting, using a 10-mL pipet.
2. Set up trays to plate no more than six dishes per tray at a time to avoid pH changes as medium is exposed to air. Plate the appropriate amount of rod-shaped myocytes in the desired culture dishes: 2 mL (containing 50,000 rod-shaped myocytes) in a laminin-coated, 35-mm dish, and 6 mL (containing 150,000 rods) in a 60-mm laminin-coated dish. This is a density of roughly 52 rod-shaped myocytes per mm^2. Use a 5-mL pipet to plate two 35-mm dishes at once, to prevent myocytes from settling in the pipet during plating. Use a 10-mL pipet to plate single 60-mm dishes. During the plating procedure, gently resuspend the myocytes constantly, to ensure

that they do not settle to the bottom of the tube, which will cause variation in plating density. Once the myocytes are plated, distribute the myocytes evenly in each dish by gently sliding the tray forward and backward and side-to-side in a cross-like pattern three to four times on the surface of the culture hood. Never swirl the medium in the dish or the myocytes will clump in the center of each dish. Immediately place finished trays in a 2% CO_2 incubator at 37°C. Incubate for 1 h to allow myocyte attachment (*see* **Notes 13–15**).

3. After 1 h, gently aspirate the plating medium with a sterile Pasteur pipet into a vacuum flask. To remove any remaining unattached myocytes and debris, wash each dish with approx 1 mL pre-equilibrated myocyte culture medium, adding medium gently to the sides of the dish, not the bottom, then aspirate the wash. When changing the medium, do not remove more than one tray from the incubator at once, and wash one dish at a time. Also minimize the time that the incubator door is open, to maintain CO_2.

4. Add myocyte culture medium to the washed cells, using 1 mL medium for 35-mm dishes, and 3 mL medium for 60-mm dishes. Immediately return myocytes to the incubator. At this stage, the myocytes are in myocyte culture medium, which is MEM containing only 1 mg/mL BSA, glutamine, and penicillin.

5. After an additional hour, count the number of rod-shaped myocytes and round myocytes under an inverted microscope with a ×20 objective. Count myocytes in three different dishes (from early, middle, and late in plating), and count at least three randomly selected fields. Average the count of the three fields for each dish, and then average the three dishes. Calculate the total number of myocytes, the number of rod-shaped myocytes, and the percent of rod-shaped myocytes. For example, in an experiment with 35-mm dishes:

   a. Total myocytes = (mean no. of myocytes counted/no. of fields) × area factor
      Example: (120 myocytes/3 fields counted) × 1052 = 42,080 myocytes/dish
      or 42,080 myocytes/962 mm^2 = 44 myocytes/mm^2.
      The "area factor" needs to be empirically determined for each microscope objective/eyepiece being used. To calculate the area factor, we use a stage micrometer to measure the diameter of the microscopic field for our ×20 objective and eyepiece. We calculate the area of the circular microscopic field, and then divide the area of the dish (supplied by the manufacturer and measured directly) by the area of the field, to give the area factor. For our microscope, the diameter of the field at ×20 is 1.08 mm, which is an area of 0.9145 mm^2, and the area of a 35-mm dish is 962 mm^2, therefore the area factor is 1052. Either total myocytes/mm^2 or total myocytes/dish can be calculated. Myocytes/mm^2 can be used to define the preparation and to calculate plating efficiency as a quality control among preparations. Total myocytes/dish allows determination of biochemical assays on a "per myocyte" basis.

   b. Total rod-shaped = (no. of rods counted/no. of fields) × area factor
      Example: (105 rods/3 fields counted) × 1052 = 36,820 rods

   c. Percent rod-shaped = (no. of rods/total no. of cells)
      Example: (36,820/42,080)*100 = 88%

A 0 hours   **B** 24 hours

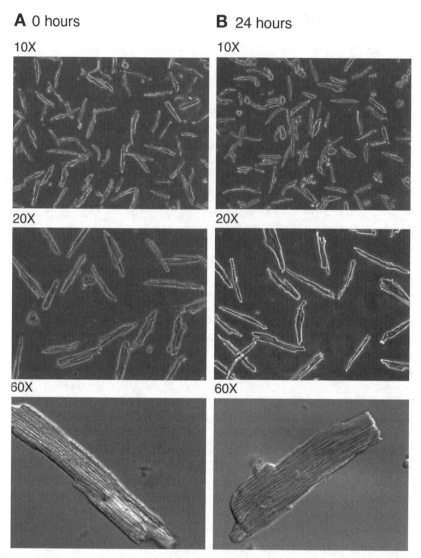

Fig. 2. Normal rod-shaped morphology of adult mouse cardiac myocytes cultured for 24 h. Myocytes were photographed under microscopy by phase contrast (×100 and ×200) and differential interference contrast (×600) after culture in myocyte culture medium for **(A)** 0 h and **(B)** 24 h.

    d. Plating efficiency 36,820 rods counted/50,000 plated = 74% plating efficiency. Record these values as the number of myocytes at T0.

6. Continue to incubate the myocytes at 37°C in 2% CO_2 until use. Count again at 24 h or at the time of use (18–24 h). **Figure 2** depicts myocytes after plating and medium change (0 h) and after 24 h in culture (*see* **Note 16**).

3.8. Long-Term Culture (72 h) of Adult Mouse Myocytes

The protocol described previously is for the short-term (24 h) culture of adult mouse cardiac myocytes. For long-term culture, we developed a separate culture medium that is supplemented with 10 mM BDM and insulin, transferring, and selenium (ITS Medium Supplement; *see* **Note 17**). **Figure 3** depicts myocytes after 0, 24, 48, and 72 h in culture with or without 10 mM BDM and ITS.

3.9. Applications of Freshly Isolated and Cultured Adult Mouse Myocytes

Recently, we have used freshly isolated and cultured myocytes in a number of experiments relevant to the study of cardiac function, hypertrophy, ischemia, and heart failure. Using freshly isolated myocytes, we measured myocyte shortening, and by loading myocytes with Fura-2AM, we also recorded calcium transients (*see* **Fig. 4**). In short-term cultured myocytes, we measured second-messenger generation (*see* **Fig. 5**) and signaling protein activation/phosphorylation by Western blot (*see* **Fig. 6**). We used freshly isolated myocytes to demonstrate that aortic constriction, a known hypertrophic stimulus, induces myocyte hypertrophy, quantified by measuring isolated myocyte size with a Coulter Multisizer, and fetal gene expression, shown by Western blot for β-myosin heavy-chain protein in isolated myocytes (*see* **Fig. 7**). In short-term cultured myocytes, we quantified myocyte apoptosis induced by hydrogen peroxide using Annexin V staining (*see* **Fig. 8**) (and TUNEL staining, not shown). Finally, we used adenoviral infection of long-term cultured myocytes to achieve high-efficiency gene transduction (*see* **Fig. 9**). These wide-ranging studies demonstrate the suitability of this culture model for studies of cardiac myocyte function, hypertrophy, signaling, and apoptosis.

3.10. Cleaning the Perfusion Rig

It is very important to have a routine protocol to clean the perfusion rig after each preparation, to minimize contamination. We use the following procedures.

1. Clean the profusion rig once a week with 1 M HCl. Perfuse and fill all parts of the rig with 1 M HCl, and incubate for 10 min, followed by three washes of sterile distilled water. Do not have the cannula attached when perfusing the rig with HCl, as the acid will corrode the metal part of the 20-G cannula.
2. Clean the rig every day with 70% ethanol before and after isolating a heart. Perfuse and fill all parts of the rig with 70% ethanol, and incubate for 10 min, followed by three washes of sterile distilled water.
3. Clean the cannula with 70% ethanol (do not use bleach) and autoclave with the surgical instruments before each isolation.

A No BDM/ITS

B BDM/ITS

0 hrs

0 hrs

24 hrs

24 hrs

48 hrs

48 hrs

72 hrs

72 hrs

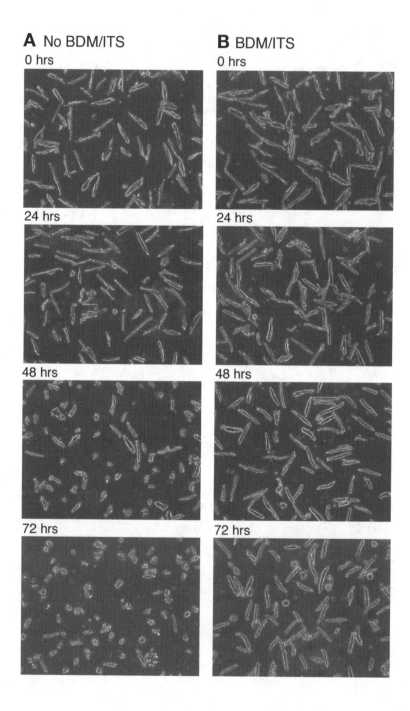

4. Notes

1. The temperature of the perfusate and the flow rate of the pump should be checked routinely. Over time, peristaltic pump tubing will fail and will need to be replaced (every 2–3 mo or 20–30 preparations) to maintain consistent flow rates. These routine checks are essential to maintain consistency in preparations.
2. In this protocol, we use isoflurane to anesthetize mice prior to removing the heart. Older and more conventional anesthetics like pentobarbital or ketamine have a much longer onset and significantly reduce respiration, which greatly increases the risk of myocardial ischemia. Isoflurane, with its rapid onset and minimal effects on respiration, avoids these problems, and we found our isolation procedure was dramatically improved by switching to isoflurane *(11)*.
3. Proper dissection of the heart is the key to a successful cannulation and perfusion. For optimal enzyme digestion of the heart and good myocyte yields, the aorta must be positioned on the cannula so that it does not pass through the aortic valve, which would prevent proper perfusion of the coronary arteries. When removing the heart, we routinely cut the transverse aorta between the carotid arteries. *See* Rokosh and Simpson for anatomy of the mouse aorta *(13)*. Too long a section of aorta will make the aorta harder to identify and to lift onto the cannula. Conversely, too short a section of aorta will make it harder to tie off the aorta on the cannula and increase the likelihood of pushing the cannula through the aortic valve, preventing good perfusion. When isolating myocytes from a heart after pressure overload by transverse aortic constriction, the proximal aorta will be quite dilated.
4. It is very highly recommended to use a magnifying lense or a dissecting microscope when cannulating the aorta, to make the heart and aorta easier to visualize and cannulate.
5. When cannulating the aorta, a rapid cannulation is essential for optimal enzyme digestion and good myocyte yields. We typically try to cannulate the heart, from removing the heart to starting perfusion, in less than 45 s.

Fig. 3. *(Opposite page)* Myocyte culture medium supplemented with 2, 3-butanedione monoxime (BDM) and insulin, transferrin and selenium (ITS) preserves normal rod-shaped morphology of adult mouse cardiac myocytes for 72 h. Myocytes were plated at a density of 50,000 rod-shaped myocytes per 35-mm culture dish and were cultured for 72 h in myocyte culture medium with or without 10 mM BDM and ITS medium supplement (10 μg/mL insulin, 5.5 μg/mL transferrin, and 5 ng/mL selenium). Myocytes cultured in (**A**) myocyte culture medium alone, or (**B**) myocyte culture medium supplemented with BDM and ITS, were photographed under phase-contrast microscopy (×100) after 0, 24, 48, and 72 h in culture. We consider rod-shaped myocytes to be the most physiological, because they mimic the cells in the intact heart. However, it should also be noted that "round" myocytes, sometimes called "meatballs" are not necessarily dead. In fact, many or most can be alive, which can be easily documented using vital cell stains.

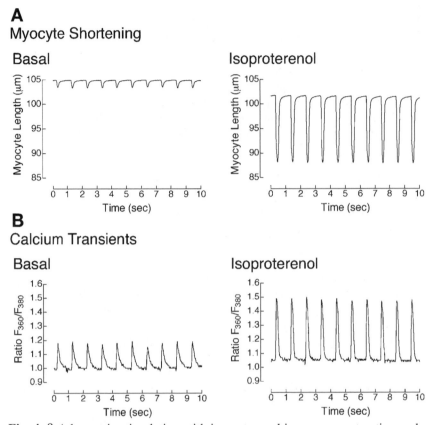

Fig. 4. β-Adrenergic stimulation with isoproterenol increases contraction and calcium transients in freshly isolated adult myocytes. Myocytes were plated and immediately loaded with the calcium indicator Fura-2AM at 25°C in HEPES buffer. For all measurements, myocytes were electrically paced (1 Hz, with 20 V pulse amplitude and 4 ms pulse duration) and basal and ligand-induced changes in length and calcium were recorded. **(A)** Contraction in a representative myocyte, measured as length change, is shown before and 5 min after treatment with isoproterenol (1 μM). The y-axis shows myocyte length in μm and the x-axis shows time in seconds. **(B)** Calcium transients are shown in a single myocyte before and 5 min after treatment with isoproterenol (1 μM). The y-axis shows Fura-2 emission ratios (360 nm/380 nm), and the x-axis shows time in seconds.

6. Note that none of the solutions are oxygenated. Other protocols use 95% O_2/5% CO_2 *(10)*, which, when used with a bicarbonate buffer, has the dual effect of oxygenating the solutions and maintaining pH. However, when the heart is arrested by perfusion with calcium free buffer, the oxygen demand of the muscle tissue is dramatically reduced. Therefore, we tested whether oxygenation of the perfusion solutions was necessary, and found that neither myocyte yields nor survival in

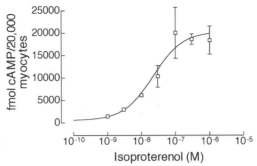

Fig. 5. β-Adrenergic stimulation with isoproterenol increases cAMP in cultured adult mouse myocytes. Myocytes were cultured for 18 h (overnight) in 35-mm dishes and then treated for 15 min with increasing concentrations of isoproterenol (1 n*M* to 1 μ*M*) in the presence of the phosphodiesterase inhibitor isobutyl-methyl-xanthine (1 μ*M*). After treatment, myocytes were lysed, and cAMP content was determined by ELISA (Amersham, Piscataway, NJ). The graph shows mean ± SEM, *n* = 3, with each measurement made in duplicate.

culture was affected if solutions were not oxygenated. The perfusion solution does contain a small amount of bicarbonate, as a metabolic requirement, but the pH is maintained by the Na-HEPES *(11)*.

7. In this protocol, we use collagenase type II (Worthington Biochemicals) to digest the heart. However, we have also used Blendzyme type I (Roche Molecular Biochemicals, Indianapolis, IN) with trypsin (Sigma), and a combination of collagenases B and D (Roche) *(11)*. In our experience, the yields using crude collagenase preparations are higher, but vary between lots, requiring testing different lots of enzyme and fine-tuning enzyme concentrations. On the other hand, Blendzyme, which is a recombinant collagenase mixture, is much more standardized, reducing variability between lots, but is more expensive and gives slightly lower yields.

8. In developing this protocol, we compared a constant flow perfusion system with a constant pressure system (70 mmHg) for myocyte isolation. Although both systems were workable, we found that the constant flow system was easier to use. This is because as the enzyme starts to digest the heart, coronary pressure increases, which significantly reduces flow and hinders digestion. We found that maintaining a constant flow at 4 mL/min reduced this problem and made the digestion process easier *(11)*.

9. After digestion the heart should appear pale and flaccid, and one might be able to see some muscle fiber dissociation on the surface of the heart. When taken down and initially cut into small pieces for further mechanical disruption with the 2-mm and 1.5-mm pipets, very little force should be needed and the heart should almost fall apart on its own.

10. The myocyte stopping buffer contains 10% CS. In the past, we tried BSA to inactivate the collagenase and other proteases, but found that serum was better *(11)*.

A ERK phosphorylation

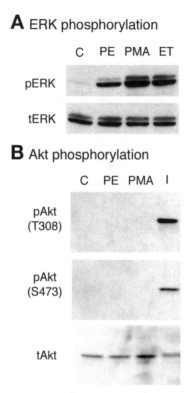

B Akt phosphorylation

Fig. 6. Hypertrophic agonists phosphorylate (activate) kinases by Western blot in cultured adult mouse myocytes. Myocytes were cultured for 18 h (overnight) in 35-mm dishes and then treated for 15 min with phenylephrine (PE; 20 μM, plus 2 μM Timolol), phorbol 12-myristate-13-acetate (PMA; 100 nM), endothelin (ET; 10 nM), or insulin (I; 6 μM). After 15 min, myocytes were washed once with PBS and lysed directly in Laemeli's sample buffer. Western blots were done to detect **(A)** phosphory-lated ERK (pERK) and total ERK (tERK), and **(B)** phosphorylated Akt (pAkt, threo-nine 308, or serine 473) and total Akt (tAkt). Antibodies were from Cell Signaling (Beverly, MA).

Collagenase is also inactivated by dilution, and a generous volume at this step additionally prevents myocyte damage during dissociation with the transfer pipets.

11. When myocytes are counted on the hemocytometer, great care must be taken when loading the counting chambers. The myocytes are very large, making it difficult to get an even dispersion in the counting chamber.

12. We found that adding ATP during the calcium reintroduction and to the plating medium improves the maintenance of rod-shaped morphology. We hypothesize that the ATP improves the metabolic state of the myocytes and helps to maintain calcium homeostasis while the cells are recovering from isolation and attaching to the culture dish.

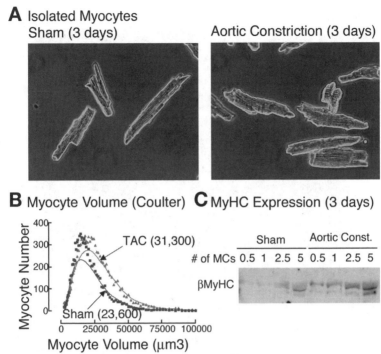

Fig. 7. Freshly isolated adult mouse myocytes document hypertrophy and fetal gene induction following pressure overload. Transverse aortic constriction (TAC) surgery was performed on mice to induce pressure overload hypertrophy *(15)*. After 3 d, myocytes were isolated and fixed immediately in 4% paraformaldehyde in phosphate-buffered saline (pH 7.4). **(A)** Myocytes from sham and TAC mice were photographed under phase microscopy at ×400. **(B)** Myocyte volumes were measured using a Coulter Multisizer *(11,12)*. The myocyte volume distribution compares myocyte volume (μm^3) vs myocyte number, demonstrating the right shift to increased cell volume after TAC. In parentheses are the median volume for sham myocytes (23,600 μm^3) and pressure-overloaded myocytes (TAC, 31,300 μm^3). **(C)** Freshly isolated myocytes were counted and lysed in 1.5X Laemeli's sample buffer, and 0.5 to 5×10^3 myocytes were used in Western blot for β-myosin heavy chain, a classical fetal hypertrophic marker gene, using the L2 antibody *(14)*, showing the increased β-myosin-heavy chain after TAC.

13. We plate the myocytes on laminin-coated dishes. However, we tried other attachment matrices, including collagen IV, fibronectin, poly-L-lysine, and gelatin. Only collagen IV performed as well as laminin, and could be substituted for laminin *(11)*.
14. To maintain rod-shaped morphology, we culture the myocytes in slightly acidic culture medium, pH approx 7.0. The myocyte culture medium base is MEM with HBSS, which contains 0.35 g/L of sodium bicarbonate. Normally, this medium is used in a 1% CO_2 incubator where the medium pH is 7.4, but when placed in a 2% CO_2 incubator, the medium pH is 6.9–7.0.

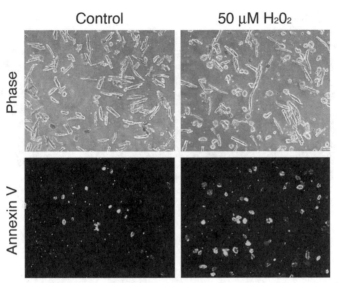

Fig. 8. Oxidative stress with H_2O_2 induces apoptosis by annexin V staining in cultured adult mouse myocytes. Myocytes were cultured for 18 h (overnight) and then treated for 2 h with hydrogen peroxide (H_2O_2, 50 μM), which is known to induce myocyte apoptosis. Myocytes were stained with FITC-Annexin V (Roche) by adding the Annexin V directly to the culture medium, to identify apoptotic cells as Annexin V-positive. Myocytes were photographed under phase and fluorescent microscopy at ×200, where Annexin V-positive myocytes are bright against the dark background (lower panels).

15. The myocytes are allowed to attach for 1 h. We have tried longer attachment times, but this did not improve the plating efficiency. We have also tried times as short as 20 min, but plating efficiency was not as good as 1 h. Plating time might require adjustment in one's own laboratory.

16. For short-term (24 h) culture of myocytes, we tested medium with and without 10 mM BDM and found that BDM caused a modest improvement in maintenance of rod-shaped morphology overnight. We found no effect of culturing myocytes in BDM on several different assays, including accumulation of cAMP, the phosphorylation of several signaling proteins including ERK and Akt, and the contractile response to β-adrenergic stimulation.

17. To maintain viable rod-shaped myocytes for more than 24 h, we tested several medium additives, including BDM, which inhibits spontaneous contractions; insulin, transferrin, and selenium medium supplement (ITS); ascorbic acid, an anti-oxidant; MnTMPyP, a superoxide dismutase inhibitor; caspase inhibitors; bongkrekic acid, an anti-apoptotic agent; verapamil, a calcium channel antagonist; and low-calcium medium. For optimal survival to 72 h, we found that the best results were obtained when myocyte culture medium was supplemented with 10 mM BDM and full-strength ITS medium supplement, which contained 1 μg/mL insulin, 0.55

A 24 hours
Control   MOI 100

B 72 hours
Control   MOI 100

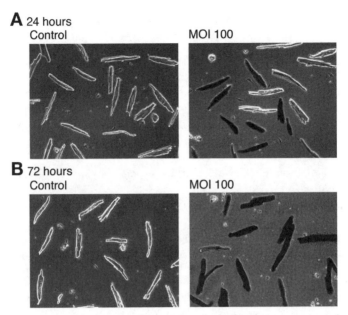

Fig. 9. High-efficiency infection of cultured rod-shaped adult mouse myocytes with adenovirus expressing β-galactosidase. Myocytes were infected at 0 h with adenovirus containing a β-galactosidase transgene at an MOI of 100, then cultured for 72 h in myocyte culture medium with 2, 3-butanedione monoxime and insulin, transferrin, and selenium. β-Galactosidase activity (dark-stained cells) was assayed at **(A)** 24 h and **(B)** 72 h. Myocytes are photographed under phase contrast microscopy (×100).

μg/mL transferrin, and 0.5 ng/mL selenium) *(11)*. We also found that 2.5 and 5 mM BDM were nearly as good as 10 mM BDM at maintaining rod-shaped morphology to 72 h, and were as good as 10 mM BDM at 48 h. In addition, the final concentration of the ITS used in our original work *(11)* could be reduced 100-fold for studies using insulin treatment. The current version of ITS (*see* **Table 6**) has slightly lower concentrations than our original *(11)*.

Acknowledgments

Marietta Paningbatan, Philip Swigart, Yuan Huang, and Luyi Li provided expert technical assistance. Keng-Mean Lin, PhD, at the Alliance for Cellular Signaling (AfCS) Laboratory at the University of Texas, Southwestern, did the experiments with myocyte contraction and calcium *(11)*. Maike Krenz and Jeff Robbins generously provided the β-myosin heavy-chain antibody (**Fig. 7**; **ref. *14***). The Department of Veterans Affairs Research Service, the National Institutes of Health, and the American Heart Association, Western States Affiliate, provided support.

References

1. Bers, D. M. (2002) Cardiac excitation-contraction coupling. *Nature* **415**, 198–205.
2. Hunter, J. J. and Chien, K. R. (1999) Signaling pathways for cardiac hypertrophy and failure. *N. Engl. J. Med.* **341**, 1276–1283.
3. Sugden, P. H. (2001) Signalling pathways in cardiac myocyte hypertrophy. *Ann. Med.* **33**, 611–622.
4. Solaro, R. J., Montgomery, D. M., Wang, L., et al. (2002) Integration of pathways that signal cardiac growth with modulation of myofilament activity. *J. Nucl. Cardiol.* **9**, 523–533.
5. Frey, N. and Olson, E. N. (2003) Cardiac hypertrophy: the good, the bad, and the ugly. *Annu. Rev. Physiol.* **65**, 45–79.
6. Dorn, G. W. 2nd and Force, T. (2005) Protein kinase cascades in the regulation of cardiac hypertrophy. *J. Clin. Invest.* **115**, 527–537.
7. Benjamin, I. J. and Schneider, M. D. (2005) Learning from failure: congestive heart failure in the postgenomic age. *J. Clin. Invest.* **115**, 495–499.
8. Wolska, B. M. and Solaro, R. J. (1996) Method for isolation of adult mouse cardiac myocytes for studies of contraction and microfluorimetry. *Am. J. Physiol.* **271**, H1250–1255.
9. Hilal-Dandan, R., Kanter, J. R., and Brunton, L. L. (2000) Characterization of G-protein signaling in ventricular myocytes from the adult mouse heart: differences from the rat. *J. Mol. Cell. Cardiol.* **32**, 1211–1221.
10. Zhou, Y. Y., Wang, S. Q., Zhu, W. Z., et al. (2000) Culture and adenoviral infection of adult mouse cardiac myocytes: methods for cellular genetic physiology. *Am. J. Physiol. Heart Circ. Physiol.* **279**, H429–436.
11. O'Connell, T. D., Ni, Y., Lin, K. M., Han, H., and Yan, Z. (2003) Isolation and culture of adult mouse cardiac myocytes for signaling studies. *AfCS Research Reports*, vol. 1. Available at www.signaling-gateway.org. Last accessed July 17, 2000.
12. O'Connell, T. D., Ishizaka, S., Nakamura, A., et al. (2003) The alpha(1A/C)- and alpha(1B)-adrenergic receptors are required for physiological cardiac hypertrophy in the double-knockout mouse. *J. Clin. Invest.* **111**, 1783–1791.
13. Rokosh, D. G. and Simpson, P. C. (2002) Knockout of the alpha 1A/C-adrenergic receptor subtype: the alpha 1A/C is expressed in resistance arteries and is required to maintain arterial blood pressure. *Proc. Natl. Acad. Sci. USA* **99**, 9474–9479.
14. Krenz, M., Sanbe, A., Bouyer-Dalloz, F., et al. (2003) Analysis of myosin heavy chain functionality in the heart. *J. Biol. Chem.* **278**, 17,466–17,474.
15. Rockman, H. A., Ross, R. S., Harris, A. N., et al. (1991) Segregation of atrial-specific and inducible expression of an atrial natriuretic factor transgene in an in vivo murine model of cardiac hypertrophy. *Proc. Natl. Acad. Sci. USA* **88**, 8277–8281.

22

Proteomic Analysis of Foam Cells

Peng-Yuan Yang, Yao-Cheng Rui, Peng-Yuan Yang, and Yan-Ling Yu

Summary

Foam cells are characteristic pathological cells in the lesions of atherosclerosis. Previous works have established macrophage-derived foam cell model to study the central role of the foam cells, and analyzed the protein expression profiles in foam cells. The reported in vitro foam cell model was established by incubating the human U937 cells with oxidized low-density lipoprotein. The global changes in protein expressions between U937 foam cell and normal U937 cells were measured with two-dimensional gel electrophoresis, and some interested proteins were tryptic-digested and then identified via mass spectrometry after capillary liquid chromatography separation. Some of the identified proteins were validated via the Internet links to the U937 proteomic map provided from the Expasy Proteomics server (http://us.expasy.org). The experimental data can provide potential markers during the inflammatory reactions for atherosclerotic studies.

Key Words: Foam cell; monocyte; oxidized low-density lipoprotein; two-dimensional gel electrophoresis; mass spectrometry.

1. Introduction

Atherosclerosis is not merely a disease in its own right, but a process that contributes principally to the pathogenesis of myocardial and cerebral infarction, gangrene, and loss of function in the extremities *(1,2)*. During the process of atherosclerosis, monocytes seem to play a central role. Once monocytes adhere to the subendothelial space and enter into the intima of the artery, oxidized low-density lipoprotein (ox-LDL) and other substances associated with atherosclerosis may participate in transformation of the monocytes in macrophage. As shown in **Fig. 1**, uptake of ox-LDL by the macrophage through scavenger receptors will lead to foam cells formation *(3,4)*.

The macrophage-derived foam cells not only result in formation of fatty streaks, which are believed to represent the earliest type of atherosclerotic plaque, but also

From: *Methods in Molecular Biology, vol. 357: Cardiovascular Proteomics: Methods and Protocols*
Edited by: F. Vivanco © Humana Press Inc., Totowa, NJ

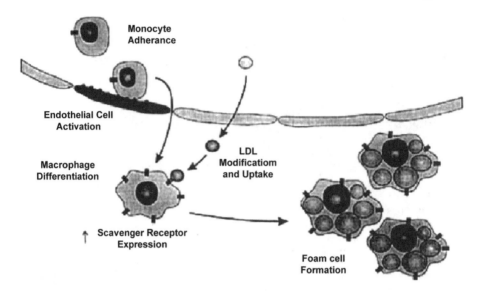

Fig. 1. The process of macrophage-derived foam cell formation.

play a role in the fibroproliferative process by their capacity to form numerous growth factors, in particular platelet-derived growth factor *(5)*, vascular endothelial growth factor *(6,7)*, interleukin (IL)-1 *(8)*, and tumor necrosis factor-α *(9)*. Therefore, a further study on the global changes in protein expression accompanying the formation of foam cells in vivo becomes an urgent mission *(10)*.

The macrophage-derived foam cell model *(11,12)* can be established through incubating the human monoblastic leukemia (U937) cells with ox-LDL. The intercellular adhesion molecule-1 and vascular endothelial growth factor expression levels have been determined to be the same as the foam cells in the atherosclerotic plaques using Western blot analysis, which confirms the successful establishment of the pathological model. The experimental data can provide potential markers during the inflammatory reactions for atherosclerotic studies.

2. Materials

2.1. Cell Culture

1. RPMI-1640 medium (Gibco/BRL, Bethesda, MD) supplemented with 10% fetal bovine serum (FBS, HyClone, Ogden, UT) (*see* **Note 1**).
2. The human monocyte U937 cell line was obtained from Cell Bank in Shanghai Institution of Biological Sciences, Chinese Academy of Science (*see* **Note 2**).
3. LDL (d = 1.019 to 1.063 g/mL; Sigma, St. Louis, MO) was sterilized by filtration through 0.45-μm Millipore membrane, and stored at 4°C.
4. Oil red O, $CuSO_4$ and EDTA (Sigma). Store at room temperature.

2.2. Protein Extract and Electrophoresis

1. Extract solution: containing 9 M urea, 2% CHAPS, 0.5% dithiothreitol, and 0.14% phenylmethyl sulfonyl fluoride.
2. Rehydration solution: containing 9 M urea, 2% CHAPS, 0.5% dithiothreitol, 0.14% phenylmethyl sulfonyl fluoride, 0.2% ampholyte (pH = 3.0–10.0, NL), and trace bromophenol blue.
3. Thirty percent acrylamide/*bis* solution (37.5:1 with 2.6% C) (this is a neurotoxin when unpolymerized and so care should be taken not to receive exposure) and (TEMED; Bio-Rad, Hercules, CA).
4. Separating buffer (4X): 1.5 M Tris-HCl, pH 8.7, 0.4% sodium dodecyl sulfate (SDS). Store at room temperature.

3. Methods

To further investigate the global changes in protein expressions between the U937 foam cells and the normal U937 cells, two-dimensional gel electrophoresis (2DE) can be applied to investigate the protein expression level of whole cell extract from monocyte, and macrophage-derived foam cell, individually. After 2DE, the protein spots of interest, especially those that changed significantly in quantity, were in-gel digested and identified by tandem mass spectrometry (MS/MS). Some protein spots were validated via the U937 proteomic map provided by links to the Expasy Proteomics Server.

3.1. Cell Culture

1. ox-LDL preparation: After EDTA was removed by dialysis, LDL was oxidized by incubating in CuSO4 10 μmol/L for 16 h at 37°C, and then dialyzed in phosphate-buffered saline (PBS) containing 0.1 mmol/L EDTA for 24 h at 4°C (*see* **Note 3**).
2. Foam cell model was established by incubating U937 cells in ox-LDL 80 mg/L for 48 h *(11,12)*. The control group was the normal U937 cells that had not been treated with ox-LDL.
3. Oil red O dyeing: the foam cells were collected and fixed with 4% paraformaldehyde for 12 h. The cells were then treated with fresh 0.3% oil red O for 20 min (**Fig. 2**).

3.2. Protein Extraction and Quantification

1. The cells were disrupted by ultrasonication (Sonics & Materials Inc., Vibra Cell™) twice using pulse mode 10 times each, i.e., 1 s on and 1 s off for one time, on ice in extract solution.
2. The resulting homogenate was centrifuged for 10 min at 12,000g. The supernatant was fractioned in aliquots and stored at −20°C until further analysis.
3. Total protein quantitation was performed as described previously *(13)*. Prepare standards and samples in duplicate to minimize some occasional error during the quantitative course.

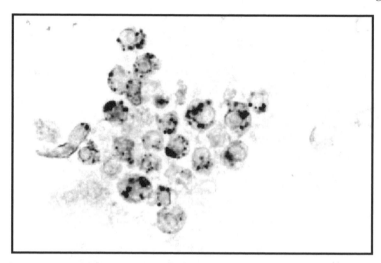

Fig. 2. The morphological form of the U937 foam cells dyed with oil red O after incubation with ox-LDL (×400).

3.4. Two-Dimensional Gel Electrophoresis

1. Each sample containing 500 μg of protein, in rehydration solution, was added to immobilized pH gradient (IPG) strips (Amersham Pharmacia, Biotech), 18 cm, nonlinearly covering a pH range of 3.0–10.0. Samples were allowed to rehydrate in 350 μL of the above solution for more than 12 h under mineral oil.
2. Isoelectric focusing electrophoresis was carried out on an IPGphor (Amersham), and the following voltage/time profile was used:
   a. Increasing voltage by gradient from 0–250 V for 30 min.
   b. 250–1000 V for 1 h.
   c. 1000–2000 V for 1 h.
   d. 2000–4000 V for 1 h.
   e. Increasing voltage by step-and-hold (rapidly) from 4000–6000 V for 2 h.
   f. 6000–8000 V for 2 h.
   g. Final phase of 8000 V for 60000 V h.
3. 2DE: the 2D (12%) was carried out using a Hoefer electrophoresis unit SE600 series (Amersham). Each gel was run at 20 mA at the beginning and 30 mA per gel until all the proteins were transferred from the strip into the gel. The run was completed when the bromophenol blue front reached the bottom of the gel.

3.5. Image Acquisition and Analysis

1. Staining: the gels were stained by colloidal Coomassie blue staining method *(14)*.
2. Images analysis: Coomassie blue-stained gels were scanned using GS-800 Calibrated Densitometer (Bio-Rad), and then processed by PDquest 2D-image-analysis software (Bio-Rad). Two separate gels were analyzed in order to minimize the contribution of experimental variations and only those spots displaying the same pattern in both samples were selected for further analysis (**Fig. 3**).

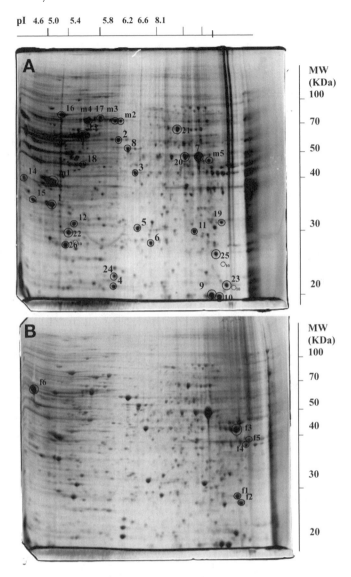

Fig. 3. (A) Proteins extracted from normal U937 cells; **(B)** proteins extracted from macrophage-derived foam cells. 2DE image of the cellular proteins extracted from the human U937 cells and macrophage-derived foam cells. The cells were disrupted by ultrasonication on ice with extraction solution containing 9 *M* urea, 2% CHAPS, 0.5% DTT, and 0.14% phenylmethyl sulfonyl fluoride. A 500-μg sample of proteins was separated by isoelectric focusing (IPG: 18 cm, nonlinearly covering a pH range of 3.0–10.0) in the first dimension and by 12% sodium dodecyl sulfate-polyacrilamide gel electrophoresis gel in the second dimension; the gel was stained with Coomassie blue. The circled protein spots are identified by MS/MS in this study.

3.6. In-Gel Digestion

1. The protein spots of interest were excised from the gel. Spots were cut into small pieces.
2. Pieces were destained twice with 60 µL 200 mmol/L NH$_4$HCO$_3$/acetonitrile (ACN) (50:50 v/v); supernatants were discarded.
3. The gel pieces were shrunk by dehydration in 60 µL of ACN twice, and completely dried at 37°C for about 20 min.
4. The gel pieces were swollen in a digestion buffer containing 100 mmol/L NH$_4$HCO$_3$ and 12.5 ng/µL trypsin (sequencing-grade, Roche Diagnostics) at 4°C.
5. After 30 min of incubation, the supernatants were removed and replaced with 20 µL 20 mmol/L NH$_4$HCO$_3$ without trypsin, to keep the gel pieces wet during enzymatic cleavage (37°C, more than 12 h).
6. Peptides were extracted by two changes of 5% TFA in 50% ACN (sonication 10 min for each change) at room temperature. The extracts were combined and concentrated to about 5 µL under the protection of a N$_2$ current.

3.7. Liquid Chromatography-MS and Liquid Chromatography-MS/MS Peptides Analysis

1. Integral high-performance liquid chromatography workstation (1100 series, Agilent) was linked to the electrospray interface of an ion trap mass spectrometer (Esquire 3000, Bruker) by an auxiliary six-port sampling valve.
2. A 0.3 × 150 mm Zorbax C18 column (Agilent) was employed for online peptide separation. The flow rate of high-performance liquid chromatography was 80 µL/min.
3. The flow was split about 20:1 via a three-port joint before the injector such that the flow rate through the column was about 4 µL /min. Mobile phase A was 0.1% trifluoroacetic acid in water, and mobile phase B was 0.1% trifluoroacetic acid in ACN.
4. Analyte was introduced into the sample loop and eluted with a gradient of 5/65/95/95% B at 5/35/45/55 min, respectively. All of the mass spectra were acquired in the full scan mode (*m/z* 150–1400).
5. MS/MS spectra were acquired automatically and the subsequent operation in MS/MS mode was on a data-dependent basis. Dynamic exclusion was enabled after one spectrum of the same parent ion and released after 0.5 min (**Fig. 4**).

3.8. Database Searching

1. Collision-activated dissociation spectra were analyzed by searching against an NCBInr database using Mascot (Matrix Science, London) search software (Bruker) via MS/MS fragments mode.
2. The *Homo sapiens* (human) subdatabase was used and one missing cleavage of trypsin was added to the database searching.

4. Notes

1. All the water used in the experiments was 18-MΩ deionized from a Milli-Q® water purification system; all other solvents used were of analytical pure grade.

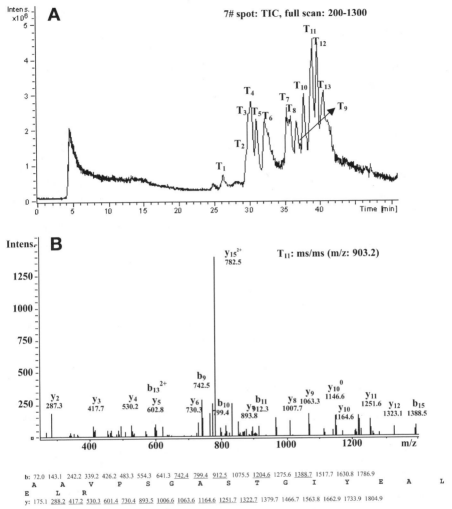

Fig. 4. Liquid chromatography-MS analysis of tryptic digests from spot no. 7. **(A)** Total ion chromatograph (TIC) showing the peptides from which 13 peptides (T1–T13) were confidently matched to 2-phosphopyruvate-hydratase α-enolase. **(B)** MS/MS spectrum of the doubly charged precursor ion with *m/z* 903.2 (T11) at the retention time of 38.6 min. A sequence is confirmed as AAVPSGASTGIYEALELR from the labeled b- and y-ions in the spectrum. Fragments observed in the spectrum are underlined and assigned.

2. Some reports have indicated that other monocyte lines, such as THP-1 *(15)*, can also be used in incubating macrophage-derived foam cells.
3. The oxidization rate of ox-LDL can be measured by quantitating the content of malondialdehyde or the rate of electrophoretic mobility of LDL.

Acknowledgments

The research was supported by National Basic Research Priorities Programme (2001CB510202), National High-Tech Programme (2001AA233031,2002AA2Z 346C), Shanghai Science and Technology Developing Programme (01JC14011), and National Natural Science Foundation of China (30300454, 39870870).

References

1. Fazio, S. and Linton, M. F. (2001) The inflamed plaque: cytokine production and cellular cholesterol balance in the vessel wall. *Am. J. Cardiol.* **88,** 12E–5E.
2. Steinberg, D. (1997) Low density lipoprotein oxidation and its pathobiological significance. *J. Biol. Chem.* **272,** 20,963–20,966.
3. Berliner, J. A. and Heinecke, J. W. (1996) The role of oxidized lipoproteins in atherogenesis. *Free Radical Biol. Med.* **20,** 707–727.
4. Russell, R. (1993) The pathogenesis of atherosclerosis: a perspective for the 1990s. *Nature* **362,** 801–809.
5. Ross, R., Masada, J., Raines, E. W., et al. (1990) Location of PDGF-β protein in macrophage on all phases of atherogenesis. *Science* **248,** 1009–1012.
6. Inoue, M., Itoh, H., Ueda, M., Naruko, T., Kojima, A., and Komatsu, R. (1998) Vascular endothelial growth factor (VEGF) expression in human coronary atherosclerotic lesions: possible pathophysiological significance of VEGF in progression of atherosclerosis. *Circulation* **98,** 2008–2016.
7. Ramos, M. A., Kuzuya, M., Esaki, T., et al. (1998) Induction of macrophage VEGF in response to oxidized LDL and VEGF accumulation in human atherosclerotic lesions. *Arterioscler. Thromb.* **18,** 1188–1196.
8. Libby, P., Friedman, G. B., and Salomon, R. N. (1989) Cytokines as modulators of cell proliferation in fibrotic diseases. *Am. Rev. Respir. Dis.* **243,** 393–396.
9. Libby, P., Ordovas, J. M., Auger, K. R., Robbins, A. H., Birinyi, L. K., and Dinarello, C. A. (1986) Endotoxin and tumour necrosis factor induce interleukin-1 gene expression in adult human vascular endothelial cells. *Am. J. Pathol.* **124,** 179–185.
10. Yu, Y. L., Huang, Z. Y., Yang, P. Y., Yang, P. Y., Rui, Y. C., and Yang, P. Y. (2003) Proteomic studies of macrophage-derived foam cell from human U937 cell line using 2D gel electrophoresis and tandem mass spectrometry. *J. Cardiovas. Pharmacol.* **42,** 782–789.
11. Yang, P. Y. and Rui, Y. C. (2003) Intercellular adhesion molecule-1 and vascular endothelial growth factor expression kinetics in macrophage-derived foam cells. *Life Sci.* **74,** 471–480.
12. Yang, P. Y., Rui, Y. C., Li, K., Huang, X. H., Jiang, J. M., and Yu, L. (2002) Expression of intercellular adhension molecule-1 in U937 foam cells and the inhibitory effect of imperatorin. *Acta Phamacol. Sin.* **23,** 327–330.
13. Neuhoff, V., Arold, N., Taube, D., and Ehrhardt, W. (1988) Improved staining of proteins in polyacrylamide gels including isoelectric focusing gels with clear background at nanogram sensitity using Coomassie brilliant blue G-250 and R-250. *Electrophoresis* **9,** 255–262.

14. Ramagli, L. S. and Rodriguez, L. V. (1985) Quantitation of microgram amounts of protein in two-dimensional polyacrylamide gel electrophoresis sample buffer. *Electrophoresis* **6,** 559–563.

15. Kawakami, A., Tani, M., Chiba, T., et al. (2005) Pitavastatin inhibits remnant lipoprotein-induced macrophage foam cell formation through ApoB48 receptor-dependent mechanism. *Arterioscler. Thromb. Vasc. Biol.* **25,** 424–429.

23

Isolation of the Platelet Releasate

Judith Coppinger, Desmond J. Fitzgerald, and Patricia B. Maguire

Summary

This chapter describes an approach to isolate, separate, and identify the contents of the platelet releasate, a fraction highly enriched for platelet granular and exosomal contents. Investigation into such a fraction will improve our understanding of platelet interactions with other cells, vascular remodeling, coagulation, and vessel growth.

Key Words: Platelet; secretion; releasate; proteomics.

1. Introduction

Mapping the complete complement of proteins expressed by a cell may provide valuable information on the identities of its biological constituents. However, such a task is technically difficult and may not provide the relevant functional insights sought after. In fact, the concentration of proteins in the cell vary by as much as eight orders of magnitude, meaning abundant housekeeping proteins overwhelm any detection system (1). If proteome analysis is to provide any truly meaningful information about cellular and regulatory processes, it must be able to penetrate to the level of regulatory proteins, which are often less abundant. Consequently, it is often preferable to prepare a set of subproteomes. Fractionation methods exploiting specific protein characteristics, such as their inherent chemical properties or differential cellular compartmentalization, may be used to select specific subproteomes (**ref. 2**; *see* Chapters 7 and 8). Focusing on specific functional compartments is particularly useful in analyzing complex structures such as platelets.

Platelets contain a number of preformed, morphologically distinguishable storage granules—α-granules, dense granules, and lysosomes—the contents of which are released upon platelet activation (3). Platelets also release two distinct

From: *Methods in Molecular Biology, vol. 357: Cardiovascular Proteomics: Methods and Protocols*
Edited by: F. Vivanco © Humana Press Inc., Totowa, NJ

membrane vesicle populations during activation: cell surface-derived micro-vesicles and exosomes of endosomal origin *(4)*. Platelet-secreted proteins play important roles in cell to cell communication and contribute to the development of diseases such as atherosclerosis; for example, platelet-derived growth factor is a regulator of smooth muscle cell proliferation, and increased levels of this protein have been observed in atherosclerotic lesions *(5)*. Here, we describe a centrifugation approach to isolate, separate, and identify the contents of the platelet releasate, a fraction highly enriched for platelet granular and exosomal contents. Proteomics analysis of this fraction using two-dimensional gel electrophoresis, multidimensional chromatography, and mass spectrometry has led to the identification of more than 300 proteins, many of which were not previously attributed to platelets and three of which, although absent in normal vasculature, were identified in human atherosclerotic plaques *(6)*. Additionally, isolation of the platelet releasate fraction has proven useful in metastasis studies, where it has been shown to investigate tumor cells adhering to the endothelial wall *(7)*. Further investigation into the extracellular functions of these novel secreted proteins will play a key role in improving our understanding of cell–cell interactions, vascular remodeling, coagulation, and vessel growth. Indeed, neutralization of platelet-derived pro-inflammatory factors may become an interesting means for therapeutic or preventative intervention in atherosclerosis.

2. Materials

1. A butterfly 19G needle and a 60-mL plastic syringe to draw blood.
2. Acid citrate dextrose (ACD) anticoagulant: 38 mM anhydrous citric acid, 75 mM sodium citrate, 124 mM D-glucose. Store at 4°C. Discard if left for more than 1 h at room temperature.
3. Plastic Pasteur pipets to transfer blood into different tubes.
4. 15- and 50-mL capped plastic tubes to centrifuge the blood.
5. Prostaglandin E1 (PGE$_1$; e.g., Sigma Aldrich). Store at −80°C.
6. Calcium chloride.
7. Modified HEPES buffer (JNL buffer): 130 mM NaCl, 10 mM trisodium citrate, 9 mM NaHCO$_3$, 6 mM dextrose, 0.9 mM MgCl$_2$, 0.81 mM KH$_2$PO$_4$, 10 mM Tris-HCl, pH to 7.4 with ACD. Store at 4°C and discard after use.
8. A benchtop centrifuge and attachable rotor with a spin capacity of at least 1000g_{max} with removable 15- and 50-mL swinging buckets (e.g., Eppendorf 5804 centrifuge).
9. A 37°C aggregometer (e.g., Biodata PAP 4).
10. Thrombin.
11. Protease inhibitor cocktail (e.g., Calbiochem).
12. A temperature-controlled microcentrifuge and attachable rotor with a spin capacity of at least 12,000g_{max} and 1.5- or 2-mL tube capacity (e.g., Eppendorf 5415 centrifuge).

13. An ultracentrifuge with a fixed-angle rotor for spinning small (2–6-mL) tubes with a spin capacity of $50,000g_{max}$ (e.g., Beckman Ultracentrifuge).
14. Protein concentrators with a low cutoff point, such as 3 kDa, to capture a wide range of proteins and remove salts.

3. Methods

1. Draw 60 mL of blood from a healthy human volunteer free from medication for 10 d into 0.15% (v/v) ACD anticoagulant using a Butterfly 19G needle (*see* **Note 1**).
2. Dispense 5 mL of blood into 12 × 15-mL tubes (*see* **Note 2**). Centrifuge the blood at $150g_{max}$ for 10 min at room temperature (*see* **Note 3**). The upper layer should contain the platelet-rich plasma (PRP), the interphase should contain the buffy coat/white cells, and the lower layer should contain the red blood cells.
3. Remove PRP and pool (*see* **Note 4**).
4. Acidify the PRP to pH 6.5 with ACD (drop by drop, pH can change rapidly) and add $1\mu M$ PGE$_1$ to prevent platelet activation during centrifugation and handling. (**Caution**: effect only lasts for 20 min.)
5. Centrifuge the PRP again at $150g_{max}$ for 10 min to remove any remaining white and red blood cell contamination.
6. Pellet platelets from PRP by centrifuging at $720g_{max}$ for 10 min at room temperature and resuspend in a modified HEPES buffer (JNL buffer) that has been supplemented with 1.8 mM CaCl$_2$ (*see* **Note 4**).
7. Record a platelet count and adjust the count to 2×10^8 platelets/mL using a JNL buffer.
8. Activate platelet at 37°C in an aggregometer (BioData PAP-4) under constant stirring (1100 rev/min), using Thrombin (0.1 or 0.5 U/mL) or an agonist of choice. It is important to thaw platelet agonists on ice and dilute from stock solutions in cold water. Record all changes in light transmission against a blank of JNL buffer (100% light transmission) (*see* **Note 5**).
9. Following 3 min of activation, platelets are immediately placed on ice and the following solutions added to both control and activated samples (*see* **Note 6**):
   a. 10 μL of a 1 mM stock solution of phenylmethylsulfonylfluoride.
   b. 10 μL of protease inhibitor cocktail (Calbiochem).
   c. 1 mM PGE$_1$.
10. All steps are now carried out on ice or at 4°C.
11. Following activation, remove all intact and clumped platelets by carefully centrifuging sequentially twice at 4°C in 1.5-mL Eppendorf tubes at $1000g_{max}$ for 10 min and harvest the supernatant (*see* **Note 7**).
12. Ultracentrifuge the supernatant for 1 h at 4°C at $50,000g_{max}$ using a 50.4 Ti rotor to remove microvesicle contamination. Remove and keep supernatant on ice for further analysis or freeze samples at −80°C for further use (*see* **Notes 8** and **9**).
13. Concentrate a sample for protein determination using centrifugal filter devices with a cutoff point of 3000 Dalton according to the manufacturer's instructions. Subsequently concentrate the other samples to the volume required to obtain the necessary concentration for proteomic analysis (*see* **Note 10**).

4. Notes

1. Draw the blood slowly in order to prevent unnecessary platelet activation.
2. Transfer blood between tubes gently again to maintain platelets in a resting state.
3. Set a break on the centrifuge so the centrifugation does not come to a halt too suddenly.
4. When separating PRP layer, be careful not to take up red cells and white cell. Take care also to remove as much plasma as possible from the platelet pellet before adding the JNL buffer.
5. Blank the aggregometer with JNL prior to activation of the platelets and make sure that you add the agonist to the bottom of the tube so correct mixing can occur. It is advisable to first set aside a platelet sample to test platelet viability with agonist.
6. This prevents protein degradation and further platelet activation, especially in control samples.
7. It is essential to perform two gentle spins when removing the platelet from the supernatant to ensure the secreted fraction is platelet-free. Ideally, use 1.5-mL tubes so the pellet can be visualized. Be extremely gentle and cautious when removing the supernatant fraction.
8. If your sample is less than the volume of your ultracentrifuge tubes, you can add paraffin oil to fill them and then you must heat-seal the tubes. Ensure all tubes are balanced properly and be very careful removing supernatant again after ultracentrifugation step because the pellet is invisible.
9. When concentrating samples in the microcentrifuge, watch the volumes because not all samples concentrate at the same level, depending on their protein content.
10. Approximately 10 µg of protein is required for a 7-cm one-dimensional Coomassie-stained gel. Approximately 80 µg and 500 µg of protein are required for an 18-cm silver-stained and Coomassie blue stained gel, respectively. Approximately 50 µg of protein is required for one multidimensional chromatography run. Releasate samples may have to be pooled to reach the concentrations required for proteomic analysis.

Acknowledgments

Supported in part by a fellowship from Enterprise Ireland, a research grant from the Health Research Board of Ireland, and funding under the Programme for Research in Third Level Institutions (PRTLI), administered by the Higher Education Authority of Ireland.

References

1. Maguire, P. B. and Fitzgerald, D. J. (2003) Platelet proteomics. *J. Thromb. Haemost.* **1,** 1593–1601.
2. Jung, E., Heller, M., Sanchez, J. C., and Hochstrasser, D. F. (2000) Proteomics meets cell biology: the establishment of subcellular proteomes. *Electrophoresis* **21,** 3369–3377.

3. Fukami, H., Holmsen, H., Kowalska, M., and Niewiarowski, S. (2001) Platelet secretion, in *Haemostasis and Thrombosis: Basic Principles and Clinical Practice* (Colman, R. W., Hirsh, J., Marder, V. J., Clowes, A. W., and George, J. N., eds.), Lippincott Williams & Wilkins, Philadelphia, PA, pp. 561–574.

4. Heijnen, H. F., Schiel, A. E., Fijnheer, R., Geuze, H. J., and Sixma, J. J. (1999) Activated platelets release two types of membrane vesicles: microvesicles by surface shedding and exosomes derived from exocytosis of multivesicular bodies and alpha-granules. *Blood* **94,** 3791–3799.

5. Raines, E. W. (2004) PDGF and cardiovascular disease. *Cytokine Growth Factor Rev.* **15,** 237–254.

6. Coppinger, J. A., Cagney, G., Toomey, S., et al. (2004) Characterization of the proteins released from activated platelets leads to localization of novel platelet proteins in human atherosclerotic lesions. *Blood* **103,** 2096–2104.

7. Lawler, K., Meade, G., O'Sullivan, G., and Kenny, D. (2004) Shear stress modulates the interaction of platelet-secreted matrix proteins with tumor cells through the integrin alphavbeta3. *Am. J. Physiol. Cell Physiol.* **287,** C1320–1327.

24

Enrichment of Phosphotyrosine Proteome of Human Platelets by Immunoprecipitation

Martina Foy, Donal F. Harney, Kieran Wynne, and Patricia B. Maguire

Summary

Proteomics offers the opportunity to comprehensively investigate the anucleate platelet. Here, we present a detailed procedure for enrichment by immunoprecipitation, using the monoclonal antibody 4G10, of the dynamic phosphotyrosine proteome of human platelets. Such an approach offers the possibility of capturing the dynamic tyrosine phosphorylation events that occur upon platelet activation and aggregation, with an aim to identify novel signaling proteins.

Key Words: Platelets; signaling; tyrosine phosphorylation; immunoprecipitation; proteomics.

1. Introduction

Platelet activation with adhesive ligands (von Willebrand factor and collagen) and/or excitatory agonists (e.g., thrombin and adenosine diphosphate) trigger a series of signaling events within the platelet that ultimately convert the integrin receptor $\alpha_{IIb}\beta_3$ into an active conformation. Fibrinogen binds to the active $\alpha_{IIb}\beta_3$ and initiates another wave of signal cascades, resulting in platelet aggregation. Such signaling pathways are regulated both positively and negatively by protein phosphorylation (1). Indeed, a multitude of lipid kinases, phosphatases, phospholipases, and GTPases can be tyrosine-phosphorylated during platelet activation. These include extracellular signal-transduction kinase 2 and the cytosolic tyrosine kinases Syk and Src (2–4). However, it is likely that additional protein phosphorylation events contribute to the dynamic changes involved in platelet activation and aggregation.

Proteomic studies offer a novel approach in providing a more comprehensive overview of these phosphorylation cascades in the anucleate platelet. Furthermore, when analyzing a complex structure such as a platelet, high-abundance

From: *Methods in Molecular Biology, vol. 357: Cardiovascular Proteomics: Methods and Protocols*
Edited by: F. Vivanco © Humana Press Inc., Totowa, NJ

proteins often mask the presence of low-abundance proteins. Therefore, sample prefractionation is essential to facilitate the identification of low-abundance signaling proteins. To date, the most comprehensive proteomic studies of tyrosine phosphorylation events in platelets employed affinity-based prefractionation prior to gel electrophoresis and mass spectrometry *(5,6)*. Using such methodology, we have investigated the dynamic phosphotyrosine changes that occur after thrombin-induced platelet activation *(5)*. Phospho-proteins were enriched by immunoprecipitation using the monoclonal antibody (MAb) 4G10 and separated by two-dimensional gel electrophoresis (2DE). Sixty-seven protein spots were reproducibly found to be unique in the thrombin-activated phospho-proteome when compared to resting platelets. Matrix-assisted laser desorption/ionization time-of-flight mass spectrometry and immunoblotting were used to positively identify several of these proteins, including focal adhesion kinase, Syk, and a mitogen-activated protein kinase *(5)*. Recently, we have extended this analysis and have identified the reversible tyrosine phosphorylation of the non-muscle myosin heavy-chain type IIA. Marcus et al. *(6)* have employed a similar methodology to identify 28 candidate tyrosine-phosphorylated proteins in thrombin-activated platelets, including filamin and tubulin. This chapter presents a detailed procedure for the enrichment of the phosphotyrosine proteome of human platelets by immunoprecipitation using a specific phosphotyrosine MAb (4G10), with extensive washing and preclearing steps. The application of proteomics to the investigation of the phosphotyrosine signaling networks in human platelets may elucidate key elements in platelet response and potential drug targets for the treatment of coronary thrombosis.

2. Materials

1. Magnetic particle concentrator (MPC).
2. End-over-end tube rotator.
3. Protein A Dynabeads (requires 25 µL per 500-µL sample) (Dynal, Wirral, UK). Store at 4°C and keep on ice during use.
4. 4G10 MAb (requires 1 µL per 500-µL sample) (Upstate Biotechnology, Buckingham, UK).
5. Bovine serum albumin (BSA). 5% BSA in JNL buffer. Always prepare a fresh solution.
6. Acid citrate dextrose anticoagulant: 38 mM anhydrous citric acid, 75 mM sodium citrate, 124 mM D-glucose. Store at 4°C. Discard if left for more than 1 h at room temperature.
7. Modified HEPES buffer (JNL buffer): 130 mM sodium chloride, 10 mM trisodium citrate, 9 mM sodium bicarbonate, 6 mM dextrose, 0.9 mM magnesium chloride, 0.81 mM potassium phosphate, 10 mM Tris-HCl, pH to 7.4 with acid citrate dextrose. Store at 4°C and discard after use.

8. 10X Phospho-lysis buffer: 1% sodium dodecyl sulphate (SDS), 5% *N*-octyl glu-copyranisidase, 10% NP-40, 20 m*M* phenylmethylsulfonylfluoride (PMSF), 1% protease inhibitor cocktail (e.g., Calbiochem), 1% tyrosine phosphatase inhibitor cocktail (e.g., Sigma Aldrich). Make up solution in JNL buffer. Vortex and store on ice prior to use (*see* **Note 1**).

9. One-dimensional (1D) electrophoresis elution buffer (2X SDS sample buffer): 3% SDS, 75 m*M* dithiothreitol, 0.05% bromophenol blue, 20% glycerol, and 125 m*M* Tris-HCl, pH 6.7. Store at −20°C in 1-mL aliquots. Thaw and keep at room temperature until use.

10. 2DE elution buffer: 9.5 *M* urea, 4% CHAPS, 40 m*M* Tris-HCl. Make up solution in dH$_2$O. Store at −80°C in 1-mL aliquots. Before use, add 1% dithiothreitol, 0.8% carrier ampholytes and 1 m*M* PMSF to a 1-mL aliquot. Thaw at room temperature and do not heat above 37°C, because it contains urea, which can cause carbamylation of proteins if heated above 37°C.

3. Methods

When preparing samples for proteomic analysis, it is important to minimize keratin contamination because large amounts of keratin can mask important proteins. Therefore, caution during immunoprecipitation is advised to reduce potential contamination.

3.1. Platelet Lysis

1. Isolate washed platelets from whole blood (*see* **Chapter 23**).
2. Resuspend washed platelets at a concentration of 1 × 10^8 platelets per 450 µL JNL.
3. Activate platelets using agonist and method of choice and terminate reaction by adding 50 µL of 10X phospho-lysis buffer directly to the 450 µL of resting or activated washed platelets.
4. To allow for complete lysis of platelets, place on ice for 1 h, vortexing intermittently.

To minimize protein degradation, *all* of the following procedures should be performed in a cold room at 4°C.

3.2. Washing of the Protein A Dynabeads

1. Resuspend the Protein A Dynabeads by thoroughly vortexing.
2. Pipet the appropriate volume of beads (25 µL per 500-µL final volume of lysed platelet sample) into a 2-mL Eppendorf tube (*see* **Note 2**).
3. Insert the tube into the MPC for 2 min. The solution clears as the beads bind to the magnet.
4. Carefully pipet off the fluid, taking care not to disturb the beads (*see* **Note 3**).
5. Add 1.5 mL of ice-cold JNL buffer to the beads.
6. Remove the tube from the MPC and vortex for 15 s.

7. Insert the tube back into the MPC, allow the beads to bind the magnet, and remove the JNL wash buffer.
8. Repeat washing **steps 4–7** twice (*see* **Note 4**).

3.3. Blocking of Nonspecific Binding Sites on the Protein A Dynabeads

1. After the final washing step, resuspend the beads in 1.5 mL of 5% BSA/JNL buffer and place on an end-over-end rotator for 1 h (*see* **Note 5**).
2. Once the nonspecific sites are blocked, wash the beads once with JNL buffer to remove excess BSA.
3. Resuspend the beads in their original volume with JNL buffer (**Subheading 3.2., step 2**). At this point, split the beads into two aliquots: one for the antibody-binding step (20 μL per sample) and the other for the preclear step (5 μL per sample).

3.4. Antibody Binding Step

1. Add 4G10 antibody (1 μL per 20 μL of beads) to the blocked beads set aside in **Subheading 3.3., step 3** for antibody binding (20 μL per sample).
2. Place on an end-over-end rotator for 3 h.
3. After the antibody has bound to the beads, wash with 1.5 mL of JNL buffer to remove any excess antibody. These beads are then used for the immunoprecipitation step (*see* **Note 6**).

3.5. Preclear Step

1. To each platelet lysate, add 5 μL of the blocked beads from **Subheading 3.3., step 3** set aside for preclearing (5 μL per sample) (*see* **Note 7**).
2. Place on an end-over-end rotator for 1 h (*see* **Note 8**).
3. Following the preclear, place the tube on to the MPC.
4. Transfer the supernatant (precleared lysate) into a new tube. Retain the preclear beads at 4°C and wash and elute as in **Subheading 3.7., steps 2–6**.

3.6. Immunoprecipitation

1. Add 20 μL of the antibody-bound beads to the precleared lysate and rotate for 4 h or overnight.

3.7. Washing of Nonspecific Binding From the Protein A Dynabeads

1. After the immunoprecipitation step, place tube on to the MPC and discard the supernatant (*see* **Note 9**).
2. Add 1.5 mL of 1X phospho-lysis buffer to the beads and place on the rotator for 10 min at 4°C (*see* **Note 10**).
3. Repeat washing step with phospho-lysis buffer twice.
4. Wash once for 10 min at 4°C in 1.5 mL of JNL buffer.
5. Wash once for 10 min at 4°C in 1.5 mL of dH_2O (*see* **Note 11**).
6. Place the tubes on the MPC and remove all the residue fluid.

3.8. Elution of Bound Proteins From the Protein A Dynabeads

3.8.1. For 1D Gels

1. Add 30 µL of 1D electrophoresis elution buffer (2X SDS sample buffer).
2. Pierce a hole in the top of the tubes and boil the samples for 10 min at 96°C, vortexing occasionally.
3. Allow the tubes to cool.
4. Centrifuge the tubes briefly to prevent any sample being lost in the lid of the tube.
5. Place in MCP and pipet off the SDS sample buffer containing the eluted proteins. Store at −20°C until ready to load onto 1D gels (*see* **Note 12**).

3.8.2. For 2D Gels

1. Add 30 µL of 2DE elution buffer.
2. Vortex the samples for 20 s and place on ice for 1 h. Continue to vortex intermittently.
3. Place in MCP and pipet off buffer containing the eluted proteins. Store at −80°C until ready to load onto 2D gels (*see* **Note 12**).

4. Notes

1. PMSF, protease, and phosphatase inhibitors must be added to the phospho-lysis buffer just prior to use.
2. Use 2-mL tubes to allow for efficient volume to wash the beads.
3. To avoid disrupting the beads, the tubes must remain on the MPC while removing the supernatant.
4. Protein A Dyna beads are supplied in an azide-containing buffer to prevent bacterial growth. Washing in JNL buffer removes any azide present on the beads.
5. Blocking of nonspecific sites with 5% BSA reduces the binding of nonphosphorylated proteins to the beads during the immunoprecipitation step.
6. The Dynabead-antibody complex is stable for 1 wk at 4°C after the addition of 0.001% azide.
7. The preclear step minimizes any nonspecific or nonphosphorylated proteins that may bind nonspecifically to the beads during immunoprecipitation.
8. To avoid the potential loss of phosphorylated protein, the preclear step should not exceed 1 h.
9. Following immunoprecipitation, the tyrosine-phosphorylated proteins/complexes in the precleared lysate are bound to the antibody-bead conjugate.
10. Washing in this buffer removes any nonspecific contaminants that may have also bound to the beads.
11. The final washing steps in dH_2O remove any residual salt from the JNL buffer wash and dilutes the SDS from the lysis buffer. SDS and salt interfere with the first dimension of 2DE by causing high resistance, resulting in poorly resolved gels.
12. From 60 mL of blood, a final yield of approx 25 µg of tyrosine phosphorylated protein is expected. Approximately 10 µg of protein is required for a 7-cm 1D Coomassie-stained gel. Approximately 80 and 500 µg of protein is required for an 18-cm silver-stained and Coomassie blue-stained gel, respectively.

Acknowledgments

Supported in part by a fellowship from Enterprise Ireland, a research grant from the Health Research Board of Ireland, and funding under the Programme for Research in Third Level Institutions (PRTLI), administered by the Higher Education Authority of Ireland.

References

1. Santos, M. T., Moscardo, A., Valles, J., et al. (2000) Participation of tyrosine phosphorylation in cytoskeletal reorganization, alpha(IIb)beta(3) integrin receptor activation, and aspirin-insensitive mechanisms of thrombin-stimulated human platelets. *Circulation* **102,** 1924–1930.
2. Kramer, R. M., Roberts, E. F., Strifler, B. A., and Johnstone, E. M. (1995) Thrombin induces activation of p38 MAP kinase in human platelets. *J. Biol. Chem.* **270,** 27,395–27,398.
3. Clark, E. A., Shattil, S. J., Ginsberg, M. H., Bolen, J., and Brugge, J. S. (1994) Regulation of the protein tyrosine kinase pp72syk by platelet agonists and the integrin alpha IIb beta 3. *J. Biol. Chem.* **269,** 28,859–28,864.
4. Saci, A., Rendu, F., and Bachelot-Loza, C. (2000) Platelet alpha IIb-beta 3 integrin engagement induces the tyrosine phosphorylation of Cbl and its association with phosphoinositide 3-kinase and Syk. *Biochem. J.* **351(3),** 669–676.
5. Maguire, P. B., Wynne, K. J., Harney, D. F., O'Donoghue, N. M., Stephens, G., and Fitzgerald, D. J. (2002) Identification of the phosphotyrosine proteome from thrombin activated platelets. *Proteomics* **2,** 642–648.
6. Marcus, K., Moebius, J., and Meyer, H. E. (2003) Differential analysis of phosphorylated proteins in resting and thrombin-stimulated human platelets. *Anal. Bioanal. Chem.* **376,** 973–993.

25

Characterization of Circulating
Human Monocytes by Proteomic Analysis

Maria G. Barderas, Verónica M. Dardé, Mari-Carmen Durán, Jesús Egido, and Fernando Vivanco

Summary

We describe a simple method for isolation of human blood monocytes with the high purity required for proteomic analysis, which avoids contamination by other blood cells (platelets and lymphocytes) and the most abundant plasma proteins (albumin and immunoglobulins). Blood monocytes were purified by gradient centrifugation followed by positive selection with monoclonal antibodies coupled to paramagnetic beads. This method is compatible with flow cytometry, which was used to assess the purity of the cell population. After solubilization of monocytes, the proteins where analyzed by two-dimensional gel electrophoresis in several pH ranges. Image analysis of gels allowed the reproducible detection and quantification of the spots present in the gel.

This method is useful for clinical studies of monocytes from a large number of patients, owing to its rapidity and reproducibility, which permits comparative analysis of normal vs pathological samples and allows follow up of the expressed proteins of monocytes from each patient.

Key Words: Human monocytes; isolation method; proteomic analysis; 2DE.

1. Introduction

Monocytes are essential cellular components of the body's host defense system, playing an essential role in the disposal of foreign agents in the immune and inflammatory responses and in the repair process following tissue injury *(1)*. Monocytes are formed in the bone marrow and are continually released into blood, where they circulate for a day and then migrate into most organs. Once they settle in tissues, they mature and differentiate into specialized macrophages, including dendritic cells *(2–4)*.

An understanding of the role of monocytes in disease would be greatly improved with the knowledge of their proteome and the identification of protein

From: *Methods in Molecular Biology, vol. 357: Cardiovascular Proteomics: Methods and Protocols*
Edited by: F. Vivanco © Humana Press Inc., Totowa, NJ

changes that occur during differentiation and maturation of the cells under different conditions and tissue location *(5,6)*.

This chapter provides protocols that facilitated the analysis of protein content in human monocytes following several considerations:

1. Isolation of monocytes requires high purity.
2. Sample preparation has to be reproducible.
3. It is necessary to solubilize the maximum number of proteins as possible.

2. Materials

2.1. Equipment

1. EDTA glass tubes (Becton Dickinson, Franklin Lakes, NJ).
2. Flow cytometer (FACS Scalibur) (Becton Dickinson).
3. Electrophoresis apparatus (Protean isoelectric focusing (IEF) cell, PROTEAN II, Bio-Rad, Hercules, CA).
4. Power supply (3500 V min; e.g., Power Pac 200, Bio-Rad).
5. ReadyStrip immobilized pH gradient (IPG) 17-cm, 4.0–7.0, 3.0–10.0, 6.0–9.0 (Bio-Rad).
6. Protean IEF cell (Bio-Rad).
7. Milli-Q® System (Millipore, Bedford, MA).
8. Shaker (Promax 1020, Heidolph).
9. Thermomixer compact (Eppendorf).
10. Ultrasonic bath.
11. Glass plates 16 × 20 cm (Bio-Rad).
12. Reswelling cassette (Bio-Rad).
13. Clamps (Bio-Rad).
14. Scanner (DUO Scan Hid, Agfa-Geavaert, Martsel, Belgium).

2.2. Reagents

1. FACS FLOW buffer (Becton Dickinson).
2. FICOLL (ICN Biomedicals, Irvine, CA).
3. Magnetic cells separation system (autoMACS, Miltenyi Biotec).
4. Fluorescein isothiocyanate (Miltenyi Biotec).
5. Benzonase (Novagen, Merck, Darmstadt, Germany).
6. Bradford (Bio-Rad).
7. Sample solubilization buffer: 2 *M* thiourea, 7 *M* urea, 2–4% CHAPS, 1% dithiothreitol (DTT). To prepare 50 mL of solubilization buffer, dissolve 7.02 g of thiourea, 21.02 g of urea, and 2 g of CHAPS in deionized water and make up to 50 mL, stir, and filter. Sample solubilization buffer can be stored at −80°C in small aliquots of 0.5 mL. Never refreeze.
8. Rehydration solution: 8 *M* urea, 0.5–4% CHAPS, 20–100 m*M* DTT, 0.5–2% Pharmalyte. To prepare 25 mL of the solution, dissolve 12 g of urea, 0.5 g of CHAPS,

7 mg of DTT, and 12.5 μL Pharmalyte, pH 3.0–10.0 in deionized water. It is important not to heat urea solutions above 37°C in order to reduce the risk of protein carbamylation.

9. Equilibration buffer: to prepare 200 mL of equilibration buffer, mix 72.07 g of urea, 6.7 mL of Tris-HCl, pH 8.8, 69 mL of glycerol (87%), and 4 g of sodium dodecyl sulfate (SDS). Add deionized water until 200 mL. Finally add bromophenol blue. During the first equilibration step, 100 mg of DTT is added to 10 mL equilibration buffer. In the second step, 250 mg of iodoacetamide is added to 10 mL equilibration buffer.

10. Acrylamide/*bis*-acrylamide solution: to prepare 100 mL of the solution, dissolve 30 g of acrylamide and 0.8 g of *bis*-acrylamide in 80 mL of deionized water. Stir for 10 min, filter, and fill up to 100 mL with deionized water.

11. Amonium persulfate solution: to prepare 1 mL of the solution, dissolve 0.1 g of ammonium per sulfate solution in 1 mL of deionized water.

12. Phosphated-buffered saline (PBS) buffer pH 7.2: to prepare 500 mL of PBS buffer, dissolve 40 g of NaCl, 1 g of KCl, 7.2 g of Na_2HPO_4, and 1.20 g of KH_2PO_4, in deionized water (500 mL final volume).

3. Methods

3.1. Isolation of Circulating Human Monocytes

For a proteomic analysis, it is important to isolate the nonactivated monocytes with the highest purity *(5)*. Blood monocytes may be isolated by different methods based on their ability to adhere to glass or plastic *(7)* or their cell size *(8)*. Other methods are based on the density of blood cells using gradient centrifugal techniques *(9)*. We describe a method that combines gradient centrifugation and positive selection using microbeads conjugated with mouse monoclonal antibody anti CD-14 *(5;* **Fig. 1**).

1. Draw 28 mL of human blood into 7 mL EDTA glass tubes and centrifuge at 600*g* for 10 min at room temperature.

2. Mix the cell pellet with an equal volume of FACS FLOW buffer over 10 mL Ficoll and then centrifuge at 700*g* for 30 min. A band over Ficoll corresponding to peripheral blood mononuclear cells (PBMCs) can be recovered at the interface.

3. Wash the PBMCs three times in FACS FLOW buffer and then with PBS buffer pH 7.2.

4. Incubate the PBMCs during 15 min at 4°C with anti-CD-14 monoclonal antibody-coated microbeads in the same buffer.

5. Isolate the monocytes by passing the PBMCs through a magnetic cell separation system (*see* **Note 1**). Magnetically labeled CD-14+ are retained in the column, while other cells run through. Obtain the magnetically retained CD-14+ cells, as a positive fraction, using PBS as elution buffer. The efficiency of magnetic cell separation can be evaluated by flow cytometry *(5)*.

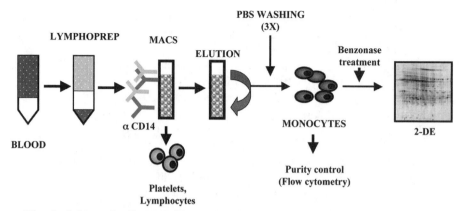

Fig. 1. Schematic diagram illustrating the methodology used for the proteomic analysis of blood monocytes.

3.2. Flow Cytometry

Incubate 3×10^4 monocytes for 15 min at 4°C in PBS buffer containing 2% AB serum and 0.01% NaN_3 with 10 μL of fluorescein-conjugated CD14 antibody. Wash and analyze the sample in a flow cytometer (**ref. *10*; Fig. 2**).

3.3. Two-Dimensional Gel Electrophoresis

3.3.1 Sample Preparation

1. Solubilize 10^7 monocytes in 200 μL of solubilization buffer and freeze at –80°C until use (*see* **Note 2**).
2. Sonicate the protein mixture for 1 min at room temperature.
3. Heat the monocytes sample at 80°C 5 min.
4. Cool the sample at room temperature.
5. Incubate the protein mixture with 200 μL of Benzonase to eliminate nucleic acids.
6. Incubate again for 50 min at room temperature.
7. Quantify the protein content using the Bradford method *(11)*.
8. Dilute 200–250 μL in rehydration solution (350 μL final volume).

3.3.2. Sample Application

1. 250 μg of proteins are mixed with 350 μL of rehydratation solution.
2. Pipet the entire sample containing rehydration solution into the groove of a swelling tray.
3. Peel off the protective cover sheets from the IPG Gel Strips (17 cm, pH 3.0–10.0, 4.0–7.0, 6.0–9.0, 6.0–11.0) and position them such that the gel of the strip is in contact with the sample (*see* **Note 3**).
4. Cover the IPG Gel Strips and the sample with mineral oil.

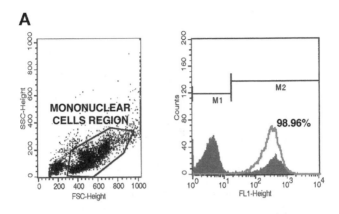

MONOCYTES CELLS PURIFICATION
TOTAL POPULATION

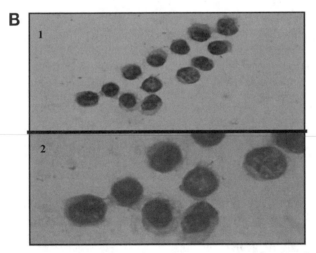

Fig. 2. **(A)** Purity control of the isolated blood monocytes using flow cytometry. **(B)** Morphological appearance of the purified monocytes, stained with MayGrünwald-Giemsa.

3.3.3. First Dimension: IEF Separation

IEF is performed in a Protean IEF Cell (BioRad). The strips are actively rehydrated for 12 h at 30 V to enhance protein uptake and then the voltage is increased according to the following program:

1. 1 h at 500 V.
2. 1 h at 1000 V.
3. 30 min at 8000 V (ramping) for a total of 50,000 Vh.

For strips in the 6.0–9.0, 6.0–11.0 pH range a different protocol is used:

1. 16 h of rehydration at 30 V.
2. 30 min at 500 V.
3. 1 h at 1000 V.
4. 30 min at 5000 V.
5. 8000 V for a total of 60,000 Vh.

3.3.4. Second Dimension: SDS-Polyacrylamide Gel Electrophoresis

A number of gel concentrations and gradients as well as different buffering systems can be used in the second dimension *(12)*. We perform SDS-polyacrylamide gel electrophoresis according the Laemmli system *(13)*, as a previously described *(14)*, using a Protean II system (Bio-Rad). Prepare a 12.5% acrylamide gel. Transfer the equilibrated IPG gel strips onto the top of the SDS gel and run it at 25 mA/gel at 4°C.

1. Place the IPG Gel Strips in a clean cassette for their equilibration. Add 5 mL equilibration buffer with DTT and rock for 15–20 min on a shaker. After that, pour off the equilibration buffer and rinse with deionized water.
2. Add 5 mL equilibration buffer with iodoacetamide and equilibrate for another 15–20 min on a shaker. Finally rinse again with deionized water and put on a piece of filter paper to remove excess of equilibration buffer (*see* **Note 4**).

3.3.5. Fixing Gels

Remove the gels from the apparatus and separate the glass plates and place them in 200 mL of 30% ethanol, 5% acetic acid overnight (fixation solution).

3.3.6. Staining

When gels are fixed, stain with Silver Staining Kit (Pharmacia Diagnostic) (*see* **Fig. 3**), to increase the reproducibility of the resultant images.

3.3.7. Processing of 2DE Gels

The protein patterns in the staining gels must be recorded as digitized images using a Desktop scanner. Evaluation and processing of two-dimensional gel electrophoresis (2DE) gels can be performed using PDQuest gel analysis software version 6.2 (Bio-Rad) or similar software of 2D gels analysis (**Fig. 4**).

3.4. In-Gel Digestion of Proteins and Sample Preparation for Mass Spectrometric Analysis

1. Select the spots to be cut from the 2DE gels.
2. Excise manually and digest using a DP protein digestion station (Bruker-Daltonics, Bremen, Germany) following the method of Schevchenko et al. *(15)*, with some variations:

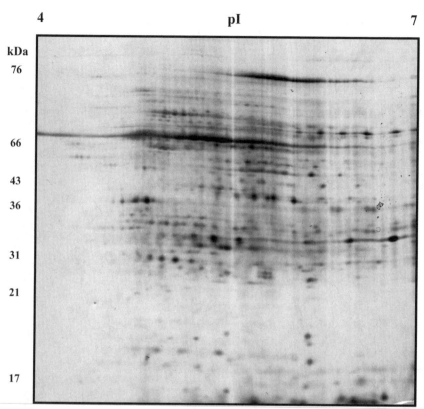

Fig. 3. Two-dimensional electrophoresis of monocyte proteins. First dimension, electrofocusing (immobilized pH gradient strip) in a pH range of 4.0–7.0. Second dimension, 12% acrylamide sodium dodecyl sulfate-polyacrylamide gel electrophoresis. The gel (silver staining) was loaded with 200 µg of protein.

1. Reduce the gels plugs with 10 mM DTT in 50 mM ammonium bicarbonate.
2. Rinse the gel pieces with 50 mM ammonium bicarbonate and acetonitrile and dry under a stream of nitrogen.
3. Add modified porcine trypsin (sequencing-grade; Promega, Madison WI) at a final concentration of 16 ng/µL in 50 mM ammonium bicarbonate to the dry gel pieces and allow the digestion to proceed at 37°C for 6 h. The reaction is stopped by adding 0.5% trifluoroacetic acid for peptide extraction.
4. An 0.4-µL aliquot of matrix solution (5 g/L 2,5-dihydroxybenzoic acid in 33% aqueous acetonitrile and 0.1% trifluoroacetic acid) followed by 0.4 µL of the above digested sample were automatically deposited onto a 400 µm AnchorChip matrix-assisted laser desorption/ionization (MALDI) probe (Bruker-Daltonics) and allowed to dry at room temperature.

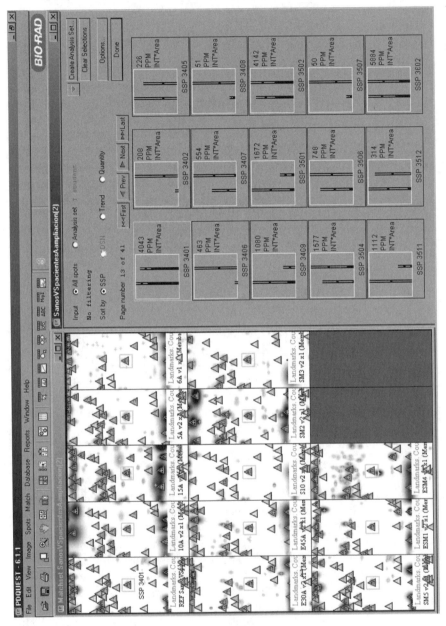

Fig. 4. PDQuest analysis illustrating the evaluation and processing of two-dimensional gel electrophoresis gels (silver staining).

3.5. MALDI Peptide Mass Fingerprinting and Database Searching

1. Peptide mass fingerprint spectra were measured on a Bruker Ultraflex TOF/TOF MALDI mass spectrometer (Bruker-Daltonics) in positive ion reflector mode using 140 ns delayed extraction and a nitrogen laser (337 nm). The laser repetition rate was 50 Hz and the ion acceleration voltage was 25 kV. Mass measurements were performed automatically through fuzzy logic-based software.

2. Each spectrum was internally calibrated with the mass signals of two trypsin auto-lysis ions: [VATVSLPR + H]$^+$ (m/z = 842.510) and [LGEHNIDVLEGNEQFINAAK + H]$^+$ (m/z = 2211.105) to reach a typical mass measurement accuracy of ±30 ppm. Known trypsin and keratin mass signals, as well as potential sodium adducts (+21.982 Da) or signals arising from methionine oxidation (+15.995 Da) were removed from the peak list.

3. The measured tryptic peptide masses were transferred through MS BioTools program (Bruker-Daltonics) as inputs to search the NCBInr database using Mascot software (Matrix Science, London, UK). The identifications were accepted if they represented the highest-ranking hit, had MOWSE scores higher than 64, and if the sequence coverage was at least 15–30% depending on protein size *(16)*. No restrictions were placed on the species of origin of the protein and no variable modifications were allowed. Up to one missed tryptic cleavage was considered and a mass accuracy of 40 ppm was generally used for tryptic-mass searches. Detailed analysis of peptide mass mapping data was performed using flexAnalysis software (Bruker-Daltonics).

4. Notes

1. Alternatively, the isolation can be carried out manually using small columns (MS Separation Columns and Minimacs; Miltenyi Biotec).

2. Sample preparation and solubilization are crucial factors for the performance of the 2DE technique. Protein complexes and aggregates should be completely disrupted in order to allow the appearance of all possible spots.

3. For the correct development of the first dimension, it is very important that when the strip is placed over the sample, there should be no bubbles present.

4. DTT is used to facilitate the conversion of disulfide bonds to free sulfhydryl groups. Sulfhydryls groups are then alkylated with iodoacetamide (carboxiamidomethylation), because the reducing reaction is reversible and the re-formation of the disulfide bridges must be prevented.

Acknowledgments

This work has been partially supported by FIS (PI02/1047) (PI 02/3093), Spanish Cardiovascular Network RECAVA (03/01), SAF-2004-06109, European Network (QLG1-CT-2003-01215), and the International Cardura Award (Pfizer, USA).

References

1. Osterud, B. and Bjorklid, E. (2003) Role of monocytes in atherogenesis. *Physiol. Rev.* **83**, 1069–1112.
2. Van Furth, R. and Cohn, Z. A. (1968) The origin and kinetics of mononuclear phagocytes. *J. Exp. Med.* **128**, 415–435.
3. Volkman, A. (1970) The origin and fate of the monocyte. *Ser. Haematol.* **3**, 62–92.
4. Richards, J., Le Naour, F., Hanash, S., and Beretta, L. (2002) Integrated genomic and proteomic analysis of signalling pathways in dendritic cell differentiation and maturation. *Ann. NY Acad. Sci.* **975**, 91–100.
5. Gonzalez-Barderas, M., Gallego-Delgado, J., Mas, S., et al. (2004) Isolation of circulating human monocytes with high purity for proteomic analysis. *Proteomics* **4**, 432–437.
6. Wang, X., Zhao, H., and Anderson, R. (2004) Proteomics and leukocytes: an approach to understanding potential molecular mechanism of inflammatory responses. *J. Proteom. Res.* **3**, 921–929.
7. Haskill, S., Johnson, C., Eirman, D., et al. (1988) Adherence induces selective mRNA expression of monocyte mediators and proto-oncogenes. *J. Immunol.* **140**, 1690–1694.
8. Boyum, A. (1983) Isolation of human blood monocytes with Nycodenz, a new non-ionic iodinated gradient medium. *Scand. J. Immunol.* **17**, 429–436.
9. Almeida, M., Silva, A., and Barral, A. (2000) A simple method for human peripheral blood monocyte isolation. *M. Mem. Inst. Oswaldo Cruz* **95**, 221–223.
10. Abendroth, A., Slobedman, B., Lee, E., et al. (2000) Modulation of major histocompatibility class II protein expression by varicella-zoster virus. *J. Virol.* **74**, 1900–1907.
11. Ausubel, F. M., Brent, R., Kingston, R. E., et al. (2005) *Current Protocols in Molecular Biology*. John Wiley & Sons, New York, NY **10**, 97–106.
12. Gorg, A., Weiss, W., and Dunn, M. J. (2004) Current two-dimensional electrophoresis technology for proteomics. *Proteomics* **4**, 3665–3685.
13. Laemmli, U. K. (1970) Cleavage of structural proteins during the assembly of the head of bacteriophage T4. *Nature* **227**, 680–685.
14. Barrio, E., Antón, L., Marques, G., Sánchez, A., and Vivanco, F. (1991) Formation of covalently linked C3-C3 dimers on IgG immune aggregates. *Eur. J. Immunol.* **21**, 343–349.
15. Schevchenko, A., Wilm, M., Vorm, O., and Mann, M. (1996) Mass spectrometric sequencing of proteins from silver stained polyacrylamide gels. *Anal. Chem.* **68**, 850–858.
16. Lefkovits, I., Kettman, J. R., and Frey, J. (2000) Global analysis of gene expression in cells of the immune system, I: analytical limitations in obtaining sequence information on polypeptides in two-dimensional gel spots. *Electrophoresis* **21**, 2688–2693.

V

CARDIOVASCULAR BIOMARKERS IN PLASMA

26

Cardiovascular Biomarker Discovery by SELDI-TOF Mass Spectrometry

Olivier Meilhac, Sandrine Delbosc, and Jean-Baptiste Michel

Summary

Surface-enhanced laser desorption/ionization time-of-flight mass spectrometry (TOF-MS) allows for rapid differential proteomic analysis of various experimental conditions. In the first step of retention chromatography, the proteome to be analyzed is fractionated based on the biochemical properties of the proteins or peptides of which it is composed. TOF-MS separates the proteins according to their molecular mass and charge (m/z in which conditions are optimized to obtain single charged proteins: $z = 1$). The most discriminating conditions are used to validate differential peaks on a larger number of samples. Once the biomarker is statistically validated, its identification is performed by reproducing the chromatographic conditions that are appropriate for its retention on spin-columns, followed by sodium dodecyl sulfate-polyacrylamide gel electrophoresis, trypsin digestion of the band of interest, and subsequent peptide mapping after MS. It may be necessary to determine part of the amino acid sequence of the protein of interest to confirm its identity.

Key Words: SELDI-TOF; mass spectrometry; biomarker; cardiovascular; protein chip; chromatography; hypertension; rat; plasma; secretome.

1. Introduction

Surface-enhanced laser desorption/ionization time of flight mass spectrometry (SELDI-TOF-MS) is an emerging technology, adapted from matrix-assisted laser desorption/ionization (MALDI)-TOF-MS for differential proteomic analysis by Ciphergen Biosystems. It is a particularly well-suited methodology for protein profiling, allowing quantitative comparison between several experimental conditions and permitting processing of an almost unlimited number of samples (1). In the first step, the proteome to be analyzed is fractionated by retention chromatography directly onto an MS-compatible ProteinChip®. Several types of surfaces allow detection of under-represented proteins: anionic or

From: *Methods in Molecular Biology, vol. 357: Cardiovascular Proteomics: Methods and Protocols*
Edited by: F. Vivanco © Humana Press Inc., Totowa, NJ

cationic exchanger, immobilized-metal affinity capture, normal, reversed-phase, or customized surfaces (i.e., phosphorylated proteins). **Figure 1** illustrates the different surfaces used with various washing conditions in order to perform a first set of experiments with a few experimental samples, which permits selection of the most discriminating conditions. After appropriate washings and application of energy absorbing molecules (the so called "matrix"), the ProteinChips® are directly placed into the SELDI-TOF-MS for reading. The resulting spectra are compared and the chromatographic/washing conditions that highlight the differences between the experimental conditions are selected for processing of a larger number of samples. Therefore, statistical validation takes place before identification of potential biomarkers.

The sampling is the most critical step for the discovery of biomarkers. In cardiovascular research, several strategies can be used. In contrast to studies using experimental animal models, the access to samples of good quality is often limited in humans. As shown in **Fig. 2**, proteomic analysis can be performed on the blood compartment (cells or serum/plasma) or on tissues (primary cell culture or tissue culture, focusing on cell/tissue proteins or secreted/released proteins). If we consider a biomarker released directly by the diseased vascular wall (triangle in **Fig. 2**), it will be diluted in the blood compartment and may be masked by the major plasma proteins (circles), or will be difficult to identify from the tissue itself because of the presence of major proteins produced by the cells or constituting the extracellular matrix. In cell culture, release of this potential biomarker may not occur because of the adaptation of the cells to a different environment. The most relevant samples to be investigated are generally the biological fluids (urine, serum, plasma, etc.), where potential biomarkers might be highly diluted and difficult to analyze using a proteomic approach. Analysis of the medium conditioned by a focal diseased cardiovascular sample is an interesting alternate strategy that allows concentration of proteins directly released by a determined area, with focus on soluble proteins that are likely to be found in plasma *(refs. 2,3* and **Fig. 3**).

2. Materials

2.1. Equipment

1. Dissection tools (sterile scalpel, forceps, and microscissors).
2. Analytical balance.
3. pH meter.
4. CO_2 incubator providing a 5% CO_2/95% air atmosphere.
5. Sonicator.
6. Homogenizer or Ribolyzer.
7. Centrifuge.
8. Bioprocessor® (Ciphergen Biosystems).

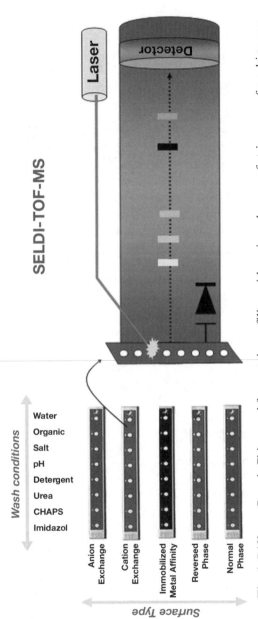

Fig. 1. Different ProteinChips used for protein profiling with various degrees of stringency of washing conditions (left). SELDI-TOF-MS principle, which illustrates laser desorption directly from the ProteinChip and subsequent flight of the proteins through the vacuum tube. The average mass of the proteins/peptides reaching the detector is calculated from their time of flight.

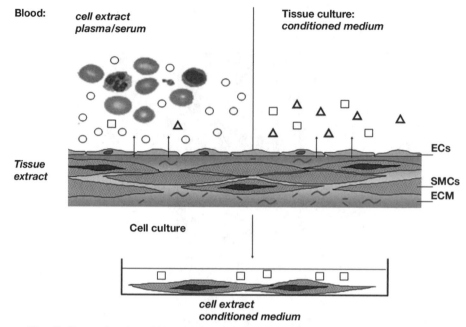

Fig. 2. Strategies for differential proteomic analysis in cardiovascular research. Analysis can be performed on blood cells or plasma/serum, on tissues or cell extracts, or on cell/tissue conditioned media. Triangles, potential biomarkers; squares, proteins released by vascular cells; circles, plasma proteins. ECs, endothelial cells; SMCs, smooth muscle cells; ECM, extracellular matrix.

9. Humid chamber.
10. Vortex.
11. ProteinChip reader (Ciphergen Biosystems).
12. Speed-vac.
13. Electrophoresis equipment (Invitrogen).

2.2. Reagents

1. Non-latex powder-free gloves (*see* **Note 1**).
2. Plastic cell culture 24-well plates.
3. Cell culture medium (RPMI without phenol red) containing penicillin, streptomycin, and amphotericin.
4. Cell/tissue lysis buffer: 9 *M* urea, 2% (w/v) CHAPS, 1% (w/v) dithiothreitol.
5. NP20 ProteinChip arrays: mimic normal-phase chromatography with silicate functionality.
6. IMAC ProteinChip arrays and corresponding spin columns (IMAC HyperCel™): Immobilized metal affinity capture.
7. H4 and H50 ProteinChip arrays, and corresponding spin columns (Methyl HyperD): Bind the proteins by reversed-phase or hydrophobic interactions.

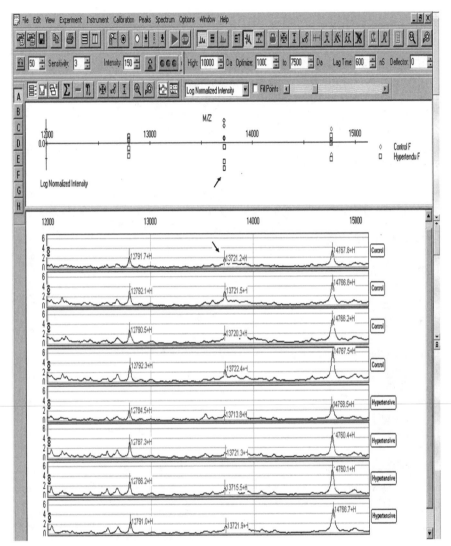

Fig. 3. Representative spectra obtained after the step of profiling on IMAC allowing comparison between hypertensive and normotensive rats. Arrows show the differentially expressed peak. ProteinChip Biomarker Wizard is a software that helps identify biomarkers by analyzing multiple sample profiles and automatically displaying differences in expression levels between sample groups (circles, control group; squares, hypertensive group).

8. CM10 ProteinChip arrays and corresponding spin columns (CM ceramic HyperD® F): weak cation exchange chromatography.
9. Q10 ProteinChip arrays and corresponding spin columns (Q ceramic HyperD F): strong anion exchange chromatography.

10. Saturated solutions of energy absorbing molecules ("matrix") are prepared in 50% acetonitrile, 0.5% trifluoroacetic acid (TFA):
    a. Sinapinic acid (SPA; molecular weight = 224.2) is used for detection of proteins.
    b. α-cyano-4-hydroxy cinnamic acid (CHCA; molecular weight = 189.2) is especially good for small proteins and peptides (<15 kDa).
11. 4–20% Gradient polyacrylamide gel.
12. Reagents for gel staining: Coomassie blue, methanol, ethanol, acetic acid, 37% formaldehyde, silver nitrate, sodium carbonate, sodium thiosulfate.
13. Reagents for trypsin digestion: Ammonium bicarbonate, acetonitrile (ACN), trypsin (Roche or Sigma, sequencing or proteomic grade).

2.3. Buffers and Solutions for Protein Profiling

2.3.1. IMAC Profiling

1. Loading buffer: 100 mM ZnCl$_2$, 100 mM CuSO$_4$.
2. Neutralization buffer: 100 mM Sodium acetate (CH$_3$COONa) pH 4.0.
3. Binding buffer: 100 mM Sodium phosphate, 500 mM sodium chloride, pH 7.0.

2.3.2. H50 Arrays

1. 5% ACN in 150 mM NaCl containing 0.1% TFA.
2. 10% ACN in 150 mM NaCl containing 0.1% TFA.
3. 20% ACN in 150 mM NaCl containing 0.1% TFA.

2.3.3. Q10 Arrays

1. 100 mM Tris-HCl, pH 7.0.
2. 100 mM Tris-HCl, pH 8.0.
3. 100 mM Tris-HCl, pH 9.0.

2.3.4. CM10 Arrays

1. 100 mM Sodium acetate, pH 4.0.
2. 100 mM Sodium acetate, pH 5.0.
3. 100 mM Sodium acetate, pH 6.0.

3. Methods

3.1. Sampling

3.1.1. Plasma/Serum

Plasma can be obtained from blood drawn on heparin, sodium citrate, or EDTA and centrifuged (3000g, 15 min). Alternatively, serum is obtained after clotting (20 min followed by centrifugation at 1000g for 15 min at room temperature; *see* **Note 2**). Before SELDI-TOF analysis, plasma or serum should be denatured. In a microfuge tube, add an equal volume of plasma/serum and denaturing buffer

(9 *M* urea, 2% CHAPS, 50 m*M* Tris-HCl, pH 9.0) and agitate for 15 min at room temperature.

3.1.2. Tissue- or Cell-Conditioned Medium

1. Dissect and weigh the tissues of interest under sterile conditions.
2. Add a ratio of 2:1 (vol/wt) of sterile RPMI without phenol red containing antibiotics and antimycotics for tissues and a standardized volume for cell culture (i.e., 100 µL/cm²).
3. Incubate for 24 h at 37°C (5% CO_2/95% air).
4. Collect the conditioned medium and centrifuge (14,000*g* for 10 min).
5. Check the protein content using Bradford quantification assay.
6. Use the supernatant for analysis.

3.1.3. Cell/Tissue Extracts

1. Add cell/tissue lysis buffer to cell pellet or tissue (100 µL/100,000 cells and 1 mL/ 100 mg tissue) containing a commercial antiprotease cocktail (i.e., Roche; *see* **Note 3**).
2. Sonicate the cells/homogenize the tissue. (Ribolyzer or mechanical homogenizer may be used.)
3. Leave at room temperature for 30 min.
4. Centrifuge at 14,000*g* (15 min at 4°C) to eliminate cell/tissue debris.
5. Collect the supernatant and check the protein content using Bradford Quantification Assay.

3.2. Protein Profiling

Profiling is usually performed using the Bioprocessor (*see* **Note 4**).

3.2.1. IMAC Profiling

1. Put 50 µL of the loading buffer in the Bioprocessor and incubate for 5 min at room temperature with gentle agitation.
2. Wash the spots with 250 µL ultrapure water.
3. Add 50 µL of the neutralization buffer to the array and incubate for 5 min at room temperature with gentle agitation.
4. Pour off the neutralization buffer and add 100 µL binding buffer to each spot and incubate for 5 min at room temperature with gentle agitation. Repeat this step once more.
5. Pour off the binding buffer and incubate the sample diluted in the binding buffer for 1–2 h at room temperature with gentle agitation (*see* **Note 5**).
6. Pour off the sample and wash the spots with 100 µL binding buffer and incubate for 5 min at room temperature with gentle agitation. Repeat this step twice.
7. Wash the array with 150 µL ultrapure water.
8. Allow to air-dry and apply 0.6 µL of the saturated SPA solution twice on each spot; air-dry before reading (*see* **Note 6**).

3.2.2. H50, CM10, and Q10 Profiling

1. Add 200 µL of the appropriate binding buffer to each spot and incubate at room temperature for 5 min with gentle agitation. Pour off the binding buffer and repeat this step once.
2. Pour off the binding buffer and incubate the sample diluted in the binding buffer for 1–2 h at room temperature with gentle agitation (*see* **Note 5**).
3. Pour off the sample and wash the spots with 100 µL binding buffer and incubate for 5 min at room temperature with gentle agitation. Repeat this step twice.
4. Wash the array with 150 µL ultrapure water.
5. Allow to air-dry and apply 0.6 µL of the saturated SPA solution twice on each spot; air-dry before reading (*see* **Note 6**).

3.3. Identification of Peaks of Interest

3.3.1. Semipreparative Purification and Electrophoresis

For the preparative purification, using spin columns is recommended. The spin columns possess the same chromatographic properties as the ProteinChip arrays. An example of purification of a biomarker detected on IMAC is given in **Subheading 3.3.1.1.** (*see* **Note 7**).

3.3.1.1. PURIFICATION ON IMAC SPIN COLUMNS

1. Use 250 µL of beads.
2. Centrifuge 30 s at 1000g to remove the storage buffer.
3. Add 500 µL ultrapure water to the columns, vortex 5 for min, and centrifuge 30 s at 1000g to remove water.
4. Add 200 µL loading buffer. Incubate at room temperature for 15 min under agitation. Centrifuge 30 s at 1000g to remove loading buffer.
5. Repeat **step 4** one more time.
6. Add 500 µL ultrapure water. Incubate at room temperature for 5 min under agitation. Centrifuge to remove water.
7. Add 200 µL binding buffer to equilibrate the beads. Incubate for 15 min under agitation. Centrifuge to remove the binding buffer.
8. Repeat this step once.
9. Add the sample diluted in the binding buffer (final volume 500 µL) to the beads and incubate at 4°C for 1 h 30 min under agitation. Centrifuge to remove the sample.
10. Wash the beads twice with 200 µL binding buffer.
11. Proceed to elution by using binding buffer (50 µL) with increasing concentrations of imidazole or by reducing the pH.
12. After the purification, the fraction containing the protein of interest is concentrated using a SpeedVac and run on an SDS-PAGE gel. Four to 20% polyacrylamide gradient gels allow one to separate the majority of proteins. For the detection of low-molecular-mass-proteins, tricine gels are recommended (Invitrogen).

For purification using other types of spin columns, *see* **Note 7**.

3.3.2. Gel Staining

3.3.2.1. COOMASSIE BLUE STAINING

1. Place the gel in a fixative solution containing 40% methanol, 10% acetic acid, 50% H_2O.
2. Place the gel in colloidal Coomassie blue solution consisting of 80% Coomassie blue solution (0.1% Coomassie Stain R-250, 50% methanol, 5% acetic acid, and 45% Milli-Q® water) and 20% methanol. Incubate between 1 h to overnight with gentle agitation.

As an alternative method, see **Note 8** for colloidal blue staining.

3.3.2.2. SILVER STAINING

1. Fixative step: Place the gel in a fixative solution consisting of 50% methanol, 12% acetic acid, and 0.5 mL/L 37% formaldehyde. Incubate at room temperature for 30 min.
2. Wash the gel in 50% ethanol and incubate for 20 min. Repeat this step twice.
3. Wash the gel for 1 min in H_2O three times.
4. Silver staining: 1 g silver nitrate, 0.75 mL 37% formaldehyde for 1 L of H_2O. Incubate for 30 min at room temperature.
5. Wash the gel for 1 min in H_2O three times.
6. Developing step: 30 g sodium carbonate, 0.25 mL 37% formaldehyde, 2 mg sodium thiosulfate up to 1 L. Incubate between 2–7 min at room temperature.
7. Wash the gel for 2 min in H_2O twice.
8. Stopping step: Place the gel in 50% methanol and 12% acetic acid for 10 min.
9. Wash the gel in 50% methanol for 20 min.
10. Gels can be stored in 10% methanol.

Silver staining can be performed using an alternative protocol (*see* **Note 9**).

3.3.3. Trypsin Digestion and Protein Identification

1. Identify the band of interest to be excised.
2. Cut the gel piece with a scalpel or a glass Pasteur pipet. Do not forget to take a piece of blank gel.
3. Transfer the bands in microfuge tubes and proceed to destaining.
4. For Coomassie blue-stained bands, add 25% methanol for at least 1 h and repeat destain step twice.
5. For silver-stained bands, Invitrogen protocol (LC6070) is preferred to homemade solutions because the bands become clear and the silver is removed more efficiently using this commercial kit.
6. Dehydrate the gel pieces with 100 µL of 100% methanol and incubate at room temperature for 5 min.

7. Rehydrate the gel with 100 µL 30% methanol and incubate at room temperature for 5 min.
8. Wash the gel with 100 µL of 100 m*M* ammonium bicarbonate containing 30% ACN for 10 min at room temperature. Repeat this step twice.
9. Dry the gel pieces in a speed-vac for 30 min.
10. Add the trypsin solution to cover gel pieces and leave at room temperature for 15 min. If the gel pieces have absorbed the solution, add ammonium bicarbonate solution.
11. Incubate at 37°C in a humid chamber for at least 16 h (overnight is recommended).
12. Put 1 or 2 µL of the digestion product on a H4 array, allow to air-dry, and apply 1 µL the 1/5 diluted CHCA matrix (50% ACN and 0.5% TFA). Dry the active spots and proceed to the reading (*see* **Note 6**).
13. Internet protein databases (i.e., Profound, ExPasy, Mascot) help to identify the protein of interest ("peptide mapping") by using the mass values of peptides obtained from trypsin digestion. Only specific peptides should be taken into account by subtracting tryspin autolysis products and the peaks resulting from digested blank gel *(4)*.

4. Notes

1. Avoid touching the spot surfaces of the ProteinChip array. Nitrile powder-free gloves are recommended for handling because latex can be detected by MS and interfere with the proteins of interest.
2. Serum is often preferred to plasma because it is an effective way of removing red blood cells and simplifying the sample in order to look at circulating proteins. However, we stress the fact that proteins of interest might be trapped within the clot and that proteases such as matrix metalloproteinase or elastase may diffuse from the clot *(5)* and proteolyze potential biomarkers. Whether plasma or serum is used for pro-filing, it is very important to standardize all the steps preceding analysis: blood sampling, centrifugation, and storage. Aliquoting is highly recommended in order to avoid freezing and thawing cycles that might damage proteins of interest.
3. Dithiothreitol may be omitted from the lysis buffer because it may interfere with binding on IMAC ProteinChips; alternative extraction buffers may be used if they do not contain ionic detergents and are compatible with MS analysis.
4. The Bioprocessor is an apparatus that allows larger volumes of sample or wash solution to be applied to individual spots of a ProteinChip array.
5. Profiling is usually performed on 2–10 µg total protein and completed to 100 µL with appropriate binding buffer. The volume of sample should not represent more that 25% of the total volume of incubation. 10–25 µL of denatured plasma/serum are incubated in the same conditions, in a Bioprocessor (Ciphergen Biosystem).
6. Before reading, a calibration in the appropriate mass range (protein or peptide) of the ProteinChip reader is necessary to ensure a reasonable mass accuracy. All-in-1 Protein Standard II (ranging from 7 to 147.3 kDa) and All-in-1 Peptide Standard (ranging from 1.08 to 7.03 kDa) from Ciphergen Biosystems can be used for calibration. Peptide calibration is performed on H4 ProteinChip array, whereas protein calibration is done on NP20 ProteinChip arrays.

7. Semipreparative purification may require chromatography columns other than IMAC.
   a. Anionic exchanger HyperQDF spin columns: binding buffer with salt concentration <100 m*M* and pH 9.0. Elution buffer: Binding buffer, pH 3.0, increasing salt concentrations.
   b. Cationic exchanger CM10 spin columns: Binding buffer with salt concentration less than 100 m*M* and pH around 4.5. Elution buffer: Binding buffer, pH 9.0, increasing salt concentration.
   c. Methyl HyperD spin columns (similar to H50 ProteinChip Array): Binding buffer is 50 m*M* phosphate buffer containing 1.5–2 *M* ammonium sulfate, pH 7.0. Elution buffer: Elution can be achieved with binding buffer without ammonium sulfate or by adding increasing concentrations of imidazole.
8. Coomassie blue staining: the use of SimplyBlue SafeStain from Invitrogen (Cat. no. LC6060) is fast and sensitive.
9. Silver staining: SilverQuest Silver Staining (Cat. no. LC6070) from Invitrogen is faster that the classical method and compatible with MS.

Acknowledgments

The SELDI-TOF-MS platform in Inserm unit 698 has been supported by MSD-Chibret Laboratories, AstraZeneca Future forum and the Leducq Foundation. We would like to thank Myriam Cubizolles and Sabine Jourdain from Ciphergen Biosystem for their help in the setup of protocols, and Mary Osborne-Pellegrin for her careful proofreading.

References

1. Issaq, H. J., Veenstra, T. D., Conrads, T. P., and Felschow, D. (2002) The SELDI-TOF MS approach to proteomics: protein profiling and biomarker identification. *Biochem. Biophys. Res. Commun.* **292**, 587–592.
2. Duran, M. C., Mas, S., Martin-Ventura, J. L., et al. (2003) Proteomic analysis of human vessels: application to atherosclerotic plaques. *Proteomics* **3**, 973–978.
3. Martin-Ventura, J. L., Duran, M. C., Blanco-Colio, L. M., et al. (2004) Identification by a differential proteomic approach of heat shock protein 27 as a potential marker of atherosclerosis. *Circulation* **110**, 2216–2219.
4. Appel, R. D., Bairoch, A., and Hochstrasser, D. F. (1999) 2-D databases on the World Wide Web, in *2-D Proteome Analysis Protocols* (Link, A. J., ed.), Humana Press, Totowa, NJ, pp. 383–391.
5. Fontaine, V., Jacob, M. P., Houard, X., et al. (2002) Involvement of the mural thrombus as a site of protease release and activation in human aortic aneurysms. *Am. J. Pathol.* **161**, 1701–1710.

27

Detection of Biomarkers of Stroke Using SELDI-TOF

Jean-Charles Sanchez, Pierre Lescuyer, Denis Hochstrasser, and Laure Allard

Summary

With a mean weight of 1500 g containing around 10 billion neurons, the adult brain represents about 2% of the total body mass, but requires 20% of the total energy produced. It consumes continuously 150 g of glucose and 72 L of oxygen every 24 h. A few minutes interruption of this supply can lead to dramatic brain damage. Manifestations and consequences of stroke depend on the location and extent of the lesions. A vascular cerebral accident, also called stroke or brain attack, is an interruption of the blood supply owing to either occlusion (ischemic stroke) or rupture (hemorrhagic stroke) of a blood vessel to any part of the brain, with an occurrence of around 80% for the ischemic type. Stroke has a devastating impact on public health and remains the third leading cause of death and the first leading cause of long-term disability in industrialized countries. An early diagnosis of the cerebral accident associated with an appropriate treatment would reduce the risk of death and enhance the chances of recovery. When the diagnosis of stroke is established, the physician needs to know the nature (ischemic or hemorrhagic), the extent, and the location of the accident in order to orient patients and to give them most suitable treatment. Because no specific and unique symptoms or early blood diagnostic markers are currently available, it was of a great interest to develop new approaches in the research and discovery area of new early diagnosis and prognosis markers of stroke.

Key Words: Plasmatic diagnostic marker; hemorrhagic stroke; ischemic stroke; SELDI-TOF; ApoCI; ApoCIII; SAA.

1. Introduction

Early diagnosis and immediate therapeutic interventions are crucial factors in reducing the extent of damage and the risk of death from stroke. Currently, the diagnosis of stroke relies on neurological assessment of the patient and neuroimaging techniques, including computed tomography (CT) and/or magnetic

From: *Methods in Molecular Biology, vol. 357: Cardiovascular Proteomics: Methods and Protocols*
Edited by: F. Vivanco © Humana Press Inc., Totowa, NJ

resonance imaging (MRI) scans. An early diagnostic marker of stroke, ideally capable of discriminating ischemic stroke from hemorrhagic stroke, would considerably improve acute management of patients.

The approach described here relied on surface-enhanced laser desorption/ionization (SELDI) protein profiling. Spectra of plasma samples from stroke patients ($n = 21$) were compared to healthy controls. Seven SELDI peaks were found differentially expressed ($p < 0.05$) and were identified as apolipoprotein (ApoCI), ApoCIII, serum amyloid A (SAA), and antithrombin-III (AT-III) fragment (1).

2. Materials

2.1. Equipment

1. Centrifuge.
2. Vortex.
3. Strong Anion Exchange arrays (SAX2 ProteinChip®, Ciphergen Biosystems, Fremont, CA).
4. ProteinChip reader system, PBS II serie (Ciphergen).
5. Ciphergen ProteinChip software (version 3.0).
6. Mini-PROTEAN® II (Bio-Rad, Hercules, CA).
7. Generator.
8. Vacuum centrifuge (HETO, Allerod, Denmark).
9. DecaXP ion trap (Thermofinnigan, San Jose, CA).
10. LC-PAL autosampler (CTC Analytics, Zwingen, Switzerland).
11. Rheos 2000 Micro-HPLC Pump (Flux Instruments, Basel, Switzerland).
12. YMS-ODS-AQ200 (Michrom Bioresources, Auburn, CA).
13. Nano-electrospray capillary (New Objective, Woburn, MA).

2.2. Reagents

1. Binding buffer: 20 mM Tris-HCl, 5 mM NaCl, pH 9.0.
2. Matrix solution: 0.5 µL sinapinic acid (SPA; Ciphergen) in 50% (v/v), acetonitrile (ACN), and 0.5% (v/v) trifluoroacetic acid (TFA).
3. Molecular-weight standard: "all-in-1" peptide molecular-weight standards (Ciphergen) diluted in the SPA matrix (1:1, v/v).
4. Laemmli's buffer: 0.125 M Tris-HCl, 4% sodium dodecyl sulfate (SDS), 40% glycerol, 0.1% bromophenol blue, pH 6.8 (2).
5. 10–20% Tris-glycine one-dimensional pre-cast gel (Bio-Rad).
6. Developing solution: sodium carbonate Na_2CO_3 30 g/L (w/v), 0.05% of 37% HCOH (v/v), 2% (v/v) of a fresh 0.2 g/L (w/v) sodium thiosulfate ($Na_2S_2O_3 \cdot 5\ H_2O$).
7. Fixing solution: 50% (v/v) methanol, 10% (v/v) acetic acid.
8. Staining solution: 2 g/L silver nitrate ($AgNO_3$).
9. Destaining solution: 30 mM K_3FeCN_6, 100 mM $Na_2S_2O_3$.
10. Drying solution: 30% ACN in 50 mM ammonium bicarbonate.

3. Method

3.1. Study Population and Sample Handling

Sixty-six consecutive patients admitted to the Geneva University Hospital emergency unit were enrolled in this study. For each patient, a blood sample was collected at the time of admission in dry heparin-containing tubes. Of the 66 patients enrolled, 21 were diagnosed with orthopedic disorders (without any known peripheral or central nervous system condition) and classified as control samples (including 12 men and 9 women, average age of 69.5 yr, range 34–94 yr) and 45 were diagnosed with stroke (including 27 men and 18 women, average age of 64.25 yr, range 27–87 yr) including 26 ischemic and 19 hemorrhagic. For the patients of the stroke group, the time interval between the neurological event and the first blood draw was 185 min (ranging from 40 min to 3 d). The diagnosis of stroke was established by a trained neurologist and was based on the sudden appearance of a focal neurological deficit and the subsequent delineation of a lesion consistent with the symptoms on brain CT or MRI images, with the exception of transient ischemic attacks where a visible lesion was not required for the diagnosis. The stroke group was separated according to the type of stroke (ischemia or hemorrhage), the location of the lesion (brainstem or hemisphere), and the clinical evolution over time (transient ischemic attacks when complete recovery occurred within 24 h, or established stroke when the neurological deficit was still present after 24 h).

1. After centrifugation at 1500g for 15 min at 4°C, plasma samples were aliquoted and stored at −70°C until analysis (*see* **Notes 1–3**).

3.2. SELDI Analysis of Stroke and Control Plasma Samples on SAX2 ProteinChip

1. SAX2 spots were first outlined with a hydrophobic pap-pen and air-dried. Chips were then equilibrated three times during 5 min with 10 µL binding buffer in a humidity chamber at room temperature (*see* **Note 4**).
2. Two microliters of binding buffer were applied on each spot and 1 µL of crude (stroke or control) plasma samples was added to each spot (*see* **Notes 5** and **6**) and incubated 30 min in a humidity chamber at room temperature (*see* **Note 6**).
3. Plasma was removed and each spot was individually washed five times for 5 min with 5 µL of binding buffer followed by two quick washes of the chip with deionized water (*see* **Note 7**).
4. Excess of H_2O was removed using absorbing paper and, while the surface was still moist, 0.5 µL of SPA in 50% (v/v) ACN and 0.5% (v/v) TFA was added twice per spot and dried.
5. The arrays were then read in a ProteinChip reader system.
6. The ionized molecules were detected and their molecular masses determined according to their TOF. TOF mass spectra, collected in the positive ion mode, were gener-

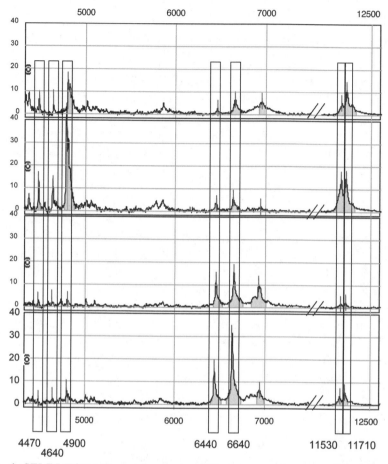

Fig. 1. SELDI-TOF-MS protein spectra of 2 representatives out of the 21 controls and 2 out of the 21 analyzed stroke plasma samples. The seven peaks found differentially expressed between controls and stroke patients ($p < 0.05$, Mann-Whitney U-test) are outlined in boxes.

ated using an average of 65 laser shots throughout the spot at a laser power set slightly above threshold (10–15% higher than the threshold).

7. Spectra were collected and analyzed using the ProteinChip software (version 3.0).
8. External calibration of the reader was performed using the "all-in-1" peptide molecular weight standards diluted in the SPA matrix (1:1, vol/vol) and directly applied onto a well of a normal phase chip.
9. Protein profile comparisons were performed after normalization on total ion current of all the spectra included in the same experiment.
10. Statistics were performed using the nonparametric Mann-Whitney U-test on the maximal intensity of each peak. Significance threshold was set at $p < 0.05$ (**Fig. 1**).

3.3. Stripping of the SAX2 ProteinChip

1. Among the 42 plasma samples analyzed on SELDI, a hemorrhagic stroke and a negative control were selected for subsequent analysis.
2. The eight spots of three SAX2 ProteinChip were loaded with control plasma sample (i.e., 24 spots) and two other chips (i.e., 16 spots) were loaded with stroke plasma sample using the protocol described in **Subheading 3.2.**
3. In order to remove and analyze specifically adsorbed proteins, chips were then stripped using Laemmli's buffer previously heated for 5 min at 90°C (*see* **Note 8**).

3.4. Mono-Dimensional Gel Electrophoresis

1. For each sample (stroke and control), the stripping products were loaded onto a 10–20% *Tris*-glycine one-dimensional pre-cast gel.
2. One microliter of crude plasma (stroke and control) sample and 7 μL of stripped plasma (stroke and control) were diluted in Laemmli's buffer up to 20 μL and heated at 95°C for 5 min (*see* **Note 9**).
3. Samples were centrifuged at 14,000g to recover evaporation in the top cover and loaded on the 10–20% SDS polyacrylamide gel.
4. Migration was performed in a *Tris*-glycine-SDS, pH 8.3, buffer.
5. The gel was then stained using MS-compatible silver staining derived from Blum *(3)*.
6. The gel was first fixed 30 min in 50% (v/v) methanol, 10% (v/v) acetic acid and then 15 min in 5% (v/v) methanol.
7. The gel was then washed three times 5 min in Milli-Q® H$_2$O and incubated 2 min in 0.2 g/L (w/v) fresh sodium thiosulfate (Na$_2$S$_2$O$_3$, 5 H$_2$O).
8. The gel was further washed three times 30 s in Milli-Q H$_2$O, and incubated 25 min in the staining solution.
9. The gel was washed three times 1 min in Milli-Q H$_2$O, and incubated in the developing solution for 10 min maximum.
10. The gel development was stopped using a 14 g/L (w/v) Na$_2$-EDTA solution for 10 min before washing in Milli-Q H$_2$O.
11. The apparent molecular masses were determined by running 2 μg of a low-molecular-weight standard (Bio-Rad).
12. The gel was scanned on a Arcus II Agfa scanner, with Agfa Fotolook version 3.6 software (**Fig. 2**).
13. Bands to be identified were immediately cut, placed in an Eppendorf tube, and stored at 4°C in 10% ethanol (v/v).

3.5. Identification of the Proteins by Nano Liquid Chromatography Electrospray Ionization Tandem Mass Spectrometry

1. MS-compatible silver-stained gel was destained as follows.
2. Each gel piece was incubated in 30 μL destaining solution with occasional vortexing until the gels were completely destained (5–10 min).
3. Gel pieces were then washed with 30 μL Milli-Q H$_2$O for 10 min. These two steps were repeated two times.

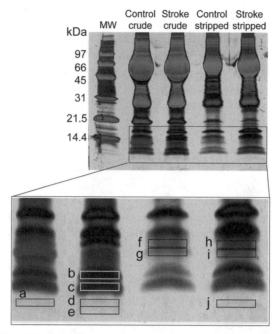

Fig. 2. Gradient 10–20% *Tris*-glycine sodium dodecyl sulfate-polyacrylamide gel electrophoresis gel of both crude and stripped plasma samples (control and stroke). Gel bands were cut and identified by tandem mass spectrometry as: antithrombin fragment (a, d, e, and j; *see* **Note 11**), apolipoprotein (Apo) CI (c, d, and e; *see* **Note 12**), ApoCIII (b, f, g, h, and i; *see* **Note 12**), and serum amyloid A (f and g; *see* **Note 10**). The Bio-Rad low-molecular-weight standards (M) are: Phosphorylase b (97.4 kDa), bovine serum albumin (66.2 kDa), ovalbumin (45 kDa), carbonic anhydrase (31 kDa), trypsin inhibitor (21.5 kDa), lysozyme (14.4 kDa).

4. Gel pieces were rehydrated with 100 µL of 50 m*M* ammonium bicarbonate, pH 8.0, for 10 min while vortexing.
5. Gel pieces were then dehydrated with 100 µL of the drying solution and dried in vacuum centrifuge for 30 min.
6. Trypsin digestion was performed as described previously *(4)*.
7. Nano liquid chromatography electrospray ionization tandem mass spectrometry was performed on a DecaXP ion trap coupled with a LC-PAL autosampler and a Rheos 2000 Micro-HPLC Pump.
8. For each experiment, 5 µL of sample in 5% ACN, 0.1% formic acid was injected on a C18 reverse phase column (75 µm inner diameter) packed in house with YMS-ODS-AQ200.
9. Peptides were eluted from the column using a CH₃CN (ACN) gradient in the presence of 0.1% formic acid.

10. For peptide elution, the ACN concentration was increased from 16–68% in 15 min.
11. A flow splitter was used to decrease the flow rate from 40 μL/min to approx 0.2 μL/min.
12. A 2-kV potential was applied on the nano-electrospray capillary.
13. Helium was used as collision gas. The collision energy was set at 35% of the maximum.
14. MS/MS spectra were acquired by automatic switching between MS and MS/MS mode.
15. The two highest peaks from each MS scan were chosen for MS/MS.
16. Spectra were converted to DTA files, regrouped using in-house software, and the database search was performed with MASCOT 1.8. A tolerance of 2 Da was chosen for the precursor and 1 Da for fragments. ESI-TRAP was selected as the instrument.
17. The combined Swiss-Prot and TrEMBL database was searched without species restriction. In these conditions, the threshold of significance was given by a score of 50 or higher by MASCOT.

4. Notes

1. Collect venipuncture blood in a syringe or tube containing a suitable anticoagulant such as heparin or EDTA. Capillary blood from a finger prick can be used provided that the skin is first cleaned thoroughly to avoid contamination with sweat. After blood collection, the syringe or tube should immediately be placed in an ice bath and brought to the laboratory for analysis. Upon arrival in the laboratory, the blood specimen must be centrifuged immediately at 2000*g* for 10 min at 5°C to avoid haemolysis, and decanted. The sample can be either processed immediately or stored at −70°C until analysis.
2. Protein degradation may occur between sample collection and solubilization owing to the presence of proteases. Protease inhibitors such as phenylmethylsulfonylfluoride (0.2 m*M*), leupeptine (50 μg/mL), benzamidine (0.8 m*M*), or a combination of these can be added to the sample.
3. Samples should be divided into several small aliquots before storage in order to avoid repeated freezing and thawing of a single sample. Solubilized samples must be frozen at −70°C or lower temperatures and kept for several years. The freezing step should be fast. Avoid freezing solubilized samples at −20°C.
4. Similar strong anionic exchange surfaces, called Q10 are now available. Q10 ProteinChips already display a hydrophobic surface all around the spots, which avoids the use of hydrophobic pap-pen.
5. The pipetting precision is essential for SELDI reproducibility. Work with appropriate pipets and tips.
6. In order to homogenize plasma sample in the buffer, several aspiration–repression should be performed with the pipet.
7. Washing is composed of two steps: first with the binding buffer by aspiration–repression several times and second with deionized water with a washing bottle.
8. The stripping was performed spot by spot for each sample (i.e., stroke or control) using always the same stripping product obtained from the previous spot.

9. It is important to work with highest quality reagents, which have been tested for electrophoresis. Commercial bulk chemicals may contain impurities. A typical example is SDS that can be sold as a mixture of different chain lengths including 12, 14, and 16 carbon chains. It binds most of the proteins in a ratio of 1.4 g SDS per gram of proteins. It has been shown that the absence of reducing agent in the solubilizing solution may decrease the SDS/polypeptides binding ratio.

10. SAA is one of the major acute-phase proteins (inflammatory marker) in humans, and may increase from 100- to 1000-fold following inflammatory stimulus. It is synthesized predominantly in the liver after induction by cytokines such as interleukin-6, and secreted as a major component of the Apos in the high-density lipoproteins particles.

11. AT-III is the most important inhibitor in the coagulation cascade. It is a serine protease inhibitor (Serpin), which inhibits the formation of thrombin. Stroke is associated with a decreased AT-III activity and an increased in thrombin–AT-III complexes.

12. ApoCI and ApoCIII belong to the Apo family. At least, nine distinct polymorphic forms of Apos are known to exist in tissues and body fluids, mainly as protein component of the lipoprotein particles. The apolipoproteins generally act as stabilizers of the intact particles. Quantitative measurement of high-, low-, and very low-density lipoprotein particles in human serum are often used to estimate an individual's relative risk of coronary heart disease. Both ApoCI and CIII are involved in triglyceride metabolism.

13. Diagnosis should probably rely on the detection of a panel of proteins. The concept of a panel of protein biomarkers is slowly replacing the ideal unique marker displaying 100% sensitivity and specificity. In this context, we are developing a robust algorithm to evidence a panel of proteins.

Acknowledgments

This work was kindly supported by Proteome Sciences plc. The authors thank all their BPRG colleagues for assistance.

References

1. Allard, L., Lescuyer, P., Burgess, J., et al. (2004) ApoC-I and ApoC-III as potential plasmatic markers to distinguish between ischemic and hemorrhagic stroke. *Proteomics* **4,** 2242–2251.
2. Laemmli, U. K. (1970) Cleavage of structural proteins during the assembly of the head of bacteriophage T4. *Nature* **227,** 680–685.
3. Blum, H., Beier, H., and Gross, H. J. (1987) Improved silver staining of plant proteins, RNA and DNA in polyacrylamide gels. *Electrophoresis* **8,** 93–99.
4. Scherl, A., Coute, Y., Deon, C., et al. (2002) Functional proteomic analysis of human nucleolus. *Mol. Biol. Cell* **13,** 4100–4109.

28

Depletion of High-Abundance Proteins in Plasma by Immunoaffinity Subtraction for Two-Dimensional Difference Gel Electrophoresis Analysis

Verónica M. Dardé, Maria G. Barderas, and Fernando Vivanco

Summary

Blood plasma is believed the most complex human-derived proteome, containing other tissue proteome subsets. Almost all body cells communicate with the plasma, either directly or through tissues or biological fluids, and many of these cells release at least a part of their content into the plasma upon damage or death. A comprehensive, systematic characterization of the plasma proteome in the healthy and diseased states will greatly facilitate the development of biomarkers for early disease detection, clinical diagnosis, and therapy. However, the characterization of human plasma proteome is a very complicated task, owing to the wide dynamic range of concentration that separates the most abundant proteins and the less common ones (10–12 orders of magnitude). The removal of its predominant proteins by affinity chromatography using an FPLC system improves the presence of low-abundance proteins in two-dimensional gel electrophoresis (2DE). The "Multiple Affinity Removal System" (Agilent Technologies) retains albumin, IgG, IgA, haptoglobin, transferrin, and antitrypsin with high specificity and reproducibility. After depletion, we have independently analyzed the flow-through (low-abundance proteins), and the retained fractions, by 2DE (4.0–7.0 pH range). Image analysis of the stained gels revealed that more than 300 spots appeared in the retained fraction and about 1800 spots appeared in the nonretained fraction. This methodology is a valuable tool for clinical proteomics, because its reproducibility allows comparative studies and quantitative analysis by 2DE or two-dimensional differential gel electrophoresis of plasma or sera samples from subjects with different pathological or physiological conditions. In addition, the method allows the comparison of experimental results from different laboratories.

Key Words: Human plasma; plasma depletion; proteomic analysis; immunoaffinity chromatography; 2DE; DIGE.

From: *Methods in Molecular Biology, vol. 357: Cardiovascular Proteomics: Methods and Protocols*
Edited by: F. Vivanco © Humana Press Inc., Totowa, NJ

1. Introduction

Plasma is one of the richest protein-containing samples, predominated by high-abundance resident proteins, such as albumin and immunoglobulins, together with proteins originated from circulating blood cells and other tissues. It also may carry tissue-leakage proteins, and thus it is expected to contain information on the physiological states of different parts of the organism (1). For this reason, it is frequently monitored for disease biomarkers and it is regularly used for diagnosis.

Despite this, human plasma has not been used often in clinical proteomics, because the unusually high concentration of a small group of proteins, including albumin and immunoglobulins, impedes the resolution of plasma proteins by conventional proteomic technology (two-dimensional [2D] gels and mass spectrometry [MS]) (2). The limited amount of protein loadable into a single 2D gel results in a profile predominated by these high-abundance proteins, leaving a large percentage of proteins with low-abundance proteins undetected. This is an important limitation, because this large percentage of protein present in lower concentrations includes physiologically relevant proteins and potential biomarkers of disease (3). However, some methods have been described lately to facilitate the analysis of plasma proteome by two-dimensional gel electrophoresis (2DE), including affinity methods for removal of high-abundance proteins (4,5).

This chapter describes a procedure to remove the six most abundant plasma proteins (albumin, IgG, IgA, haptoglobin, transferrin, and antitrypsin), which constitute 85–90% of total plasma protein, using a commercially available affinity system (Multiple Affinity Removal System, Agilent Technologies) in order to raise the concentration of low-abundance proteins in our 2D gels, improving analysis of plasma by 2DE. In addition, this method allows downstream quantitative analysis by 2D difference gel electrophoresis (DIGE) (6,7) and makes it possible to perform differential analysis of plasma or sera samples from healthy people (controls) and diseased patients by comparison of the expression profiles of their low-abundance plasma proteins.

2. Materials

2.1. Equipment

1. EDTA glass tubes (Becton Dickinson, Franklin Lakes, NJ).
2. Electrophoresis apparatus (Ettan IPGphor Unit, Ettan DALT*six* Electrophoresis Unit, Amersham Biosciences, Uppsala, Sweden).
3. Ettan IPGphor ceramic strip holders (Amersham).
4. Electrophoresis Power supply (EPS 601, Amersham).
5. Immobiline DryStrip 24 cm, pH 4.0–7.0, 3.0–10.0 NL (Amersham).
6. Milli-Q® System (Millipore, Bedford, MA).

7. Shaker (Promax 1020, Heidolph Instruments GmbH & Co., Schwabach, Germany).
8. Vortex mixer (Promax Reax Top, Heidolph Instruments GmbH & Co).
9. Thermomixer compact (Eppendorf AG, Hamburg, Germany).
10. DALT Glass plates (26 × 20 cm, Amersham).
11. Reswelling cassette (Bio-Rad Laboratories, Inc., Hercules, CA).
12. DALTsix Gel Caster (Amersham).
13. GS-800 Calibrated Densitometer (Bio-Rad).
14. HITACHI U-1100 Spectrophotometer (Hitachi, Ltd., Tokyo, Japan).
15. Freezemobile 12SL Freeze Dryer (VirTis Company Inc., Gardiner, NY).
16. Allegra X-12R Centrifuge (Beckman Coulter, Inc., Fullerton, CA).
17. DALTsix Gradient Maker (Amersham).
18. Micro Tweezers MTW7 (FONTAX Technologie SA, Lausanne, Switzerland).
19. FPLC System (ÄKTApurifier, Amersham).
20. Multiple Affinity Removal Column, 4.6 mm × 50 mm (Agilent Technologies, Inc., Palo Alto, CA).
21. Vivaspin 5KDa cut-off spin concentrators (Vivascience AG, Hannover, Germany).
22. Spin filters, 0.22-μm cellulose acetate (Agilent).
23. PDQuest gel analysis software, version 7.1 (Bio-Rad).
24. DeCyder Differential Analysis Software, version 5.0 (Amersham).
25. Typhoon 9400 fluorescence scanner (Amersham).

2.2. Reagents

1. Buffer A for loading, washing, and equilibrating Multiple Affinity Removal Column and buffer B for elution of bound proteins from Multiple Affinity Removal Column (Agilent).
2. Agarose LM2 (low melting point [LMP], 65°C) for analytical and preparatory electrophoresis (Pronadisa, Laboratorios Conda, S.A., Madrid, Spain, Ref. 8050).
3. HCl 37% PA-ACS-ISO (Ref. 131020) and glycerol (RFE, USP, BP, Ph. Eur.) PRS-CODEX (Ref. 141339, Panreac Química S.A., Barcelona, Spain).
4. Acetic acid reagent grade, ACS, ISO (Ref. AC0344) and methanol, extra pure, Ph Eur, NF (Ref. ME0301, Scharlau Chemie S.A., Barcelona, Spain).
5. CyDye DIGE Fluor minimal labeling kit, *N*,*N*-dimethyl formamide, DeStreak reagent, immobilized pH gradient (IPG) buffers (4.0–7.0 and 3.0–10.0 NL), and PlusOne Silver Staining Kit (Amersham).
6. Urea, CHAPS, Tris-HCl (electrophoresis purity reagent), sodium dodecyl sulfate (SDS), dithiothreitol (DTT), iodoacetamide, bromophenol blue, mineral oil, acrylamide, *N*,*N*'-methylene-bisacrylamide, ammonium persulfate solution (APS), TEMED, SYPRO Ruby, and Bio-Rad Protein Assay Dye Reagent Concentrate from Bio-Rad.
7. Ethanol and ammonium bicarbonate (Merck, Whitehouse Station, NJ).

2.3. Solutions

1. Ammonium bicarbonate solution (25 m*M*): Dissolve 99 mg of ammonium bicarbonate in double-distilled water (50 mL final volume). Prepare just prior to use.

2. DIGE labeling buffer: 7 *M* urea, 2 *M* thiourea, 4% (w/v) CHAPS, 30 m*M* Tris-HCl. Dissolve 42 g of urea, 15.2 g of thiourea, 4 g of CHAPS, and 360 mg of Tris base in 50 mL of double-distilled water. Adjust to pH 8.5 with diluted HCl. Make up to a final volume of 100 mL. Store in 1-mL aliquots at −20°C. This buffer is a cell lysis buffer compatible with the labeling method. In this case it is used to resuspend lyophilized samples before labeling.

3. CyDye DIGE Fluor minimal dye stock solution (1 nmol/μL). Reconstitute each CyDye DIGE Fluor minimal dye in the specified volume of *N,N*-dimethyl formamide (*see* specification sheet supplied with the CyDye DIGE Fluor minimal dye).

4. Rehydration stock solution: 9 *M* urea, 4% (w/v) CHAPS, 10 m*M* Tris-HCl. Dissolve 10.8 g of urea, 800 mg of CHAPS, and 24 mg of Tris base in double-distilled water to a final volume of 20 mL and add a trace of bromophenol blue. Store in 1-mL aliquots at −20°C. 1.2% (w/v) DeStreak reagent and 1% (v/v) IPG buffer (pH 3.0–10.0 NL or 4.0–7.0) must be added just prior to use (*see* **Note 1**).

5. SDS equilibration stock solution: 6 *M* urea, 50 m*M* Tris-HCl, pH 8.8, 2% (w/v) SDS, 30% (v/v) glycerol. Mix 72.07 g of urea, 6.7 mL of Tris-HCl 1.5 *M*, pH 8.8, 60 mL of glycerol (87%), and 4 g of SDS. Add double-distilled water up to 200 mL. Finally add a trace of bromophenol blue. Store at −20°C (*see* **Note 1**).

   Prior to the first equilibration step, add 100 mg of DTT per 10 mL of equilibration stock solution. Prior to the second equilibration step, add 250 mg of iodoacetamide per 10 mL of equilibration stock solution.

6. Acrylamide stock solution: 30% (w/v) acrylamide, 0.8% (w/v) *N,N'*-methylenebisacrylamide. To prepare 100 mL, dissolve 30 g of acrylamide and 0.8 g of bisacrylamide in 80 mL of double-distilled water. Stir for 10 min, filter through a 0.45-μm filter, and fill up to 100 mL with double-distilled water. Store at 4°C in the dark.

7. APS (10% [w/v]): dissolve 1 g of APS in 10 mL of double-distilled water. Prepare just prior to use.

8. SDS electrophoresis buffer (1X): 25 m*M* Tris, 192 m*M* glycine, 0.1% (w/v) SDS. Mix 3.03 g of Tris base, 14.4 g of glycine, and 1g of SDS. Add double-distilled water up to 1 L. Store at room temperature.

9. Agarose sealing solution: to prepare 100 mL, suspend 0.5 g of LMP agarose and a trace of bromophenol blue to 100 mL of SDS electrophoresis buffer. Store in 2-mL aliquots at room temperature. To dissolve the agarose, heat aliquots in a thermomixer at 65°C and cool the solution below 37°C just prior to apply, to avoid decomposition of urea and protein carbamylation (*see* **Note 1**).

10. Fixation solution (for silver staining): 30% (v/v) ethanol, 5% (v/v) acetic. Mix 60 mL of ethanol and 1 mL of acetic acid. Add double-distilled water up to 200 mL. Prepare just prior to use.

3. Methods

3.1. Depletion of High-Abundance Proteins

To visualize low-abundance proteins of plasma or serum by 2DE, removal of high-abundance proteins is required (*1*). Different methods for removal of albumin have been described through years (**refs.** *3–11*; *see* Chapter 29). Among these methods, it was recently found that antibody-based methods are more effective than those based on ion-exchange chromatography *(7)*. In 2003, Piepper et al. *(12)* developed a strategy to generate multicomponent immunoaffinity subtraction chromatography matrices capable of specifically depleting albumin and other high-abundance plasma proteins. A similar approach has been applied by Agilent Technologies to produce the Agilent Multiple Affinity Removal System, which comprises an affinity column, with binding capacity for 20 µL of human serum or plasma, and optimized mobile phases (Buffers A and B). The affinity column is packed with immobilized affinity-purified polyclonal antibody resins for removing albumin, IgG, IgA, haptoglobin, transferrin, and antitrypsin with high specificity. The ratio of immobilized antibodies coupled to the affinity column reflects the ratio of target proteins in human serum or plasma. We have tweaked and expanded the manufacturer's basic protocol to perform a comparative analysis of plasma samples by 2DE (also valid for sera), assessing the two main fractions obtained by using this commercially available system (*see* **Note 2**).

To avoid column saturation, dilute up to 15–20 µL of human plasma five times in Buffer A (15 µL of plasma and 60 µL of buffer) and spun through a 0.22-µm spin filter tube at maximum speed (about 16,000*g*) on a tabletop microfuge for 1 min (*see* **Note 3**). Alternatively, if enough plasma is available, add 400 µL of buffer A to 100 µL of plasma. Distribute the 500 µL in five aliquots of 100 µL and use 90 µL for the analysis (this corresponds to 18 µL of plasma). Then, the sample is ready for injection into a fast protein liquid chromatography System (ÄKTApurifier, Amersham Biosciences) coupled to the Multiple Affinity Removal Column. Buffer A and Buffer B must be set up as the only mobile phases. The chromatographic steps must be performed according to the instruction manual supplied with the column:

1. Inject 75–100 µL of the diluted plasma with Buffer A at a flow rate of 0.25 mL/min for 9 min.
2. Monitor the absorbance at 280 nm and the conductivity, which reflects the linear concentration increase for Buffer B (% gradient).
3. Collect flow-through fraction when the absorbance more than 0.02 absorbance units (it usually appears between 1.5 and 4.5 min) (*see* **Fig. 1**) and store collected fractions at −20°C if not analyzed immediately.
4. Elute bound proteins from the column with Buffer B at a flow rate of 1 mL/min for 3.5 min. Collect the elution fraction when the absorbance more than 0.02 absorbance units. Store collected fractions at −20°C if not analyzed immediately.
5. Regenerate column by equilibrating it with Buffer A for 7.5 min at a flow rate of 1 mL/min, with a total run cycle of 20 min (*see* **Note 4**).
6. Control the pressure limit of the column (120 bar) (*see* **Note 3**).

Similar chromatograms are obtained when overlaying several chromatographic runs (**Fig. 1**). The great reproducibility of this method allows the combination of flow-through fractions after several chromatographic cycles if high
amounts of protein are necessary for downstream applications (in this case, 2DE).
Normally, 100–300 µg of nonretained proteins, in 1 mL (two chromatographic
fractions; *see* **Fig. 1**), are recovered after every injection (*see* **Note 5**). The flow-
through fractions can be also used for SELDI-TOF analysis.

Transfer the flow-through fractions (4 × 1 mL) into 5-KDa cut-off spin concentrators and centrifuge at 3000*g* in a swinging-bucket rotor for 30 min (the
volume is reduced to about 0.5 mL). Then, equilibrate fractions in a 25 m*M*
ammonium bicarbonate solution by addition of 3–4 mL, and concentrate them,
again to about 0.5 mL. Quantify the protein content using the Bradford method
(13) and pipet the amount of protein required for a single 2D gel into 1.5-mL
tubes (*see* **Note 6**). Freeze aliquots at −80°C and lyophilize them for 15–18 h.

3.2. Sample Labeling

2D-DIGE is an approach that allows a quantitative analysis of differential
protein expression between biological samples. It is commercialized as the Ettan
DIGE proteomics system (Amersham) and it is based on the specific properties of the three spectrally resolvable dyes: CyDye DIGE Fluor minimal dyes
(Cy2, Cy3, and Cy5). This fluorescent compounds are matched for mass and
charge, and they are used to covalently label lysine residues of proteins in different samples prior to mixing and separation on the same 2D gel. Labeled proteins can be detected at appropriate excitation and emission wavelengths using
a fluorescence detection device (Typhoon 9400, Amersham). The signals may
be then compared using appropriated software applications.

This methodology has several advantages, because all samples are subjected
to the same handling procedures and, as a fluorescence method *(14)*, it provides higher sensitivity and linear dynamic range of detection than many other

Fig. 1. (*Opposite page*) Schematic representation of the depletion method using the
Multiple Affinity Removal System. Three peaks appear in the resulting chromatograms.
The first peak corresponds to the flow-through fraction containing the low-abundance
proteins, whereas the second one corresponds to the retained fraction. There is a third
small peak with no protein content that might reflect changes in buffers from B to A.
Similar chromatograms are obtained when overlaying several chromatographic runs,
confirming the high reproducibility of this method. Removal of the six most abundant
plasma proteins (retained fraction) allows the detection of many low-abundance proteins (flow-through fraction) using high-resolution two-dimensional gels (first dimension, 24 cm immobilized pH gradient, pH 4.0–7.0 range; second dimension 9–16%
acrylamide gradient). More than 1800 and 300 spot proteins were visualized by silver
staining of the nonretained (left) and retained (right) plasma fractions, respectively.

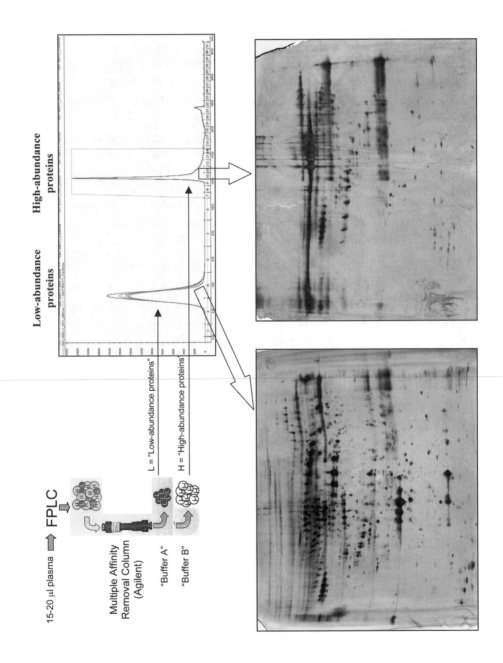

High-abundance proteins

Low-abundance proteins

15–20 µl plasma → FPLC

Multiple Affinity Removal Column (Agilent)

"Buffer A"

"Buffer B"

L = "Low-abundance proteins"

H = "High-abundance proteins"

staining techniques. Thus, protein changes can be detected and quantified with higher confidence.

Prior to performing a 2D-DIGE experiment, a meticulous experimental design is required. Guidelines for the design of 2D-DIGE experiments are freely available from Amersham Biosciences (Ettan DIGE System User Manual; http://www5. amershambiosciences.com/). In this case, an experiment is described in which a plasma sample from a healthy subject is compared with a diseased one (*see* **Fig. 2**).

1. Resuspend lyophilized samples in labeling buffer to a final concentration of 5–10 mg/mL (*see* **Note 7**).
2. Check pH for each sample. Make sure it is between pH 8.0–9.0; otherwise, the labeling reaction will fail.
3. Aliquot 50 µg each sample into separated tubes. To create an internal standard, mix 25 µg of Sample 1 with 25 µg of Sample 2 in a third tube.
4. Add 400 pmol of the appropriate CyDye to each sample (usually, Cy3 to Sample 1, Cy5 to Sample 2, and Cy2 to the pooled internal standard) and incubate the samples on ice for 30 min in the dark.
5. Add 1 µL of 10 m*M* lysine and incubate for 10 min on ice in the dark to quench the labeling reaction.

Once the labeling reaction is finished, the labeled samples can be processed immediately or stored for 3 mo at −70°C in the dark. Before proceeding with the sample preparation for isoelectric focusing (IEF) (*see* **Subheading 3.3.1.**), the three labeled samples must be combined into a single tube. Further steps (IEF, equilibration, and SDS-PAGE) should be performed in the dark.

3.3. Two-Dimensional Electrophoresis

3.3.1. Sample Preparation for the First Dimension: IEF

Sample preparation is essential for the success of any 2DE experiment (including 2D-DIGE). Prior to IEF, the protein mixture must be placed in a denaturing environment capable of solubilizing proteins, avoiding protein modification and/or aggregation, and maintaining their native electrostatic properties.

1. Dilute samples in rehydration solution (up to 450 µL).
2. Vortex sample to mix well.
3. Incubate the sample with the rehydration solution at room temperature for 20 min before sample application.

Fig. 2. (*Opposite page*) Example of a two-dimensional-difference gel electrophoresis experiment in which a plasma sample from a healthy subject is compared to a diseased one. (**A**) Flow-through fraction of plasma from a healthy volunteer, labeled with Cy3. (**B**) Flow-through fraction of plasma from a diseased patient, labeled with Cy5. (**C**) Images A and B overlaid.

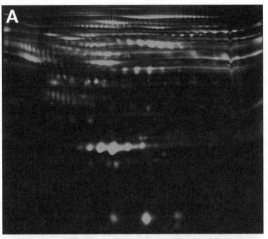

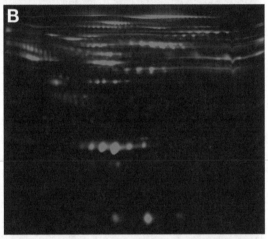

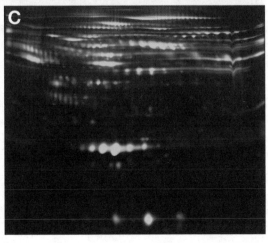

3.3.2. Sample Application

Samples are applied into the Immobiline DryStrips (IPG strips) by rehydration loading (*see* **Note 8**).

1. Position the reswelling cassette on a level surface on the bench top and ensure it is clean and dry.
2. Pipet 450 µL of the rehydration solution, including sample, and deliver this volume along one slot of reswelling cassette. If any bubbles are formed, remove them.
3. Retire the protective plastic cover from the Immobiline DryStrip (24 cm, pH 3.0–10.0 NL or 4.0–7.0) and put the gel side in contact with the rehydration solution carefully. The entire strip must be coated with the solution and no bubbles must be trapped under it.
4. Overlay the IPG strip with mineral oil, to avoid water loss and urea crystallization.
5. Repeat this process, using a new strip for each sample.
6. Allow the IPG strips to rehydrate overnight at room temperature.

3.3.3. IEF Separation

Proteins are amphoteric molecules that may be positively or negatively charged depending on the pH of the environment. IEF separates proteins according to their isoelectric points (pI), defined as the specific pH at which net charge of a protein is zero. If an electric field is applied on a pH gradient, the proteins inside it migrate until they reach the pH equivalent to their pI, where they lose any charge and stop moving. Thus, to perform IEF we need a pH gradient (IPG strips) and a power supply.

After passive rehydration, each strip is placed in an independent Ettan IPG phor ceramic strip holder. Then, all the strip holders (up to 12) are positioned on the Ettan IPGphor Unit where IEF is performed at 20°C, according to the following program:

1. 30 min at 300 V 1 h.
2. 3 h at 3500 V (gradient).
3. 3 h at 3500 V.
4. 1.5 h at 6000 V (gradient).
5. 6000 V for a total of 100,000 V/h.

After focusing, strips can be stored at −70°C for up to 3 mo or they can be run immediately on the second dimension.

3.4. Second Dimension: SDS-PAGE

3.4.1. IPG Strips Equilibration

Prior to SDS-PAGE, IPG strips must be saturated in a SDS buffer system. This equilibration buffer contains several reagents that are essential for the second-dimension separation, like SDS, which denatures proteins and forms negatively charged SDS–protein complexes that migrate towards the anode during electrophoresis. The equilibration solution also contains urea and glycerol, to diminish electroendosmotic effects *(15)*. The introduction of DTT in the first equilibration step keeps the reduced state of the denatured proteins. The addition of iodoacetamide during the second equilibration step is used to alkylate reduced proteins and residual DTT, in order to avoid protein reoxidation and point streaking, respectively (**ref. *16***; *see* **Note 9**).

1. Place each IPG strip in one channel of a clean cassette, acrylamide side up.
2. Add 5 mL of equilibration buffer containing 1% (w/v) DTT to each strip. Position the cassette on a shaker and stir for 15–20 min.
3. Pour off the equilibration buffer and rinse with double-distilled water.
4. Add 5 mL of equilibration buffer containing 2.5% (w/v) iodoacetamide to each strip and shake again for 15–20 min.
5. Finally rinse again with double-distilled water and drain excess of equilibration buffer on a clean filter paper before applying the strips to the SDS-PAGE gel.

3.4.2. SDS-PAGE Separation

Focused proteins in the IPG strip are further separated on the basis of molecular mass using SDS-PAGE in the second-dimension. We perform SDS-PAGE according to Laemmli *(17)*, using the Ettan DALT*six* Electrophoresis System. The choice for the percentage of acrylamide and the type of gel (gradient vs homogeneous gels) depends on the molecular weight range we want to resolve. For analysis of complex protein mixtures, gradient gels are normally used. For serum or plasma, very good results are obtained using a 9–18% acrylamide gradient.

After equilibration, IPG strips are placed on the top of the vertical gel and overlaid with agarose sealing solution to ensure good contact between the gel and the strip (*see* **Note 10**). Then gels are run at 20°C, setting the power supply at 1 W/gel overnight or at 18 W/gel for 5–6 h.

3.5. Gel Staining

Once SDS-PAGE run has finished, protein spots must be visualized. There are several staining techniques, with different characteristics (sensitivity, linearity, homogeneity, reproducibility). The most sensitive method is silver staining, although the linear range of detection in fluorescent techniques (e.g., SYPRO Ruby or DIGE) is wider.

When performing 2D DIGE, staining is not required. Because proteins are covalently labeled with fluorescent compounds, gels can be immediately visualized in an appropriate fluorescence detection device (Typhoon 9400). Nevertheless, colloidal Coomassie brilliant blue and SYPRO Ruby protein stains can be used in combination with CyDye labeling. When staining with SYPRO Ruby protein stain, gels must be fixed for at least 1 h in a mixture of 10% (v/v) methanol and 7% (v/v) acetic acid and then incubated in SYPRO Ruby protein stain overnight. Prior to imaging in a fluorescence scanner, to reduce background, gels may be washed for 30 min in double-distilled water.

In order to use PlusOne Silver Staining Kit (high reproducibility and compatibility with MS), gels must be incubated in fixation solution for at least 1 h. Then, perform staining steps according to manufacturer's indications. This silver-staining protocol is based on the methodology of Heukeshoven and Dernick *(18)*.

3.6. Image Analysis of the 2DE Gels

Gels must be digitized before image analysis can be performed. Silver-stained gels may be recorded as digitized images using a desktop scanner or a densitometer like GS-800 Calibrated Densitometer (BioRad). SYPRO Ruby stained gels and 2D DIGE gels must be digitized in a fluorescence scanner, like Typhoon 9400 (Amersham). Then, digital images can be imported to a gel analysis software (e.g., PDQuest, DeCyder) to perform the evaluation, processing, and analysis of the 2D gels.

4. Notes

1. Solutions containing urea must not be heated over 37°C, in order to prevent urea decomposition into isocyanate, which may cause protein modification by carbamylation.
2. Columns of higher capacity are available from Agilent Technologies (4.6 × 100 mm; 1.66 mL), for 30–40 μL of human serum/plasma, or the most recent one (Agilent High Capacity Multiple Affinity Removal System for the Depletion of High-Abundant Proteins from Human Proteomic Samples, Ref. 5188-5921), with at least twice the loading capacity and 300 injections in the fast protein liquid chromatography with high reproducibility. This column is made with a novel attachment process for antibodies and the same buffers and protocols are used.
3. Do not load the column directly with crude serum or plasma. It is essential to dilute it in order to prevent clogging of the column. If the inlet frit is clogged, replace both the inlet and outlet frits simultaneously. Clogged inlet frits may increase backpressure, which affects negatively to column lifetime. Remove particulate materials that are sometimes present in serum or plasma by quick spin using the 0.22-μm spin filters.
4. When not in use, store the column after equilibration with Buffer A at 2–8°C in a refrigerator to minimize loss in column capacity. Be sure that the end-caps are tightly sealed. Do not expose column to organic solvents or reducing agents (which affect the structure of the antibodies). Do not freeze the column.
5. After the first elution cycle of a plasma sample, the binding capacity of the column may drop a 10–25%. This is a phenomenon frequently observed, known as "first cycle effect." It is probably caused by high-avidity binding of some plasma proteins to a fraction of the immobilized antibodies *(11)*.

6. Protein estimation is very important. The amount of protein that can be loaded to a single IPG strip depends on its length. Usually 50–500 µg of protein are applied to 17 cm or 24 cm IPG strips.
7. Before labeling is finished, avoid addition of any primary amine compounds to the protein mixture, because these compounds may compete with the proteins for dye.
8. Use gloves and tweezers when handling IPG strips to avoid protein contamination.
9. DTT must be added just prior to use. It is used to break disulfide bonds by reducing sulfhydryl groups. In the second equilibration step, iodoacetamide is added to alkylate free sulfhydryl groups, preventing reoxidation and resurgence of disulfide bonds.
10. After adding agarose, bubbles may be formed around the IPG strip. These bubbles must be removed because they may interfere in protein migration from the strip to the gel.

Acknowledgments

This work was partially supported by grants from Ministerio Ciencia y Tecnología (BMC2002-02596), Fundación MMA, FIS (Cardiovascular Network, RECAVA 03/01).

References

1. Vivanco, F., Martín-Ventura, J. L., Duran, M. C., et al. (2005) The quest for novel cardiovascular biomarkers by proteomic analysis. *J. Proteome Res.* **4,** 1181–1191.
2. Anderson, N. L. and Anderson, N. G. (2002) The human plasma proteome: history, character, and diagnostic prospects. *Mol. Cell. Proteomics* **1,** 845–867.
3. Ahmed, N. and Rice, G. E. (2005) Strategies for revealing lower abundance proteins in two-dimensional protein maps. *J. Chromatogr. B.* **815,** 39–50.
4. Yang, Z. and Hancock, W. S. (2004) Approach to the comprehensive analysis of glycoproteins isolated from human serum using a multi-lectin affinity column. *J. Chromatogr. A.* **1052,** 79–88.
5. Marshall, J., Jankowski, A., Furesz, S., et al. (2004) Human serum proteins preseparated by electrophoresis or chromatography followed by tandem mass spectrometry. *J. Proteome Res.* **3,** 364–382.
6. Morita, A. and Szafranski, C. (2004) Differential Analysis of Ovarian Cancer Patient Serum Using the Multiple Affinity Removal System. Application, Agilent Technologies, 5989-1839EN, http://www.chem.agilent.com/. Last accessed July 2005.
7. Chromy, B. A., Gonzales, A. D., Perkins, J., et al. (2004) Proteomic analysis of human serum by two-dimensional differential gel electrophoresis after depletion of high-abundant proteins. *J. Proteome Res.* **3,** 1120–1127.
8. Gianazza, E. and Arnaud, P. (1982) A general method for fractionation of plasma proteins. Dye-ligand affinity chromatography on immobilized Cibacron blue F3-GA. *Biochem. J.* **201,** 129–136.
9. Tirumalai, R. S., Chan, K. C., Prieto, D. A., Issaq, H. J., Conrads, T. P., and Veenstra, T. D. (2003) Characterization of the low molecular weight human serum proteome. *Mol. Cell. Proteomics* **2,** 1096–1103.

10. Rothemund, D. L., Locke, V. L., Liew, A., Thomas, T. M., Wasinger, V., and Rylatt, D. B. (2003) Depletion of the highly abundant protein albumin from human plasma using the Gradiflow. *Proteomics* **3,** 279–287.
11. Steel, L. F., Trotter, M. G., Nakajima, P. B., Mattu, T. S., Gonye, G., and Block, T. (2003) Efficient and specific removal of albumin from human serum samples. *Mol. Cell. Proteomics* **2,** 262–270.
12. Pieper, R., Su, Q., Gatlin, C. L., Huang, S. T., Anderson, N. L., and Steiner, S. (2003) Multi-component immunoaffinity subtraction chromatography: an innovative step towards a comprehensive survey of the human plasma proteome. *Proteomics* **3,** 422–432.
13. Ausubel, F. M., Brent, R., Kingston, R. E. (2005) *Current Protocols in Molecular Biology.* John Wiley & Sons, 10.I.4. Supplement 35.
14. Patton, W. F. (2000) A thousand points of light: the application of fluorescence detection technologies to two-dimensional gel electrophoresis and proteomics. *Electrophoresis* **21,** 1123–1144.
15. Görg, A., Weiss, W., and Dunn, M. J. (2004) Current two dimensional electrophoresis technology for proteomics. *Proteomics* **4,** 3665–3685.
16. Görg, A., Postel, W., Günter, S., et al. (1987) Elimination of point streaking on silver stained two dimensional gels by addition of iodoacetamide to equilibration buffer. *Electrophoresis* **8,** 122–124.
17. Laemmli, U. K. (1970) Cleavage of structural proteins during the assembly of the head of bacteriophage T4. *Nature* **227,** 680–685.
18. Heukeshoven, R. and Dernick, R. (1985) Simplified method for silver staining of proteins in polyacrylamide gels and the mechanism of silver staining. *Electrophoresis* **6,** 103–112.

29

A Rapid, Economical, and Reproducible Method for Human Serum Delipidation and Albumin and IgG Removal for Proteomic Analysis

Qin Fu, Diane E. Bovenkamp, and Jennifer E. Van Eyk

Summary

Serum is a readily available source for diagnostic assays, but the identification of disease-specific serum biomarkers has been impeded by the dominance of human serum albumin (HSA) and immunoglobulin G (IgG) in the serum proteome. Therefore, in order to observe lower-abundance serum proteins, removal or depletion of at least these two proteins is required. However, the depletion method needs to be inexpensive and reproducible. We describe such a protocol that combines delipidation by centrifugation, IgG removal with Protein G Sepharose™, and HSA depletion with sodium chloride/ethanol precipitation. The protocol is streamlined to increase reproducibility and is compatible with many proteomic platforms, including two-dimensional gel electrophoresis, and high-performance liquid chromatography either offline or coupled online with a mass spectrometer. The reproducible depletion of lipids, IgG, and HSA permits a higher load of the remaining serum proteins, facilitating the identification of disease biomarkers.

Key Words: Serum; immunoglobulin; albumin; IgG; HSA; lipid; biomarker; proteomics; disease; depletion.

1. Introduction

There is immense potential for proteomic analysis of human serum to improve disease diagnosis and to monitor drug effects. However, the wide range of protein concentrations within the serum proteome presents an exceptional challenge for proteomic analysis (1). For most discoveries, the low-abundance proteins need to be enriched and separated from major serum proteins, including IgG and human serum albumin (HSA), prior to proteomic analysis. Often sample preparation is hindered by the lipid content of serum samples, which can vary dramatically between patients. Because biomarker discovery requires the analysis of a large number of patient samples, it is critical to develop reproducible sample processing procedures that minimize technical variation and accentuate biologically relevant differences.

Albumin can be depleted with affinity chromatography-based methods, such as Cibracon Blue F3G-A (Affi-Gel Blue™) *(2)* and antibody-based immunoaffinity matrices (POROS®). However, affinity-based chromatography experiments are traditionally problematic owing to potential nonspecific interactions, carry-over between runs, and poor reproducibility *(3)* (for Cibacron Blue example; *see* **Fig. 1A**). Recently, newer columns (e.g., anti-albumin POROS) report low nonspecific binding *(4,5)*. Unfortunately, the new columns are prohibitively expensive for routine academic laboratory usage, even if they can be recycled *(4)*.

The requirement for large patient cohort analyses in biomarker discovery stresses the importance of developing a simple, reliable, and economical method to remove abundant serum proteins. This type of sample processing is an essential first step in any proteomic analysis of serum, and it can be combined with downstream protein separation technologies to maximize the number of lower-abundance serum proteins and isoforms that can be analyzed. The method described here (**Fig. 2**) is a rapid, simple, inexpensive, and reproducible way to remove lipids and highly abundant serum proteins such as IgG and HSA *(6)*.

2. Materials

2.1. Human Serum

1. Pooled normal human sera (Serological Corporation, Norcross, GA, cat. no. 27,000).
2. Pooled normal human male sera (Sigma, St. Louis, MO, cat. no. S7023).
3. Age-specified individual healthy male and female donor sera (BioReclaimation Inc., Hicksville, NY). (Note: any serum sample of interest can be used.)
4. Serum was separated into smaller aliquots upon receipt and stored at −80°C. Aliquots were thawed at room temperature for use in a single experiment.
5. 1.5-mL Polypropylene tubes, noncoated.

2.2. IgG Depletion

1. Protein G Sepharose beads (Amersham, Piscataway, NJ).
2. Handee Mini-spin column (Pierce, Rockford, IL); 1.5-mL polypropylene tubes, noncoated.
3. 100 mM NaCl, 10 mM HEPES, pH 7.4.
4. 0.15 M NaCl.
5. End-over-end rotator (Thermolyne* LabQuake™ Tube Rotator; Barnstead International, Dubuque, IA).

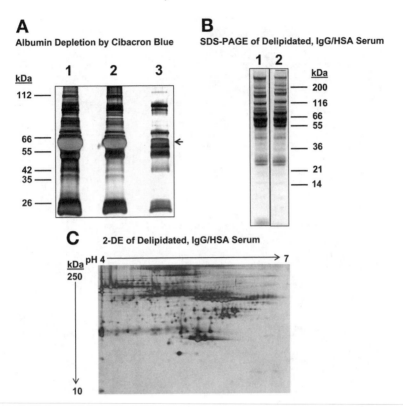

Fig. 1. (A) Serum albumin depletion by Cibacron Blue F3G-A beads. Proteins were separated on a 12% Tris-glycine gel and silver-stained *(10)*. Serum sample is from a healthy 35-yr-old male, with equal loading of serum equivalents in each lane. Lane 1, whole serum; lane 2, proteins bound to beads; lane 3, depleted serum; arrow points to albumin; **(B)** SDS-PAGE analysis of IgG/HSA depleted serum proteins. IgG/HSA depleted proteins (20 µg) were separated on 4–12% Bis-Tris gels and stained with Coomassie blue. Serum samples are from healthy individuals: lane 1, 30-yr-old male; lane 2, 26-yr-old male. **(C)** 2DE separation of IgG/HSA depleted serum. Proteins (200 µg) are separated by isoelectric point in the first dimension (horizontal) and molecular weight in the second dimension (vertical) and silver-stained *(7,10)*. Serum sample is from a healthy 30-yr-old male.

2.3. Albumin Depletion/Fractionation

1. 95% Ethanol (Sigma).
2. Vortex mixer, with foam insert set for microfuge tubes.
3. 1% Sodium dodecyl sulfate (SDS), 10 m*M* HEPES, pH 7.4.
4. 55°C heat block.
5. BCA™ Protein Assay Kit (Pierce).

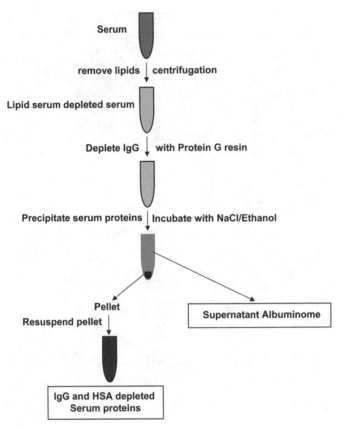

Fig. 2. A schematic diagram of delipidation, IgG/HSA depletion.

2.4. SDS-Polyacrylamide Gel Electrophoresis

1. 4–12% NuPAGE *Bis-Tris* gels (Invitrogen, Carlsbad, CA).
2. NuPAGE Running Buffer (MOPS or MES); NuPAGE sample buffer and reducing agent (Invitrogen).
3. Coomasie Brilliant blue R-250 Stain (Pierce).
4. 5% Acetic acid, 40% methanol destaining buffer.

3. Methods

3.1. Delipidation

1. Thaw serum at room temperature and centrifuge for 15 min at 15,000*g* at room temperature.
2. Transfer the clear yellow serum to a new microfuge tube for further processing.
3. Discard the insoluble matter in the pellet and the top lipid layer. The centrifugation removes mostly triglyceride-rich chylomicrons and very low-density lipo-

proteins (VLDL) *(8,9)*. This procedure is sufficient to eliminate the confounding influence of lipids in downstream protein separation methods and reduces the protein streaks on electrophoresed gels.

4. After centrifugation, the top lipid layer can be easily disrupted and diffuses quickly. Therefore, handle samples with extreme care and we recommend processing only four to six samples for each centrifugation (*see* **Note 1**).

3.2. IgG Depletion

All steps are performed at room temperature.

1. Transfer 1 mL of Protein G beads to a 2-mL tube and add wash buffer (100 m*M* NaCl, 10 m*M* HEPES, pH 7.4) to a total volume of 1.5 mL. Transfer an equal amount of bead suspension (300 μL) to each Handee Mini-spin column (5 spin columns per 1 mL of beads). Make one spin column per sample.
2. For each serum sample, transfer 115 μL of delipidated serum to a clean microfuge tube and add 230 μL of 0.15 *M* NaCl. Gently mix the tube using a vortex mixer and spin briefly.
3. Centrifuge all of the Handee Mini-spin columns containing the Protein G bead suspension at 3300*g* for 3 min and remove excess wash buffer. Add all of one serum/ NaCl mixture from **step 3** into one spin column and close the lid of the tube. Repeat application of the remaining serum/NaCl samples into separate spin columns.
4. Rotate the tubes end-over-end for 1 h at room temperature (*see* **Note 2**).
5. Place the spin column in a new tube and collect the IgG-depleted serum by centrifugation (3300*g* for 3 min). Wash the column with 200 μL of 100 m*M* NaCl, 10 m*M* HEPES, pH 7.4, and centrifuge again into the same tube (this combines the wash with the supernatant).

3.3. Albumin Depletion/Fractionation

All steps are performed at 4°C (*see* **Note 3**).

1. Transfer 500 μL of each delipidated and IgG depleted serum sample into a new microfuge tube and equilibrate to 4°C by gentle mixing (continuously on a vortex mixer at a low setting) for 1 h.
2. During this 1 h equilibration, place a 15-mL tube of 95% ethanol on ice (in a conical polypropylene tube).
3. Add 396.2 μL of cold 95% ethanol, to a final concentration of 42%, and incubate the mixture for an additional hour with gentle mixing on the vortex mixer.
4. Centrifuge all of the samples at 16,000*g* for 45 min at 4°C.
5. Remove the supernatant (albumin-enriched serum fraction) and store it at −80°C until ready for analysis (this is considered the "albuminome," an albumin-enriched fraction that consists of approx 95% albumin).
6. Briefly re-spin the microfuge tube containing the pellet (1 min at 16,000*g* at 4°C) and discard any residual supernatant.
7. Store the pellet at −80°C until ready for analysis (delipidated and IgG/HSA depleted fraction; *see* **Note 4**).

3.4. Protein Concentration Estimation and SDS-Polyacrylamide Gel Electrophoresis

All steps are performed at room temperature.

1. Thaw frozen pellets at room temperature for 15 min prior (*see* **Note 5**) to resuspension in 300 μL SDS solubilization buffer (1% SDS, 10 m*M* HEPES, pH 7.4) and place tubes in a 55°C heating block.
2. Incubate for 3–4 h at 55°C, with occasional vortex-mixing, until sample is completely solubilized (as judged by the absence of a pellet after centrifugation for 15 min at 16,000*g*). If a pellet is still visible after centrifugation, then vortex and re-solubilize for another hour at 55°C (*see* **Note 6**).
3. After pellets are completely dissolved, spin the tubes for 15 min at 16,000*g* at room temperature.
4. After centrifugation, avoid the cloudy white lipid layer on top of the clear serum sample while you transfer each serum supernatant to a new microfuge tube.
5. Dilute 10 μL of supernatant with 90 μL deionized distilled water (a 1:10 dilution) prior to using the BCA™ Protein Assay Kit.
6. Protein concentrations are determined in duplicate according to the manufacturer's Microplate Procedure.
7. Mix serum proteins (20 μg) with NuPAGE lithium dodecyl sulfate sample buffer and reducing agent and heat for 10 min at 70°C prior to loading samples onto pre-cast 4–12% *Bis-Tris* NuPAGE gels. Electrophorese with NuPAGE MOPS or MES running buffer.
8. After separation, visualize proteins by staining with Coomasie Brilliant blue R-250 and destaining with 40% methanol/5% acetic acid (*see* **Fig. 1B**) (or stain proteins with your method of choice, such as silver stain *[10]*).
9. Alternatively, visualize the proteins in the depleted serum with two-dimensional gel electrophoresis, where they are separated by isoelectric point in the first dimension and molecular weight in the second dimension (*see* **Fig. 1C**). For two-dimensional gel electrophoresis separation of serum, *see* **ref.** *(6)* for more details.

When starting a search for disease biomarkers in human serum, it is essential to treat all samples in the same way, with a method that will give consistent results. Some methods remove confounding high-abundant proteins but also non-specifically reduce other serum proteins, making it difficult to identify genuine biomarkers. This protocol will effectively and reproducibly reduce lipids, IgG, and HSA from human serum samples, leaving high-quality protein mixtures for analysis. The development of this protocol has been published in Colantonio et al. *(3)* and Fu et al. *(6)*.

4. Notes

1. If the top lipid layer diffuses, a second centrifugation is recommended.
2. Gently tap the tube to re-suspend Protein G beads prior to end-over-end rotation.
3. We perform all 4°C steps in cold room.

4. Protein pellets are the preferred format for storing. No significant difference is observed after pellets are stored at −80°C for a few months.
5. Bringing the tubes with depleted protein pellets to room temperature helps resuspend pellets completely in SDS-HEPES solution.
6. It is extremely important to dissolve a pellet completely. If a pellet is not completely re-suspended in SDS-HEPES solution, reproducibility of samples preparation is reduced and pre-analytical variations will be introduced.

Acknowledgments

This work was supported by grants from the NHLBI (contract N0-HV-28180), the Donald W. Reynolds Foundation, and funds from the Daniel P. Amos Family Foundation supporting the Johns Hopkins Bayview Proteomic Center.

References

1. Anderson, N. L., Polanski, M., Pieper, R., et al. (2004) The human plasma proteome: a nonredundant list developed by combination of four separate sources. *Mol. Cell. Proteomics* **3,** 311–326.
2. Ahmed, N., Barker, G., Oliva, K., et al. (2003) An approach to remove albumin for the proteomic analysis of low abundance biomarkers in human serum. *Proteomics* **3,** 1980–1987.
3. Colantonio, D. A., Dunkinson, C., Bovenkamp, D. E., and Van Eyk, J. E. (2005) Effective removal of albumin from serum. *Proteomics* **5,** 3831–3835.
4. Pieper, R., Su, Q., Gatlin, C. L., Huang, S. T., Anderson, N. L., and Steiner, S. (2003) Multi-component immunoaffinity subtraction chromatography: an innovative step towards a comprehensive survey of the human plasma proteome. *Proteomics* **3,** 422–432.
5. Anderson, N. L., Anderson, N. G., Haines, L. R., et al. (2004) Mass spectrometric quantitation of peptides and proteins using Stable Isotope Standards and Capture by Anti-Peptide Antibodies (SISCAPA). *J. Proteome Res.* **3,** 235–244.
6. Fu, Q., Garnham, C. P., Elliott, S. T., Bovenkamp, D. E., and Van Eyk, J. E. (2005) A robust, streamlined and reproducible method for proteomic analysis of serum by delipidation, albumin and IgG depletion, and 2-dimensional gel electrophoresis. *Proteomics* **5,** 2656–2664.
7. Gorg, A., Weiss, W., and Dunn, M. J. (2004) Current two-dimensional electrophoresis technology for proteomics. *Proteomics* **4,** 3665–3685.
8. Terpstra, A. H. (1985) Isolation of serum chylomicrons prior to density gradient ultracentrifugation of other serum lipoprotein classes. *Anal. Biochem.* **150,** 221–227.
9. Rodriguez-Oquendo, A. and Kwiterovich, P. O. Jr. (2000) Dyslipidemias, in *Inborn Metabolic Diseases* (Fernandes, J., Saudubray, J. M., and en Berghe, G., eds.), Springer-Verlag, Berlin, pp. 319–336.
10. Shevchenko, A., Wilm, M., Vorm, O., and Mann, M. (1996) Mass spectrometric sequencing of proteins silver-stained polyacrylamide gels. *Anal. Chem.* **68(5),** 850–858.

VI

Bioinformatics: Emerging Technologies

30

Mitoproteome

Human Heart Mitochondrial Protein Sequence Database

Purnima Guda, Shankar Subramaniam, and Chittibabu Guda

Summary

The human mitochondrial proteome database has been developed by deriving data from a combination of public repositories and experimental and computational prediction methods. The experimental data is derived from highly purified mitochondria from human heart tissue, whereas predictions have been performed by MITOPRED, a genome-scale method for the prediction of nucleus-encoded mitochondrial proteins. Mitochondrial protein sequences from different sources have been clustered to generate a nonredundant dataset. Annotations related to the protein function, structure, disease association, pathways, and so on are collected from a number of public databases using commonly used UNIX and Perl scripts. This chapter provides a detailed description of various data sources and methods used to download, curate, parse, and generate meaningful annotations from primary as well as derived databases.

Key Words: Mitochondria; mitoproteome; human heart; biological database; MITO-PRED.

1. Introduction

Mitochondria are semi-autonomous intracellular organelles that play a crucial role in many metabolic pathways operating in eukaryotic cells. Hence, genetic and/or metabolic alterations in this organelle are associated with more than 100 known human diseases, including many cardiac disease conditions such as ischemia, reperfusion, aging, ischemic preconditioning, and cardiomyopathy *(1)*. Human mitochondrial genome is a circular DNA molecule approx 16.5 Kb in size, encoding only 37 gene products, including 22 tRNAs, 2 rRNAs, and 13 polypeptides. However, an estimated 1500 proteins functioning inside mitochondria (mitoproteome) are encoded by the nuclear genome and imported into mitochondria by various transport mechanisms. Recently, we developed a comprehensive

From: *Methods in Molecular Biology, vol. 357: Cardiovascular Proteomics: Methods and Protocols*
Edited by: F. Vivanco © Humana Press Inc., Totowa, NJ

database for human mitoproteome (*2*; http://www.mitoproteome.org) where the data was primarily obtained from highly purified human heart mitochondria by mass spectrometric analysis (*3,4*). This database also includes data based on the experimental annotations from public databases such as SwissProt, GenBank, and so on, and from computational prediction methods such as MITOPRED (*5,6*). This chapter provides a detailed description of various data sources and methods used to download, curate, parse, and generate meaningful annotations from primary as well as derived databases.

2. Materials

2.1. Hardware

1. Workstation: Sun Microsystems® SPARC workstation, 4 processors (CPU), 4 GB of random access memory (RAM) with Solaris 5.9 operating system. Hard disks are mounted via network file system (NFS) with automatic backup and archival facilities (*see* **Note 1**). However, any workstation with equivalent or better configuration can be used.
2. Compute cluster: 16-CPU or better compute cluster, 32 GB of RAM with Solaris or Linux operating system (*see* **Note 2**) and hard disks mounted via NFS.

2.2. Software and Data Sources

1. *See* **Table 1** for a list of software and data resources used in this work and the URL addresses for accessing or downloading software or data.

3. Methods

3.1. Data Collection

1. Download human protein records from the SwissProt (SP) database using an ftp connection on a UNIX terminal (*see* **Fig. 1**). Connect to ftp server at "ftp.ebi.ac.uk," use "anonymous" as username and "your_email_id" as password. After changing to appropriate directory, use "get filename" command to retrieve files and "gunzip filename" command to uncompress (*see* **Note 3**) files. Similarly, download data or software from the other database resources using appropriate ftp addresses (**Table 1**).
2. SP database records contain experimental annotations for proteins, including available information on the "SUBCELLULAR LOCATION" under the CC field (comments). Use regex patterns with UNIX and Perl scripts to identify all the mitochondrial proteins from the SP or GenBank records (*see* **Note 4**).
3. Download 615 mitochondrial proteins obtained from human heart tissue by mass spectrometric analysis (*4*). This data is available as supplemental information to **ref.** (*4*) at http://www.nature.com/nbt/journal/v21/n3/extref/nbt793-S5.pdf.
4. Predict human mitochondrial proteins using the MITOPRED web server (http://bioinformatics.albany.edu/~mitopred), a computational method used for the prediction of nucleus-encoded proteins targeted to mitochondria. Predictions are possible

Table 1
Data Sources and the URLs for Accessing and Downloading Data

Name	Description	URL
Software		
BioPerl	BioPerl module	http://www.bioperl.org
CD-HIT	Clustering software	http://bioinformatics.ljcrf.edu/cd-hi
HMMER	Software to search Pfam database	http://hmmer.wustl.edu/
NCBI Toolkit	BLAST software	http://www.ncbi.nlm.nih.gov/BLAST/
Perl	Perl interpreter	http://www.perl.com/perl
Swissknife	Software to process SP data	http://swissknife.sf.net
Databases		
DIP	Database of Interacting Proteins	http://dip.doe-mbi.ucla.edu/
Ensembl	Ensembl genome browser	http://www.ensembl.org/
ENZYME	Enzyme database	http://us.expasy.org/enzyme/
GenBank	GenBank	http://www.ncbi.nlm.nih.gov/
GO	Gene Ontology	http://www.geneontology.org/
InterPro	Protein families database	http://www.ebi.ac.uk/interpro
LOCUSLINK	Locuslink database	http://www.ncbi.nlm.nih.gov/LocusLink/
MINT	Molecular INTeraction database	http://mint.bio.uniroma2.it/mint
MITOP	Mitochondrial Protein database	http://ihg.gsf.de/mitop2
MITOPRED	Prediction method for mitochondrial proteins	http://bioinformatics.albany.edu/~mitopred
Mitoproteome	Mitoproteome database	http://www.mitoproteome.org
OMIM	Online Mendelian Inheritance in Man	http://www.ncbi.nlm.nih.gov/entrez/query.fcgi?db=OMIM
PDB	Protein Data Bank	http://www.rcsb.org/pdb/
Pfam	Protein families database	http://pfam.wustl.edu/
PRINTS	Compendium of protein fingerprints	http://umber.sbs.man.ac.uk/dbbrowser/PRINTS/
RefSeq	Reference Sequence database	http://www.ncbi.nih.gov/RefSeq/
SP	Swiss-Prot	http://www.ebi.ac.uk/swissprot

```
pathway (3) % ftp ftp.ebi.ac.uk
Connected to alpha4.ebi.ac.uk.
220 ftp1.ebi.ac.uk FTP server (Version wu-2.6.2(2) Wed Aug 20 08:58:45 BST 2003) ready.
Name (ftp.ebi.ac.uk:babu): anonymous
331 Guest login ok, send your complete e-mail address as password.
Password:
230-Welcome anonymous@pathway.sdsc.edu
Remote system type is UNIX.
Using binary mode to transfer files.
ftp> cd pub/databases/swissprot/special_selections/
250 CWD command successful.
ftp> get human.seq.gz
200 PORT command successful.
150 Opening BINARY mode data connection for human.seq.gz (19766620 bytes).
226 Transfer complete.
local: human.seq.gz remote: human.seq.gz
19766620 bytes received in 68 seconds (282.76 Kbytes/s)
ftp> bye
221-You have transferred 19766620 bytes in 1 files.
221-Total traffic for this session was 19767706 bytes in 1 transfers.
221-Thank you for using the FTP service on ftp1.ebi.ac.uk.
221 Goodbye.
pathway (4) % gunzip human.seq.gz
pathway (5) % ▊
```

Fig. 1. An ftp session for downloading data from Swiss-Prot database.

at different accuracy thresholds, and we used 100% threshold to obtain only the most accurately predicted mitochondrial proteins.

3.2. Data Curation

1. Annotation on the subcellular location of proteins from SP records often contains ambiguous and/or uncertain information that needs to be cleaned using a combination of automatic and manual methods. For example, the following UNIX command will remove all the proteins in the list that contain the words such as "by similarity," "possible," "probable," "potential," and so on in their annotations (*see* **Note 5**).

    (1)% egrep –iv "similarity|probable|possible|potential" input_file >output_file

2. Generate the final list of trusted mitochondrial proteins after manually removing the ambiguously annotated proteins from the SP and GenBank entries and generate a FASTA formatted sequence file for these entries (*see* **Note 6**).
3. Combine the FASTA formatted sequence files for all mitochondrial proteins, i.e., those derived from public databases, mass spectrometric analysis, and those predicted by MITOPRED program.
4. Remove redundant proteins in the combined dataset by clustering at 99% identity with CD-HIT program using the following command (*see* **Note 6**).

    (1)% cd-hit –i input_file –c 0.99 –o output_file

3.3. Annotation of Mitochondrial Proteins

Mitochondrial proteins have been annotated using a number of public databases; however, we used SP records as the primary source because they contain

Table 2
Summary of Protein Annotations
and the Fields Parsed Under Each Data Category

Data category	Source	Data
Primary annotations	Swiss-Prot (SP)	Accession number, sequence, function, subcellular localization, posttranslational modification, tissue specificity, splice-variants, pathway, cross-references to other databases
	GenBank	GenBank ID, RefSeq ID, LocusLink ID
	Mass spectrometric data	Peptides, molecular weight, isoelectric point
Diseases	OMIM	OMIM number, Disease, Gene symbol
Domains and Protein families	Pfam	ID, model name, domain description
	Interpro	ID, GO ID, GO classification, signatures
	PRINTS	ID, fingerprint description
Genes	LOCUS LINK	Locus ID, gene symbol, alias symbols and names, chromosome number and location, RefSeq, and Unigene information
	ENSEMBL	ID, Chromosome Information
Protein–protein Interactions	DIP and MINT	Homologous node, interacting partners, method, database IDs, species of origin
Structures	PDB	ID, Structural motifs
Homology	NCBI NR (Nonredundant) database	Top hits based on pair-wise comparisons
Pathways	KEGG	Association to metabolic and disease pathways, enzyme reaction, substrates, and products
EC numbers	ENZYME	EC number, cross-links to SP IDs
Classification	Manual	Based on the primary function, subfunction, subcellular localization

rich and comprehensive annotation at one place. Hyperlinks to all public databases are provided in **Table 1** and annotated features have been summarized in **Table 2**. Here, we provide various methods and thresholds used to retrieve annotations from each database.

1. Annotation of protein function: For all proteins with SP accession numbers, obtain functional annotation by parsing SP records using Swissknife software (*see* **Note 4**). This includes information on the physical properties, amino acid sequence, primary

```
pathway (2) % ls
mitoproteome.fa    pdb.fa
pathway (3) % formatdb -i pdb.fa -o T -b T
pathway (4) % ls
formatdb.log       pdb.fa            pdb.fa.pin       pdb.fa.pni      pdb.fa.psi
mitoproteome.fa    pdb.fa.phr        pdb.fa.pnd       pdb.fa.psd      pdb.fa.psq
pathway (5) % blastall -p blastp -i mitoproteome.fa -d pdb.fa -e 1e-5 -m 8 -o out_file
pathway (6) % cut -f1-3 out_file |sort +2 -nr >out_file2
pathway (7) % cat out_file2
gi|106140|pir||S18490    gi|2554808|pdb|1AFO|A    100.00
gi|106221|pir||S21403    gi|229913|pdb|1FDH|G     100.00
gi|1082428|pir||A49678   gi|2624718|pdb|1RGP|     100.00
gi|1082428|pir||A49678   gi|3402094|pdb|1AM4|A    100.00
gi|1082428|pir||A49678   gi|6730013|pdb|1GRN|B    100.00
pathway (8) % cut -d'|' -f2,8 out_file2
106140|1AFO
106221|1FDH
1082428|1RGP
1082428|1AM4
1082428|1GRN
pathway (9) % █
```

Fig. 2. Using blast program and UNIX commands to search and extract related information from databases.

function, subcellular localization, enzymatic properties, posttranslational modification, pathway, and so on, and cross-references to other major databases such as EMBL, GO, InterPro, PDB, PRINTS, PROSITE, PFAM, SMART, and so on. For those proteins without SP accession numbers, functional annotation can be obtained from GenBank records or by directly searching against the aforementioned databases as described in **step 4**.

2. Protein structure annotation: Search mitoproteome sequences against Protein Data Bank (PDB) sequences to identify structures with sequence homology to mitochondrial proteins using the programs from Blast software suite. The "blastall" (blastp) program is useful for searching protein query sequence(s) against protein databases. Before using "blastp," the protein database (in FASTA format) needs to be converted into blast-readable format using "formatdb" program, which generates seven blastp-readable files. **Figure 2** shows an example of the blast session on UNIX machine (*see* **Note 7**). Formatted databases can be searched with "blastp" using an E-value threshold of 1e-5 and writing the output in a tabular format (*see* **Note 8**) to "out_file." Using UNIX commands, cut the required columns from outfile and sort them based on the percent identity between the query and the target sequences. Separate blast hits with more than or equal to 90% identity into a separate file (out_file2) and cut only those columns containing the ID of query sequence and ID of target database (PDB IDs in this case). E-value and percent identity could be adjusted to recover structures with sequence homology at various thresholds.

3. Disease information: Similar to PDB, use protein sequences from Online Mendelian Inheritance in Man (OMIM) as the target database to search against mitoproteome sequences and extract disease-related information from the matching hits.

4. Gene annotation: Gene annotation could be obtained primarily from the GenBank records. There are several ways to map the accession numbers between SP and

GenBank databases, however: one direct way is using International Protein Index (IPI) cross-reference tables (*see* **Note 9**) to obtain corresponding RefSeq ID from SP accession number. Using RefSeq IDs, cross-references to GenBank ID, locus link, and other gene-related annotations could be obtained. Whenever RefSeq IDs are not available, direct string matching between protein sequences could be done using Perl scripts for mapping IDs across various databases.

5. Enzyme annotation: Each protein may contain one or more enzyme classification (EC) numbers and each EC number may include one or more proteins. Also, all mitochondrial proteins are not enzymes. Where applicable, obtain EC number(s) for each protein from SP or GenBank records and identify unique EC numbers corresponding to all mitochondrial proteins. For each EC number, parse the substrates, reaction, and the products from the enzyme database provided by Kyoto Encyclopedia of Genes and Genomes, using Perl scripts.

6. Protein–protein interactions: Use protein sequences from Database of Interacting Proteins (DIP) as a target database to search against mitoproteome sequences with "blastp." Identify the DIP nodes (protein sequences) that have sequence identity of more than or equal to 90% with mitoproteome sequences. Interacting partners for each DIP node could be identified from the DIP database.

7. Pathway annotation: Obtain available pathway information from SP records and for the missing cases, perform a blast search against the human protein sequences in Kyoto Encyclopedia of Genes and Genomes database and extract pathway information based on the matching hits.

8. Tissue specificity: Majority of the proteins in the mitoproteome (615) have been derived from the human heart tissue by mass spectrometric analysis *(4)*. For the remaining proteins, available annotation on the tissue specific expression could be obtained from the SP records using the "TISSUE SPECIFICITY" keyword under the comments (CC) field with UNIX and Perl scripts.

4. Notes

1. NFS-mounted hard disk is not an absolute requirement; however, it is a very useful feature for maintaining updated versions of large databases such as GenBank, Swiss-Prot, and so on at a central location and make them accessible to several groups in an institutional setting.

2. We used mostly the SUN SPARC machines with sun4u kernel architecture; however, the latest breed of CPUs with x86_64 (Intel Xeon) or ppc64 (IBM) architecture running under Linux OS are proven to be more efficient for performing compute-intense jobs in this project.

3. Sometimes several files or directories are bundled into one file with ".tar" extension. In such cases, use the command "tar –xvf filename.tar" to extract the files.

4. Common tasks such as keyword searches could be performed using "regex patterns" (regular expressions) with "grep" or "egrep" commands in UNIX (see an example in **Subheading 3.2.**). Basic UNIX and Perl programming experience is required for handling data in this project. However, open source Perl modules are available from several sources such as BioPerl, Swissknife, and so on (**Table 1**),

and can be easily integrated to develop customized Perl scripts for parsing a variety of data from primary database records.

5. The expression grep utility searches files for a pattern of characters and prints all lines that contain the pattern. The "-i" option denotes ignoring the upper/lower case distinction in pattern matching and the "-v" denotes printing all lines except those that contain matched patterns. In other words, the "-v" option eliminates the lines with matched patterns from the output file.

6. CD-HIT program expects input file in FASTA format and generates the output file in the same format. FASTA format is very concise and the most commonly used, for representing genomes and proteomes. An example is as follows:

>ACD8_HUMAN (Accession number) Description in the first line
MLWSGCRRFGARLGCLPGGLRVLVQTGHRSLTSCIDPSMGLNEEQKEFQ
KVAFDFAAREMAPNMAEWDQKELFPVDVMRKAAQLGFGGVYIQTDVGG
>68MP_BOVIN
MLQSLIKKVWIPMKPYYTQAYQEIWVGTGLMAYIVYKIRSADKRSKALKA
SSAAPAHGHH
>ACON_CANAL
VGLLGSCTNSSYEDMTRYTVSPGSVQQR

7. To get blast programs work on your UNIX terminal, you need to create an ".ncbirc" file in your root directory that contains appropriate path for blast substitution matrices. For more details, please read "Setting up standalone Blast for UNIX" at the URL http://www.ncbi.nlm.nih.gov/Class/BLAST/README.bls

8. Programs in the standalone blast suite are empowered with a number of options to manipulate the quality and format of the output data. Because it is difficult to remember all these options, they can easily be retrieved by typing the program name followed by "-," as an argument at the command line. Tabular output format can be selected with "-m 8" option, which is easy to manipulate with simple UNIX commands and extract required information from the output files (**Fig. 2**).

9. Mapping accession numbers between databases often gets confusing because there may be multiple accession numbers for the same protein, which is especially true in the case of GenBank. However, the IPI resource provides cross-references between SP accession number and RefSeq ID. This information for human sequences can be downloaded from the IPI ftp site at URL ftp://ftp.ebi.ac.uk/pub/databases/IPI/current

Acknowledgments

We would like to thank Dawn Cotter for developing mitoproteome web resource and Dr. Brian Saunders for maintaining parsed data from major public resources such as Swiss-Prot, GenBank, and so on at the San Diego Supercomputer Center, UCSD.

References

1. Lesnefsky, E. J., Moghaddas, S., Tandler, B., Kerner, J., and Hoppel, C. L. (2001) Mitochondrial dysfunction in cardiac disease: ischemia–reperfusion, aging, and heart failure. *J. Mol. Cell. Cardiol.* **33,** 1065–1089.
2. Cotter, D., Guda, P., Fahy, E., and Subramaniam, S. (2004) MitoProteome: mitochondrial protein sequence database and annotation system. *Nucleic Acid Res.* **32,** D463–D467.
3. Taylor, S. W., Fahy, E., and Ghosh, S. S. (2003) Global organellar proteomics. *Trends Biotechnol.* **21,** 82–88.
4. Taylor, S. W., Fahy, E., Zhang, B., et al. (2003) The mitochondrial proteome of normal human heart muscle. *Nature Biotechnol.* **21,** 281–286.
5. Guda, C., Fahy, E., and Subramaniam, S. (2004) MITOPRED: A genome-scale method for prediction of nucleus-encoded mitochondrial proteins. *Bioinformatics* **20,** 1784–1794.
6. Guda, C., Guda, P., Fahy, E., and Subramaniam, S. (2004) MITOPRED: A web server for genome-scale prediction of mitochondrial proteins. *Nucleic Acid Res.* **32,** W372–W374.

31

Techniques to Decipher
Molecular Diversity by Phage Display

Dawn R. Christianson, Michael G. Ozawa,
Renata Pasqualini, and Wadih Arap

Summary

Combinatorial phage display technology may be applied to decipher the molecular diversity of peptide binding specificity to isolated proteins, purified antibodies, cell surfaces, intracellular/cyto-domains, and blood vessels in vivo. The application of such a strategy ranges from identifying receptor–ligand pairs and antigen binding sites to understanding the progression of diseases by their differential expression patterns and developing therapeutic targeting strategies. Different strategies can be used to isolate peptides from diverse libraries displayed on the surface of bacteriophage by exposing the library to a target molecule or organ, washing away nonbinding phage, eluting and amplifying the bound phage for multiple round use, and then analyzing the peptide sequences of the enriched phage. The following methods first outline the construction of a phage library and then delineate various in vitro and in vivo biopanning applications to probe isolated integrins, purified antibodies, cell surface molecules, and vascular endothelial cells.

Key Words: Phage display; vascular targeting, molecular markers; antibody fingerprinting; angiogenesis; biopanning; BRASIL; protein.

1. Introduction

Phage display random peptide libraries were originally designed to define binding sites of antibodies in isolated immunoglobulins *(1)*. Combinatorial phage display technology has since expanded in its application to decipher the molecular diversity of peptide binding specificity to isolated proteins, purified antibodies, cell surfaces, intracellular/cyto-domains, and blood vessels in vivo *(2–5)*. Ultimately, this approach allows for the selection of peptides in an unbiased functional assay without any preconceived notions about the nature of their

From: *Methods in Molecular Biology, vol. 357: Cardiovascular Proteomics: Methods and Protocols*
Edited by: F. Vivanco © Humana Press Inc., Totowa, NJ

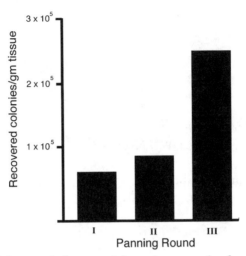

Fig. 1. Enrichment of phage pool from three rounds of panning *(38)*.

target. The application of such a strategy ranges from identifying receptor–ligand pairs and antigen binding sites to understanding the progression of diseases by differential molecular expression patterns and developing therapeutic targeting strategies. For example, probing the molecular diversity of blood vessels will help localize novel markers that can help in the understanding of diseases such as atherosclerosis, arthritism, diabetic retinopathy, and tumor formation where angiogenesis and vascular remodeling occur *(2,4,5)*.

Different strategies can be used to isolate peptides from diverse libraries displayed on the surface of bacteriophage. In brief, phages are propagated in pilus-positive bacteria that are not lysed by the phage but rather secrete multiple copies of phage that display a particular insert. The number of transducing units (TU) for a phage preparation is calculated from its efficiency to infect bacteria and produce colonies. A phage library is then exposed to a target of interest. Non-binding phage are washed away and phage bound to a target molecule or within an organ are eluted and then amplified by growth in host bacteria. Multiple rounds of panning are performed to enrich the pool of phage until a population of selective binders is obtained (*see* **Fig. 1**; ref. *6*). The amino acid sequence of the recovered peptides is determined by sequencing the DNA corresponding to the insert in the phage genome. Sequences can be analyzed to monitor enriched peptide frequencies compared to the unselected library. Statistically significant enriched sequences can be further analyzed by comparing them to a current peptide database (Basic Local Alignment Search Tool) to identify proteins, or ligands, with which the motifs share sequence homology *(3,7–11)*.

1.1. In Vitro Panning and Cell Targeting

1.1.1. Cell-Free Screening on Isolated Receptors and Antibodies

In vitro phage display panning provides a straightforward method to characterize the peptide binding specificity for proteins such as immunoglobulins *(12–15)*, integrins that mediate cellular adherence *(16)*, and membrane proteins that can discriminate between diseased and normal states. Panning on immobilized receptors has helped identify unknown ligands, including cell adhesion molecules, proteases, cell-cycle regulators, viruses, oncoproteins, and tumor-suppressor proteins, among others *(17,18)*.

The humoral immune response is very specific to foreign peptides and self-expressed proteins in tumors as well as other autoimmune diseases. Selection and isolation of phage clones for a desired antibody pool involves immobilization of antibodies on Protein G agarose beads, pre-clearing the phage library of nonspecific binders, and then exposing the nonbinding phage from the pre-clearing step to the desired antibody pool using a series of buffers for elution and neutralization (**Fig. 2**). Identification of specific binding peptides has helped gain insight into the state/progression of disease and may be further applied to targeted treatment strategies *(14,15)*.

1.1.2. Profiling Cell Surface Markers

Cells express molecularly recognizable surface markers that can be probed with phage libraries by an effective cell panning method termed biopanning and rapid analysis of selective interactive ligands (BRASIL) *(19)*. BRASIL selects for specific receptors binding high- and low-affinity ligands *(20,21)* in the context of whole cells. Because the affinity and activation states of cell surface proteins, such as integrins, are highly dependent on molecules that associate with their cytoplasmic domain, probing cell surface molecules in the context of whole cells is important. In addition, membrane-bound proteins are more likely to preserve their functional conformation, which can be lost upon purification and immobilization outside the context of intact cells. Through a single differentiation centrifugation cycle that drives cells from a hydrophilic environment into a nonmiscible organic phase, BRASIL separates phage–cell complexes from the remaining unbound phage (**Fig. 3**). Because the organic phase is hydrophobic, it excludes water-soluble materials surrounding cell surfaces. Bound phage are recovered from the cell pellet, whereas the unbound phage remain soluble in the upper aqueous phase, eliminating the need for repeated washes.

1.2. Vascular Targeting and In Vivo Combinatorial Applications

The organ and tissue microenvironment greatly influences the expression of cell surface molecules by endothelial cells. Identification of biochemical recognition markers in vitro is often difficult because of the dramatic phenotypic

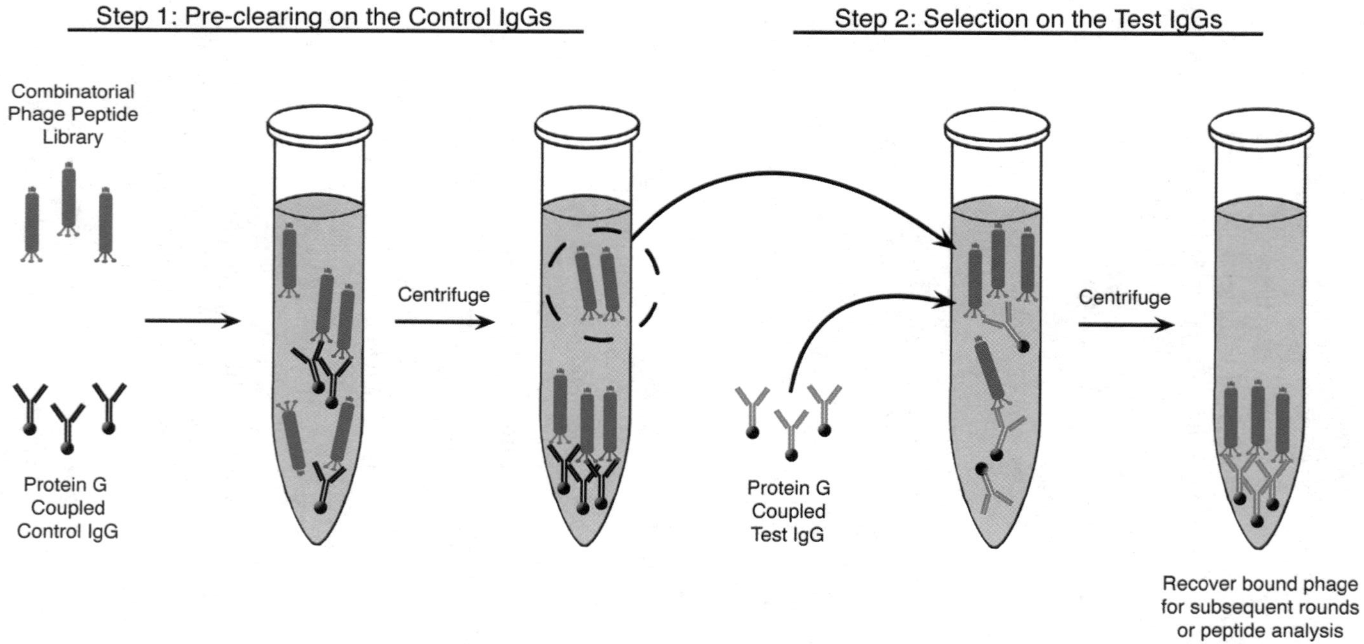

Fig. 2. Phage display peptide library screening on antibodies. The following steps are performed to generate a specific enriched sequence: Step 1, pre-clearing the library of nonspecific binders; Step 2, incubating the unbound phage from the pre-clearing step with the test antibodies, recovering the bound phage and repeating Step 2 to enrich the selective binders or amplification and peptide sequencing of the recovered phage.

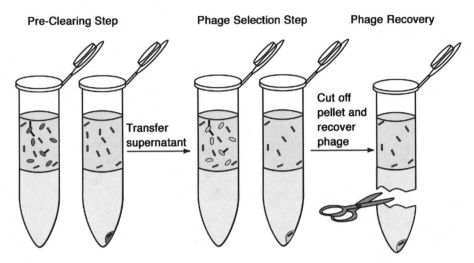

Fig. 3. BRASIL. Separation of phage bound cells from unbound phage by centrifugation through an organic phase.

changes endothelial cells undergo in culture and of the cellular context often necessary for expression. Thus, the application of phage display in vivo circumvents these challenges and has accelerated the discovery of novel peptide sequences targeting specific vascular beds. In brief, a phage display library is injected intravenously, allowed to circulate, and then the bound phage are recovered from the dissected organ and analyzed. To date, numerous peptide sequences have been successfully identified that target vascular receptors expressed in normal organs as well as in diseased tissue, such as tumors *(9,22–28)*. Characterization of these peptide sequences and their spatially regulated receptors has brought valuable insights into the biology of the microvasculature as well as the mechanistic basis of blood vessel proliferation in angiogenesis-related diseases *(2,29–33)*. In addition, selective homing peptides have been used to direct therapies to desired targets in mice and thus have led to the development of more effective and less toxic therapeutic agents *(4,8,34–37)*.

Moreover, integration of in vivo phage display with other technologies has also improved the understanding of vascular diversity at the suborgan level. Several organs, including the heart, pancreas, and kidney, contain functionally distinct regions that also exhibit unique vascular organization *(35,38)*. Isolation of spatially regulated ligand–receptor pairs within these regions is often difficult owing to the mechanical manipulation necessary for conventional in vivo phage display. Fortunately, the combination of in vivo phage display with laser pressure catapult microdissection has helped minimize tissue manipulation and allow for identification of unique peptide sequences corresponding to differentially expressed receptors.

2. Materials

2.1. Construction of Phage Display Peptide Libraries

1. fUSE5 plasmid.
2. Electrocompetent *Escherichia coli* strain MC1061.
3. Luria Bertani (LB)-tetracycline (tet.)/streptomycin (strep.) medium (LB medium): 10 g bacto-tryptone, 10 g NaCl, 5 g yeast extract in 1 L. Adjust pH to 7.4 and autoclave. Add tet. to 20 µg/mL and strep. to 100 µg/mL (*see* **Note 1**).
4. LB-tet. plates (LB agar plates containing 40 µg/mL tet.).
5. Double-distilled (dd)H$_2$O.
6. SOC medium (Invitrogen, Carlsbad, CA).
7. Plasmid purification kit (Qiagen Inc., Valencia, CA; Promega, Madison, WI; or equivalent).
8. Cesium chloride (CsCl).
9. Ethidium bromide (EtBr).
10. TE buffered *N*-butanol: 10 m*M* Tris-HCl, 1 m*M* EDTA with ddH$_2$O saturated butanol.
11. 3 *M* Sodium acetate (NaOAc), pH 5.2.
12. 100% Ethanol.
13. 0.8% Agarose gel.
14. *Sfi*I and *Bgl*I restriction enzymes with corresponding buffers (Roche Applied Sciences, Indianapolis, IN or equivalent).
15. Taq polymerase and polymerase buffer (Promega, Roche Applied Sciences, or equivalent).
16. Synthetic oligonucleotides:
    5'-CACTCGGCCGACGGGGCT (NNK)$_x$ GGGGCCGCTGGGGCCGAA-3'
    library sense: 5'-CACTCGGCCGACG-3'
    library antisense: 5'-TTCGGCCCCAGCGGC-3'
17. Dimethyl sulfoxide (DMSO) (Fisher Scientific, Pittsburgh, PA).
18. dNTP (Promega, Fisher, or equivalent).
19. Nucleotide Removal Kit (Qiagen, Promega, or equivalent).
20. T4 DNA ligase and buffer (Promega, Invitrogen, or equivalent).
21. Commercial buffer exchange kit (Qiagen, Promega, or equivalent).

2.2. In Vitro Panning

2.2.1. Coating Receptors on Plates

1. Recombinant or purified protein of interest.
2. Flat bottom 96-well microtiter plates (Nalge-Nunc, Rochester, NY).
3. Phosphate-buffered saline (PBS): 2 g potassium chloride, 2 g potassium phosphate monobasic, 80 g sodium chloride, 11.4 g sodium phosphate dibasic anhydrous in 1 L. Adjust pH to 7.4 with diluted HCl.

4. Tris-buffered saline (TBS): 50 mM Tris-HCl, pH 7.5, 100 mM NaCl, dissolve in ddH$_2$O autoclave.
5. 25 mM N-octyl-β-D-glucopyranoside in TBS (*see* **Note 2**).
6. 3% Bovine serum albumin (BSA)/PBS.
7. 1 mM MnCl$_2$ in TBS (*see* **Note 3**).

2.2.2. Panning

1. Phage display library of choice.
2. 1% BSA, 1 mM MnCl$_2$ in TBS.
3. 0.5% Tween-20 in TBS.
4. Elution buffer: 0.1% BSA, 0.05% phenol red, 0.1 M glycine-HCl, pH 2.2 in ddH$_2$O.
5. Neutralization buffer: 1 M Tris-HCl, pH 9.0 in ddH$_2$O.
6. LB-tet./kanamycin (kan.) medium (20 μg/mL tet. and 100 μg/mL kan.).
7. *E. coli* K91/kan. bacteria.

2.2.3. Phage Recovery

1. LB-tet./kan. medium (0.2 μg/mL tet. and 100 μg/mL kan.; *see* **Note 4**).
2. LB-tet./kan. plates (40 μg/mL tet. and 100 μg/mL kan.).
3. 16.7% polyethelyne glycol (PEG), 3.3 M NaCl, dissolve in ddH$_2$O.
4. PBS.

2.2.4. TU Titer Assay for Phage Preparations

1. PBS.
2. LB-tet./kan. plates (40 μg/mL tet. and 100 μg/mL kan.).
3. *E. coli* K91/kan. bacteria.

2.2.5. Antibody Fingerprinting

1. Phage display library of interest.
2. Antibody samples.
3. Protein G agarose beads (Amersham Pharmacia Biotech, Piscataway, NJ).
4. Binding buffer: 0.5% Tween-20 in PBS.
5. Elution buffer: 0.1 M glycine buffer, pH 2.2 in ddH$_2$O.
6. Neutralization buffer: 1 M Tris-HCl, pH 9.0 in ddH$_2$O.
7. LB-tet./kan. medium (20 μg/mL tet. and 100 μg/mL kan.).
8. *E.coli* K91/kan. bacteria.

2.2.6. Cell Surface Binding (BRASIL)

1. Cells of interest.
2. Phage display library of interest.
3. PBS.
4. 400-μL Ultracentrifuge Eppendorf tubes (Fisher Scientific).
5. 2.5–5 mM EDTA (Sigma, St. Louis, MO).
6. 1% BSA, Dulbecco's modified Eagle's medium (DMEM) containing Earle salts filtered through a 0.22-μm filter.

7. Organic combination: nine parts dibutyl phthalate (Sigma); one part cyclohexane (Sigma) (v:v). Always wear gloves when handling either component (*see* **Note 5**).
8. LB-tet./kan. medium (20 µg/mL tet. and 100 µg/mL kan.).
9. LB-tet./kan. agar plates (40 µg/mL tet. and 100 µg/mL kan.).
10. *E. coli* K91/kan. bacteria.

2.3. In Vivo Panning

2.3.1. Mouse Organ Targeting

1. Phage display library of interest.
2. Mouse of interest for targeting.
3. DMEM.
4. Avertin®: 2.5 g 2,2,2-tribromoethanol, 5 mL 2-methyl-2-butanol. Heat to dissolve. Add 200 mL ddH$_2$O. Filter-sterilize, and store away from light at 4°C.
5. Grinding buffer with protease inhibitors: DMEM containing Earle's salts supplemented with 1% BSA, 1 mM 4-(2-aminoethyl) benzenesulfonyl fluoride hydrochloride (AEBSF), 10 µg/mL aprotinin, and 1 µg/mL leupeptin.
6. Glass homogenizers (VWR, Suwanee, GA).
7. *E. coli* K91/kan bacteria.

2.3.2. Intra-Organ Vasculature Targeting

1. Phage display library of interest.
2. Laser pressure catapult microdissection system (P.A.L.M. system or equivalent).
3. Avertin®: 2.5 g 2,2,2-tribromoethanol and 5 mL 2-methyl-2-butanol. Heat to dissolve. Add 200 mL ddH$_2$O. Filter-sterilize, and store away from light at 4°C.
4. DMEM.
5. Protease inhibitor cocktail: 1 mM AEBSF, 20 µg/mL aprotinin, 10 µg/µL leupeptin, 1 mM elastase inhibitor I, 0.1 mM Na-tosyl-Phe chloromethyl ketone, and 1 nM pepstatin A; (Calbiochem, San Diego, CA). Wear gloves at all times when handling protease cocktail reagents.
6. Glass homogenizers.
7. Fluorescein conjugated *Lycopersicon esculentum* lectin (Vector Laboratories Inc., Burlingame, CA).
8. LB-tet./kan. plates (40 µg/mL tet. and 100 µg/mL kan.).
9. LB-tet./kan. medium (0.2 µg/mL tet. and 100 µg/mL kan.).
10. Tissue Tek O.C.T. compound (Sakura, Torrance, CA).
11. fUSE5 primers:
    Forward:
    5'-TAATACGACTCACTATAGGGCAAGCTGATAAACCGATACAATT-3'
    Reverse: 5'-CCCTCATAGTTAGCGTAACGATCT-3'
    Nested primers:
    Sense: 5'-CACTCGGCCGACG-3'
    Antisense: 5'-TTCGGCCCCAGCGGC-3'

2.4. Sequencing of Phage Insert

1. fUSE5 primer:
   Reverse: 5'-CCCTCATAGTTAGCGTAACGATCT-3'
2. PBS.
3. dNTPs (Promega, Roche, or equivalent).
4. Taq polymerase and polymerase buffer (Promega, Roche, or equivalent).
5. ddH$_2$O.
6. DMSO.
7. StrataClean® resin (Stratagene, La Jolla, CA).
8. Glycogen.
9. TE: 10 m*M* Tris, 1 m*M* EDTA.
10. 3 *M* NaOAc.
11. 100%, 70% Ethanol.
12. 2% Agarose gel (E-gels; Invitrogen).

3. Methods

3.1. Construction of Phage Display Peptide Libraries

Filamentous phage display of peptide libraries is based on cloning DNA fragments encoding a peptide sequences into the phage genome fused to the PIII coat protein gene. Incorporation and expression of the gene fusion product will result in the presentation of the peptide on the phage surface, where it can interact and bind with an intended target. A phage library can consist of up to 10^9 phage clones, each displaying a different peptide *(39)*. The size of the peptide insert and cyclic or linear expression orientation are options to be considered before constructing a library. However, it is generally difficult to insert or obtain a reasonable expression yield from peptides longer than 12 amino acids (6–8 amino acids is ideal). The success of the screening is highly dependent on how well the library is constructed *(40)*. Although insertless phage cannot be avoided and will be eliminated in the phage selection biopanning procedures, it is important to limit the number of insertless phage so as to maximize the diversity of the library. The following procedure outlines the steps for isolating, purifying, and replacing the fUSE5 vector "stuffer" sequence with the desired peptide library insert.

3.1.1. Preparation of fUSE5 Vector

The fUSE5 plasmid *(39)* is propagated in F'-minus host bacteria *E. coli* MC1061. Grow and maintain these bacteria in LB- tet./strep. medium.

1. Electroporate about 100 ng of fUSE5 plasmid into 25 μL of *E. coli* MC1061.
2. Add the sample to 1 mL SOC medium and shake at 225 rpm for 1 h at 37°C.
3. Plate serial dilutions on LB-tet. plates and leave overnight at 37°C.

4. The next day inoculate a starter culture from one colony in 5 mL LB-tet./strep. media for 2 h in a 225 rpm shaker at 37°C.
5. Add the starter culture to 1 L LB-tet./strep. media and shake overnight at 37°C (*see* **Note 6**).
6. The next day, centrifuge the culture at 6000*g* for 10 min at 4°C.
7. Purify the plasmid with two commercially available plasmid preparation purification kits.
8. Purify the plasmid further by using a CsCl-gradient. Add 3.2 g CsCl and 400 μL EtBr (10 mg/mL) per 3 mL DNA/TE.
9. Centrifuge the samples overnight at 100,000*g* at 25°C (remove the tube from the centrifuge very carefully so as to not disturb the bands).
10. Extract the DNA band with an 18G needle (*see* **Note 7**) and place in a 1.5 mL Eppendorf tube (400 μL max volume per tube).
11. Remove the EtBr from the DNA by adding 600 μL of the organic phase (upper layer) of TE buffered *N*-butanol to each tube. Mix by inversion, then remove and discard the upper phase once the layers separate.
12. Repeat **step 11** three times. The last time, remove the bottom aqueous DNA phase carefully with a pipet, being careful not to extract any of the organic layer (*see* **Note 8**).
13. Ethanol precipitate the fUSE5 plasmid. Based on the volume of DNA sample, add 1/10 the volume of 3 *M* NaOAc and two times the volume of 100% ethanol. Place at −20°C for 2 h (the DNA can be left frozen for an extended period of time).
14. Centrifuge at 10,000*g* for 30 min at 4°C.
15. Wash the pellet with 70% ethanol and centrifuge for 5 min at 4°C.
16. Air-dry the pellet, then resuspend in 500 μL ddH$_2$O.
17. Calculate the concentration and purity of the DNA by measuring the optical density at 260 nm. The concentration should be about 50 μg/mL and have an A$_{260}$/A$_{280}$ around 1.8.

3.1.2. Digestion of fUSE5 Vector

The fUSE5 vector was engineered to be noninfective by disrupting the gene III reading frame with of a 14-bp "stuffer" *(39)*. Infectivity is restored when the "stuffer" sequences is replaced with an in-frame insertion. Removal of the fUSE5 "stuffer" sequence within gene III is achieved with the restriction enzyme *Sfi*I (*see* **Note 9**). This leaves the overhanging sites incompatible with each other and allows for unidirectional cloning of a *Bgl*I digested DNA insert *(39,41)*.

1. Digest 50 μg of the fUSE5 vector with *Sfi*I enzyme (use at least 30 μg to allow for loss during the preparation steps) in a 100 μL total reaction volume for 2 h at 50°C.
2. Separate the *Sfi*I-digested fUSE5 vector from the "stuffer" using a commercial kit. Final elution volume should be 200 μL.
3. Confirm that the approx 9.5-kb vector has been delinearized by running an aliquot on a 0.8% agarose gel.

3.1.3. Preparation of Insert

The synthetic inserts should be purchased or synthesized as single-stranded degenerate oligonucleotides. The sequence of the template is:

5'CACTCGGCCGACGGGGCT (NNK)$_X$GGGGCCGCTGGGGCCGAA3',

where N indicates an equimolar mixture of all four nucleotides; K indicates an equimolar mixture of G and T, preventing the introduction of a stop codon into the sequence; and X represents the number of repeats. Perform at least 6 reactions/ library insert.

1. Convert the oligonuceotides to double-stranded DNA according to the polymerase chain reaction settings below using the following sample volumes per reaction (20 µL total volume):
   a. 500 ng Oligonucleotide.
   b. 2 µg Library antisense.
   c. 2 µg Library sense.
   d. 10.6 µL ddH$_2$O.
   e. 0.4 µL DMSO.
   f. 2 µL 0 mM dNTP.
   g. 2 µL Taq buffer.
   h. 0.5 µL Taq polymerase.

(PCR) settings:
   a. 2 min 94°C.
   b. 30 s 94°C.
   c. 30 s 60°C.
   d. 30 s 72°C.
   e. 5 min 72°C.
   f. Hold at 4°C.
   g. 35 Cycles.

2. Purify the reaction with any nucleotide removal kit. Final elution volume should be 100 µL per reaction.
3. Digest the double-stranded oligonuceotide with *Bgl*I enzyme in a 100 µL total reaction volume for 2 h at 37°C. This produces overhanging ends that are compatible to those on the *Sfi*I-digested fUSE5 vector. Perform at least three digestions.
   a. 10 µL Buffer H.
   b. 4 µL *Bgl*I (10 U/µL).
   c. 86 µL Purified dsDNA insert.

3.1.4. Test Ligations

Before performing the final ligation between the *Sfi*I-digested fUSE5 vector and the *Bgl*I-digested insert, test ligations should be performed, including a negative control (without adding the *Bgl*I-digested insert fragment) to determine the ideal molar ratio of vector to insert and to determine ligation background and efficiency (different insert amounts and insert to vector ratio; *see* **Note 10**). To expedite the test, ligations may be carried out at room temperature for 3–4 h, and then ethanol precipitated for at least 1 h at −20°C.

1. Try the following molar ratios: 1:1000, 1:300, 1:100, 1:30, 1:10, 1:3, 1:1 (vector: insert).
2. Incubate 2 μL 10X ligase buffer, 1 μL T4 DNA ligase, molar ratio of vector (500 ng vector) and insert and H_2O to a final 20 μL reaction volume.
3. Incubate 4 h at room temperature (RT).
4. Ethanol precipitate the samples by adding 2X the sample volume of ethanol plus 1/10 the volume of 3 *M* sodium acetate, pH 5.2 for 1 h at −20°C.
5. Centrifuge at 1600*g* for 30 min at 4°C.
6. Air-dry the pellet and resuspend in 50 μL ddH$_2$O.
7. Electroporate 1 μL of DNA solution and 25 μL of electrocompetent F'-minus bacteria (MC1061).
8. Add each sample to 1 mL of SOC medium and shake at 225 rpm for 1 h at 37°C.
9. Plate serial dilutions (e.g., 2 μL and 200 μL) on LB-tet. plates.
10. Count the number of colonies and determine the optimal vector-to-insert molar ratio, transformation efficiency, and background from the negative control ligation.

3.1.5. Library Ligation

After determining the optimal molar ratio for the ligation and ensuring that ligation of the vector alone yields a low or no background, the final library ligation may be performed. Keep all the samples and bacteria on ice at all times.

1. Ligate at least 10 μg of *Sfi*I-digested fUSE5 vector for production of the library according to the best-determined test ligation in **Subheading 3.1.4.**
2. Carry out the ligation (**Subheading 3.1.4., step 2**) overnight at 16°C in a final volume of 500 μL.
3. Exchange the ligation buffer to ddH$_2$O by using a commercial buffer exchange kit or ethanol precipitation. The final volume should then be approx 200 μL.
4. Electroporate 1 μL of DNA solution and 25 μL of electrocompetent F'-minus bacteria (MC1061). Incubate on ice for at least 1 min before electroporation.
5. Pool every 25 electroporations into 25 mL of SOC medium. Shake at 225 rpm for 1 h at 37°C.
6. Add each 25 mL to 1 L of LB-tet./strep. medium. Up to 100 electroporations (100 mL SOC medium) can be added to 1 L LB-tet./strep. medium.
7. From each 5-L flask, plate serial dilutions (for example, 0.5 and 50 μL) on LB-tet. plates and incubate at 37°C overnight. To amplify the library, go to **step 9**.
8. The next day, count the colonies from the plates to determine the diversity of the library (in TUs). Ideally, it should be approx 10^9 TU.
9. To amplify the phage library, incubate the bacteria in 37°C shaker at 225 rpm overnight (final volume of 2 L).
10. Proceed to **Subheading 3.2.4., step 6** to precipitate the phage.

3.2. In Vitro Panning

3.2.1. Coating Receptors on Microtiter Plates

1. To a 96-well microtiter plate, add 25 μL of purified desired receptor at various concentrations (from 10–500 ng per well) in 25 mM N-octyl-β-D-glucopyranoside, TBS. (Coat wells in triplicate; coat additional wells with 3% BSA/PBS to be used as negative controls.)
2. Add 200 μL of 1 mM MnCl$_2$/TBS.
3. Incubate at 4°C for 16–24 h. Most proteins will bind well after 1–2 h at RT.
4. Wash wells twice with 200 μL PBS.
5. Saturate the wells by incubating with 200 μL 3% BSA, 1 mM MnCl$_2$/TBS for 1 h at room temperature.
6. Wash the wells three times with 200 μL 1 mM MnCl$_2$/TBS, and blot on paper towel to remove most of the liquid.

3.2.2. First Round of Panning

1. Add 80 μL of 1% BSA, 1 mM MnCl$_2$/TBS to each receptor-coated well.
2. Add 20 μL of the phage library constructed in **Subheading 3.1.** containing 10^{10}–10^{11} TU.
3. Incubate at 4°C for 16–24 h with gentle shaking.
4. Wash the wells 10 times with 400 μL 0.5% Tween-20/TBS. Allow each wash to stand for 1 min.
5. Invert plate on paper towel to remove most of the washing buffer.
6. Elute the phage bound to the receptor with 100 μL of the elution buffer for 10 min with gentle shaking.
7. Neutralize the low-pH elution buffer by adding it to a sterile 50 mL tube containing 8 μL of the neutralization buffer. Note neutralization of the solution by color change to red. Do not discard the pipet tip (*see* **Note 11**).
8. Add 100 μL of log phase K91/kan. bacteria to the 50-mL tube. Also add 100 μL of bacteria directly to the well to harvest the phage from the well and the pipet tip.
9. Continue to **Subheading 3.2.4.** for phage recovery.

3.2.3. Successive Rounds of Panning

1. Follow the procedure in **Subheading 3.2.2.**, using the selected and amplified phage from the previous panning.
2. At least two or three rounds of panning should be performed.
3. Sequence the DNA inserts starting from round 2 as described in **Subheading 3.4.**

3.2.4. Phage Recovery

1. Gently mix the bacteria with the phage and let the tubes stand for at least 20 min (but no more than 1 h) at RT. In the meantime, warm LB-tet./kan. media to 37°C.
2. Add 18 mL of the pre-warmed LB-tet./kan. media to the phage–bacteria mixture and shake the tubes at 37°C for 40–60 min.

3. Adjust tetracycline to a final concentration of 20 µg/mL. Spread dilution (2 µL to 200 µL) aliquots of the bacterial culture onto LB-tet./kan. plates to determine the amount of phage clones recovered from the panning procedure.
4. Incubate the plates overnight at 37°C.
5. Amplify the phage by growing the remaining of the bacterial culture in 500 mL LB-tet./kan. media in a 37°C shaker for 16–24 h.
6. After overnight culture bacteria infection, centrifuge the culture at 6000*g* for 10 min at 4°C and transfer the supernatant to a new tube.
7. Add 1.5 mL PEG/NaCl per 10 mL of phage solution. Mix by inversion 10 times and incubate on ice for 2–4 h.
8. Centrifuge the sample at 10,000*g* for 30 min at 4°C.
9. Aspirate the supernatant without disturbing the pellet and re-centrifuge at 10,000*g* for 5 min at 4°C to remove the remaining PEG supernatant.
10. Re-suspend the phage in 1 mL PBS.
11. Completely solubilize the pellet and then centrifuge at 10,000*g* for 10 min.
12. Transfer the supernatant to a sterile tube and add 150 µL PEG/NaCl/mL.
13. Leave on ice for 1 h for a second phage precipitation.
14. Centrifuge at 10,000*g* for 20 min at 4°C.
15. Discard the supernatant. Centrifuge at 10,000*g* for 5 min at 4°C.
16. Re-suspend the phage pellet in 50 µL of PBS.
17. Centrifuge at 10,000*g* for 5 min to remove any bacterial contaminants and store at 4°C.
18. Titer the phage as described in **Subheading 3.2.5.**

3.2.5. TU Titer Assay for Phage Preparations

1. Prepare serial dilutions of 10^{-6}, 10^{-7}, and 10^{-8} for each phage preparation in PBS.
2. Resuspend 4 µL of each dilution in 400 µL log phase *E. coli* K91/kan. and allow infection for 20–30 min at room temperature.
3. Plate 100 µL of each dilution in triplicates on LB-tet./kan. plates and incubate at 37°C overnight.
4. The next day, count the colonies on the plates and calculate the TU. For example, an average of 86 colonies from a 10^{-6} dilution plate will equal 86×10^6 TU/µL.

3.2.6. Antibody Fingerprinting

1. Immobilize 50 µL of purified IgGs from the desired source (e.g., cell-free ascites, serum) with 50 µL of commercially available Protein G-agarose beads. Follow protocol provided with commercial beads.
2. If a comparison between multiple antibody pools is desired, a preclearing step should be performed in order to eliminate nonspecific phage binding.
3. Incubate 10^9 TU of the phage library constructed in **Subheading 3.1.** with the control coupled IgG for 1 h at 4°C (pre-clearing the phage library).
4. Centrifuge the samples at 1000*g* for 1 min.

5. Remove the supernatant containing the unbound phage (pre-cleared library) and incubate it with 50 μL of the second sample of immobilized IgGs for 2 h at 4°C. (If no pre-clearing step is desired, then incubate the original 10^9 TU random phage peptide library with the desired immobilized IgGs.)
6. Centrifuge the samples at 1000g for 1 min.
7. Discard the supernatant and wash the pellet two times each with 500 μL binding buffer to remove unbound phage.
8. Elute the antibody-bound phage complexes from the IgGs with 50 μL elution buffer and immediately neutralize with 10 μL of the neutralization buffer.
9. Add 1 mL of *E. coli* K91/kan. in log phase and incubate 20 min at RT.
10. Transfer everything to a 50-mL tube containing LB-tet./kan. and incubate in a 37°C shaker for 16–24 h.
11. Recover the amplified phage as described in **Subheading 3.2.4.**
12. For successive rounds of panning, repeat **steps 5–11** in this section.
13. If clone sequencing is desired, refer to **Subheading 3.4.**

3.2.7. Cell Surface Binding

The following outlines the steps for screening, selection, and sorting cell surface binding peptides from phage libraries through the BRASIL procedure (*see* **Note 12**).

1. Wash cell culture flask three times with sterile PBS. If using suspension cells, wash cells three times with cold PBS by centrifuging the cells at 500g for 5 min and carefully aspirating the media.
2. Detach adherent cells with 4 mL (75 cm² flask) cold 5 mM EDTA/PBS, and incubate on ice to release the cells. If cells are tightly bound to the flask, use a scraper to detach the cells.
3. Collect the cells into a 15-mL tube, add 1% BSA/DMEM to a final volume of 15 mL and centrifuge 500g for 5 min.
4. Aspirate the media carefully and resuspend the cell pellet in 15 mL fresh 1% BSA/DMEM and repeat the centrifugation.
5. Carefully aspirate the media and resuspend the cells at 10^6 cell/mL in 1% BSA/DMEM.
6. Incubate 10^9 TU phage display peptide library constructed in **Subheading 3.1.** with 1×10^6 cells in a 1.5-mL Eppendorf tube for 2–4 h on ice (*see* **Notes 13** and **14**).
7. Place 200 μL of the organic combination into a 400-μL ultracentrifuge Eppendorf tube.
8. Gently transfer 100 μL of the cell-phage suspension to the top of the nonmiscible organic combination of the 400-μL Eppendorf tube.
9. Centrifuge 10,000g for 10 min at 4°C (*see* **Note 15**).
10. Place the tube at −80°C for 10–15 min to freeze the upper phase.
11. Slice off the bottom of the tube, being careful not to cross-contaminate with the phage remaining in the aqueous phase.

12. Remove the excess oil from the bottom of the tube with a pipet and place the bottom of the tube containing the pellet into a 1.5-mL tube.
13. Add 200 μL of *E. coli* K91/kan host bacteria in log phase for 20 min at RT (gently pipet up and down to mix).
14. Add 800 μL of LB-tet./kan. media and incubate for 1 h at 37°C with shaking.
15. Transfer all the contents of the 1.5-mL tube to a sterile 50-mL tube with 18-mL LB-tet./kan. media.
16. Wash the 1.5-mL tube with 1 mL LB-tet./kan. media and add to the 50-mL tube.
17. Plate 1, 10, and 100 μL from the 50 mL tube in triplicate to LB-tet./kan. plates for colony counting to compare multiple rounds of panning.
18. Incubate the 20 mL culture in the 50-mL tube for 16–18 h at 37°C in a 250–300 rpm shaker to amplify the phage.
19. The next day, count the colonies on the 1, 10, and 100 μL plates and calculate the total recovered phage. Example: If 30 colonies appear on the 10 μL plate then the total TU recovered from that round of panning is 6×10^4. (*See* **Note 16** and **Fig. 1** for details and an example of panning enrichment.)
20. Selected colonies may be sequenced as described in **Subheading 3.3.**
21. Double-precipitate the 20 mL amplified phage as described in **Subheading 3.2.4., steps 6–18**.
22. The amplified recovered phage can be used for subsequent rounds of BRASIL.

3.3. In Vivo Panning

The development of in vivo phage display technology has enabled the selection of peptides from phage display random peptide libraries capable of homing to different vascular beds in vivo *(26)*. In the in vivo selection method, peptides that home to specific vascular beds can be selected after intravenous administration of a phage display random peptide library *(26,42)*.

3.3.1. Individual Organ

Using the in vivo phage display technique, one can isolate peptide sequences that preferentially target the vasculature of an organ of interest.

1. Anesthetize a single mouse with 0.015–0.017 mg/g Avertin® injected intraperitoneally.
2. Once deeply anesthetized, intravenously inject via the tail vein 10^9 TU of the phage library prepared in **Subheading 3.1.** Dilute the library in DMEM to a final volume of 200 μL.
3. Allow the phage library to circulate for 6 min while maintaining the mouse body temperature at 37°C on a heating pad.
4. Systemically perfuse the mouse through the left ventricle with 5 mL 37°C DMEM to remove unbound phage, cutting the vena cava as the outlet.
5. Remove the organ of interest, place in small beaker containing DMEM, and weigh the recovered tissue.

6. Homogenize the tissue in 1 mL ice-cold grinding buffer (*see* **Note 17**).
7. Centrifuge the homogenate at 5000–6000*g* for 5 min at 4°C.
8. Discard the supernatant and resuspend the tissue homogenate in the grinding buffer. Repeat wash and grinding two additional times.
9. Recover the bound phage by infecting the homogenate with 1 mL log phase *E. coli* K91/kan host bacteria for 30 min at RT. Continue with recovery and amplification as described in **Subheading 3.2.4.**
10. For subsequent rounds of panning, repeat **steps 1–9** using the amplified phage recovered from the previous round in place of the library. Recovered phage can be sequenced as described in **Subheading 3.4.**

3.3.2. Intra-Organ Targeting

Several organs have functionally distinct regions that often have unique vascular architectures. Combining in vivo phage display with laser capture microdissection, one can identify phage that selectively target vascular beds embedded within an organ.

1. After performing several rounds of phage selection on the organ of interest, intravenously inject 10^9 TU of the amplified phage from the last round of panning into an Avertin anesthetized mouse.
2. During the final 2 min of circulation, intravenously inject 50 μg of fluorescein-conjugated lectin diluted to a total volume of 100 μL in PBS.
3. Infuse the mouse through the left ventricle with 3 mL of 37°C DMEM, cutting the inferior vena cava.
4. Remove the organ of interest and quickly freeze in Tissue Tek O.C.T. compound at −80°C.
5. Cut 14-μm sections of the frozen organ and place on suitable microscope slides for laser pressure catapult microdissection.
6. Excise region of interest using laser pressure catapult microdissection and catapult into 0.75-mL tubes containing 30 μL of either 1 m*M* EDTA (proceed to **step 7**) or protease inhibitor cocktail in PBS (proceed to **step 8**).
7. For regions catapulted into EDTA, amplify peptide insert sequences by PCR using the fUSE5 forward and reverse primers. Perform a second PCR amplification to incorporate *Sfi*I restriction sites using the nested sense and antisense primers. Ligate products to *Sfi*I digested fUSE5 vector and electroporate into MC1061 bacterial host. Plate bacteria on LB-tet./kan. plates and select single colonies for sequence analyses as described in **Subheading 3.4.**
8. For regions catapulted into the protease inhibitor cocktail, incubate tissue with 1 mL of log phase K91/kan. bacteria at room temperature. Transfer mixture to 1.2 mL of LB-tet./kan. media and incubate in the dark for 40 min.
9. Increase tetracycline concentration to 40 μg/mL and incubate overnight at 37°C with agitation.
10. The next day, plate onto LB-tet./kan. plates and select single colonies for PCR and sequencing as described in **Subheading 3.4.**

3.4. Sequencing of Phage Insert

The amino acid sequence of the recovered peptides is determined by sequencing the DNA corresponding to the insert in the phage genome.

1. Pick well-separated colonies from the agar plates and transfer them to 5 mL of LB-tet./kan. media. Grow 14 mL of the bacterial culture in a 50-mL tube. Incubate in a 37°C shaker for 16 h.
2. Centrifuge the 50-mL tube at 6000g for 15 min to pellet the bacteria.
3. Transfer the supernatant to an Eppendorf tube containing 750 μL PEG/NaCl, vortex for 1 min and incubate the tubes on ice for at least 1 h.
4. Centrifuge at 6000g for 15 min. Aspirate the supernatants, centrifuge briefly, and remove the rest of the liquid.
5. Resuspend phage in 1 mL of PBS by vortexing. Store the phage at −20°C.
6. For preparation of single-stranded DNA, transfer 200 μL of the phage to an Eppendorf tube containing 10 μL StrataClean® resin. Vortex for 1 min. Centrifuge at 2000g for 4 min.
7. Transfer 180 μL of the supernatant to a new Eppendorf tube containing 10 μL of StrataClean resin. Repeat vortexing and centrifugation.
8. Transfer 150 μL to an Eppendorf tube containing 1 μL of glycogen, 150 μL of TE, and 40 μL of 3 M NaOAc. Add 1 mL of 100% ethanol. Incubate the tubes on ice for 1 h.
9. Centrifuge at 8000g for 30 min at 4°C. Aspirate the supernatant.
10. Re-centrifuge for 10 min and remove the rest of the liquid.
11. Wash with 1 mL of 70% ethanol. Centrifuge at 18,000g for 15 min and aspirate the supernatant.
12. Re-centrifuge for 10 min and remove the rest of the liquid.
13. Resuspend the DNA in 5 μL of ddH$_2$O. The dissolved DNA is ready for cycle sequencing.
14. Sequence the DNA directly using the 5'CCCTCATAGTTAGCGTAACGATCT 3' primer. An alternative method is to first prepare a double-stranded PCR product containing the insert site and then sequence the double-stranded DNA product.
15. If you use an automatic Perkin Elmer (Norwalk, CT) ABI PRISM sequencing system, the following protocols apply.

3.5. PCR Mix

1. 100 ng Template single-stranded DNA.
2. 8 μL Terminator Ready Reaction Mix.
3. 3.2 pmoL reverse primer (5'CCCTCATAGTTAGCGTAACGATCT 3') at 20 ng/reaction.
4. Add ddH$_2$O to 20 μL.

3.5.1. PCR Setting

1. 96°C for 10 s, 50°C for 5 s, 60°C for 4 min. Repeat for 25 cycles.
2. Precipitate the DNA, wash with 70% ethanol, and air-dry.

4. Notes

1. Check the genotype of the host bacteria for other drug resistance and add additional antibiotics to the media and plates to prevent contamination (e.g., streptomycin, kanamycin).
2. The detergent, 25 mM N-octyl-β-D-glucopyranoside, helps keep the integrins solubilized and is important for attachment.
3. Mn^{+2} cations activate integrins and enhance the binding of peptides to purified integrins.
4. A starting tetracycline concentration of 0.2 μg/mL has been suggested to induce the promoter of the tetracycline resistant gene of fUSE5. After initial incubation, the concentration should be increased to 20 μg/mL in order to prevent contamination.
5. Other organic solvents will work and should be tested with each cell line prior to panning.
6. Grow at least a 1 L culture because fUSE5 is replication deficient and gives low yield *(39)*. Maintain the fUSE5 vector in F pilus-negative bacteria such as MC1061. Keep a glycerol stock of the bacterial clone and restreak plates every 2 wk in minimal media plates. When restreaking, be sure to streak a negative-control LB-tet. plate. If tet.-resistant colonies are present on the negative-control plate, fUSE5 cross-contamination is a possibility.
7. Insert one needle into the top of the tube to release pressure and insert a second needle into the wide band to extract the DNA. Do not extract the small upper band, if present. This band contains nicked and small genomic DNA.
8. The EtBr and CsCl can optionally be removed by dialyzing twice against TE (5 L) in a 10,000-molecular weight cutoff dialysis cassette.
9. Be sure to read the manufacturer guidelines and troubleshooting notes with each restriction enzyme kit for optimal time and temperature use.
10. A 1:3 ratio of vector to insert is recommended as to avoid the possibility of insert ends ligating to each other. However, varying the ratio and total amount of insert during the test ligation will allow optimization of the reaction.
11. In the first round of panning, it is important to recover as much bound phage as possible so that it will be represented in sequential pannings.
12. Prior to panning on cells of interest, a pre-clearing of nonspecific phage may be performed as shown in the first step of **Fig. 3**. This optional step may be performed to select differential binding phage to one cell line over another or to eliminate phage that bind to commonly expressed cell surface proteins.
13. Centrifuge the phage library 16,000g for 5 min before adding to the cells.
14. Incubation time will be dependent on the cell line in use as well as the phage library. Generally 2 h works, but longer incubation times may be necessary if the phage recovery is very low.
15. Oil density is temperature sensitive and some cell lines do not enter the organic phase as well as others when centrifuged at 4°C. Testing a small aliquot of cells at first 4°C and then RT with the organic combination is recommended prior to centrifuging the test samples.

16. The total recovered phage is representative of each phage clone selected during panning (each clone is not necessarily a different phage). Amplifying the recovered phage for subsequent pannings allows the enrichment of specific binding clones.
17. Wash the homogenizer with up to 500 µL grinding buffer to maximize phage recovery. Be sure to keep the amount of grinding buffer constant in order to accurately calculate the phage recovery per gram of tissue.

Acknowledgments

The authors would like to thank Drs. Ricardo Giordano and Paul Mintz for their helpful comments on the manuscript.

References

1. Scott, J. K. and Smith, G. P. (1990) Searching for peptide ligands with an epitope library. *Science* **249,** 386–390.
2. Pasqualini, R., Koivunen, E., Kain, R., et al. (2000) Aminopeptidase N is a receptor for tumor-homing peptides and a target for inhibiting angiogenesis. *Cancer Res.* **60,** 722–727.
3. Kolonin, M. G., Saha, P. K., Chan, L., Pasqualini, R., and Arap, W. (2004) Reversal of obesity by targeted ablation of adipose tissue. *Nat. Med.* **10,** 625–632.
4. Marchio, S., Lahdenranta, J., Schlingemann, R. O., et al. (2004) Aminopeptidase A is a functional target in angiogenic blood vessels. *Cancer Cell* **5,** 151–162.
5. Pasqualini, R., Arap, W., and McDonald, D. M. (2002) Probing the structural and molecular diversity of tumor vasculature. *Trends Mol. Med.* **8,** 563–571.
6. Koivunen, E., Wang, B., and Ruoslahti, E. (1995) Phage libraries displaying cyclic peptides with different ring sizes: ligand specificities of the RGD-directed integrins. *Biotechnology* **13,** 265–270.
7. Rajotte, D., Arap, W., Hagedorn, M., Koivunen, E., Pasqualini, R., and Ruoslahti, E. (1998) Molecular heterogeneity of the vascular endothelium revealed by in vivo phage display. *J. Clin. Invest.* **102,** 430–437.
8. Arap, W., Pasqualini, R., and Ruoslahti, E. (1998) Cancer treatment by targeted drug delivery to tumor vasculature in a mouse model. *Science* **279,** 377–380.
9. Arap, W., Kolonin, M. G., Trepel, M., et al. (2002) Steps toward mapping the human vasculature by phage display. *Nat. Med.* **8,** 121–127.
10. Koivunen, E., Arap, W., Rajotte, D., Lahdenranta, J., and Pasqualini, R. (1999) Identification of receptor ligands with phage display peptide libraries. *J. Nucl. Med.* **40,** 883–888.
11. Marchio, S., Alfano, M., Primo, L., et al. (2005) Cell surface-associated Tat modulates HIV-1 infection and spreading through a specific interaction with gp120 viral envelope protein. *Blood* **105,** 2802–2811.
12. Cortese, R., Monaci, P., Luzzago, A., et al. (1996) Selection of biologically active peptides by phage display of random peptide libraries. *Curr. Opin. Biotechnol.* **7,** 616–621.
13. Burritt, J. B., Bond, C. W., Doss, K. W., and Jesaitis, A. J. (1996) Filamentous phage display of oligopeptide libraries. *Anal. Biochem.* **238,** 1–13.

14. Mintz, P. J., Kim, J., Do, K. A., et al. (2003) Fingerprinting the circulating repertoire of antibodies from cancer patients. *Nat. Biotechnol.* **21**, 57–63.
15. Vidal, C. I., Mintz, P. J., Lu, K., et al. (2004) An HSP90-mimic peptide revealed by fingerprinting the pool of antibodies from ovarian cancer patients. *Oncogene* **23**, 8859–8867.
16. Ruoslahti, E. (1996) RGD and other recognition sequences for integrins. *Ann. Rev. Cell Dev. Biol.* **12**, 697–715.
17. Koivunen, E., Arap, W., Valtanen, H., et al. (1999) Tumor targeting with a selective gelatinase inhibitor. *Nat. Biotechnol.* **17**, 768–774.
18. Cardo-Vila, M., Arap, W., and Pasqualini, R. (2003) avb5 integrin-dependent programmed cell death triggered by a peptide mimic of annexin V. *Mol. Cell* **11**, 1151–1162.
19. Giordano, R. J., Cardo-Vila, M., Lahdenranta, J., Pasqualini, R., and Arap, W. (2001) Biopanning and rapid analysis of selective interactive ligands. *Nat. Med.* **7**, 1249–1253.
20. Giordano, R., Chammas, R., Veiga, S. S., Colli, W., and Alves, M. J. (1994) An acidic component of the heterogeneous Tc-85 protein family from the surface of Trypanosoma cruzi is a laminin binding glycoprotein. *Mol. Biochem. Parasitol.* **65**, 85–94.
21. Levesque, J. P., Hatzfeld, A., and Hatzfeld, J. (1985) A method to measure receptor binding of ligands with low affinity. Application to plasma proteins binding assay with hemopoietic cells. *Exp. Cell Res.* **156**, 558–562.
22. Kolonin, M. G., Pasqualini, R., and Arap, W. (2002) Teratogenicity induced by targeting a placental immunoglobulin transporter. *Proc. Natl. Acad. Sci. USA* **99**, 13,055–13,060.
23. Pasqualini, R., Arap, W., and McDonald, D. M. (2002) Probing the structural and molecular diversity of tumor vasculature. *Trends Mol. Med.* **8**, 563–571.
24. Essler, M. and Ruoslahti, E. (2002) Molecular specialization of breast vasculature: a breast-homing phage-displayed peptide binds to aminopeptidase P in breast vasculature. *Proc. Natl. Acad. Sci. USA* **99**, 2252–2257.
25. Rajotte, D. and Ruoslahti, E. (1999) Membrane dipeptidase is the receptor for a lung-targeting peptide identified by in vivo phage display. *J. Biol. Chem.* **274**, 11,593–11,598.
26. Pasqualini, R. and Ruoslahti, E. (1996) Organ targeting in vivo using phage display peptide libraries. *Nature* **380**, 364–366.
27. Trepel, M., Arap, W., and Pasqualini, R. (2002) In vivo phage display and vascular heterogeneity: implications for targeted medicine. *Curr. Opin. Chem. Biol.* **6**, 399–404.
28. George, A. J., Lee, L., and Pitzalis, C. (2003) Isolating ligands specific for human vasculature using in vivo phage selection. *Trends Biotechnol.* **21**, 199–203.
29. Koivunen, E., Gay, D. A., and Ruoslahti, E. (1993) Selection of peptides binding to the alpha 5 beta 1 integrin from phage display library. *J. Biol. Chem.* **268**, 20,205–20,210.
30. Pasqualini, R., Koivunen, E., and Ruoslahti, E. (1997) Alpha v integrins as receptors for tumor targeting by circulating ligands. *Nat. Biotechnol.* **15**, 542–546.

31. Burg, M. A., Pasqualini, R., Arap, W., Ruoslahti, E., and Stallcup, W. B. (1999) NG2 proteoglycan-binding peptides target tumor neovasculature. *Cancer Res.* **59,** 2869–2874.
32. Hoffman, J. A., Giraudo, E., Singh, M., et al. (2003) Progressive vascular changes in a transgenic mouse model of squamous cell carcinoma. *Cancer Cell* **4,** 383–391.
33. Joyce, J. A., Laakkonen, P., Bernasconi, M., Bergers, G., Ruoslahti, E., and Hanahan, D. (2003) Stage-specific vascular markers revealed by phage display in a mouse model of pancreatic islet tumorigenesis. *Cancer Cell* **4,** 393–403.
34. Kolonin, M. G., Pasqualini, R., and Arap, W. (2001) Molecular addresses in blood vessels as targets for therapy. *Curr. Opin. Chem. Biol.* **5,** 308–313.
35. Pasqualini, R. and Arap, W. (2002) Translation of vascular proteomics into individualized therapeutics, in *Pharmacogenomics: The Search for Individualized Therapies* (Licino, J. and Won, M . L., eds.). Wiley-VCH, Weinheim, Germany, pp. 525–530.
36. Ellerby, H. M., Arap, W., Ellerby, L. M., et al. (1999) Anti-cancer activity of targeted pro-apoptotic peptides. *Nat. Med.* **5,** 1032–1038.
37. Curnis, F., Sacchi, A., Borgna, L., Magni, F., Gasparri, A., and Corti, A. (2000) Enhancement of tumor necrosis factor alpha antitumor immunotherapeutic properties by targeted delivery to aminopeptidase N (CD13). *Nat. Biotechnol.* **18,** 1185–1190.
38. Yao, V. J., Ozawa, M. G., Trepel, M., Arap, W., McDonald, D. M., and Pasqualini, R. (2005) Targeting pancreatic islets with phage display assisted by laser pressure catapult microdissection. *Am. J. Pathol.* **166,** 625–636.
39. Smith, G. P. and Scott, J. K. (1993) Libraries of peptides and proteins displayed on filamentous phage. *Methods Enzymol.* **217,** 228–257.
40. Barbas, C. F., III, Burton, D. R., Scott, J. K., and Silverman, G. J. (2001) *Phage Display: A Laboratory Manual.* Cold Spring Harbor Laboratory Press, Cold Spring Harbor, New York.
41. Smith, G. P. (1985) Filamentous fusion phage: novel expression vectors that display cloned antigens on the virion surface. *Science* **228,** 1315–1317.
42. Kolonin, M., Pasqualini, R., and Arap, W. (2001) Molecular addresses in blood vessels as targets for therapy. *Curr. Opin. Chem. Biol.* **5,** 308–313.

Index

A

ABCA1, 16
Acetonitrile, 17, 19
Achromobacter I, protease, 42
ACEi, 46
Acetonitrile (ACN), 17, 19, 36, 107,
    113, 117, 121, 129, 130, 132,
    134, 146, 153–155, 186, 192,
    201, 209, 239, 244, 262, 340,
    344, 348, 349
Acrylamide tricine gel, 106
Actin, 87, 89, 92, 152, 208
β-Actin, 173
F-Actin, 182
Acute myocardial injury (AMI), 91
Adenosine diphosphate, 313
Adenovirus, 295
Adhesion proteins, 182, 183
β-Adrenergic stimulators, 92, 271, 290, 291
Affinity beads, IDM, SELDI-TOF, 91,
    93, 95
Agarose, 36, 256, 353
Agarose, low melting point, 209, 353, 354
Aggregates, cytoplasmic, 83
Aggregometer, 308–310
Akt, 292
Albumin, bovine (BSA), 75, 314, 316
Albumin depletion, 67, 68, 70, 351,
    365, 367, 369
Albumin, human, 351, 352, 355, 365
Albuminome, 368, 369
Alkylation, cysteine, 111, 211
AMI, 91
Amberlite MB-1, 105
Amidosulfobetaine (ASB-14), 60, 62,
    107, 108, 110, 121, 149
Ampholyte, 47, 51, 62, 64, 242, 320, 321

Anchor chips, 325
Angina pectoris, 33
Angiogenesis, 182, 385, 386
Angioplasty, 33, 253
Angiotensin II, 46, 55, 182
Angiotensin converting enzyme
    inhibitor, 46
Annexin V, 294
Annotation proteins, 25, 378–381
Antibodies, monoclonal, 165, 166, 170,
    175, 176, 316, 388
Antibody fingerprinting, 385, 391, 398
Antibody-HRP, 167
Anticoagulant, acid citrate dextrose
    (ACD), 308, 314
Antihypertensive drugs, 45
Anti-thrombin III, 344, 350
Anti-thrombin fragments, 348
α1-Antitrypsin, 351, 352
Apo CI, CIII, 343, 344, 348, 350
ApoE, 170, 173
Apolipoprotein CI, CIII, 343, 344,
    348, 350
Apoptosis, 253, 257, 271, 294
Aprotinin, 90, 92, 240, 392
Aquapore RP-300, 107, 113, 121
Argon, 159
Arrays (H50, Q-10), 94, 336
Arteria
    proteome, 173, 225
    secretome, 225
Arteriosclerosis, 46
Ascorbate, 216, 217, 294
Asthma, 235
Atheroma, 141, 144, 151
Atheroma plaque, 151, 152
Atherosclerosis, 46, 141, 151, 165, 253,
    297, 308, 386

407